Particle and Bulk Powder Characterization 2018

Topical at the 8th World Congress on Particle Technology

Orlando, Florida, USA
22 - 26 April 2018

ISBN: 978-1-5108-6966-0

Printed from e-media with permission by:

Curran Associates, Inc.
57 Morehouse Lane
Red Hook, NY 12571

CURRAN ASSOCIATES INC. proceedings .com

Some format issues inherent in the e-media version may also appear in this print version.

Printed by Curran Associates, Inc. (2018)

AIChE
120 Wall Street, FL 23
New York, NY 10005-4020

Phone: (800) 242-4363
Fax: (203) 775-5177

www.aiche.org

Additional copies of this publication are available from:

Curran Associates, Inc.
57 Morehouse Lane
Red Hook, NY 12571 USA
Phone: 845-758-0400
Fax: 845-758-2633
Email: curran@proceedings.com
Web: www.proceedings.com

TABLE OF CONTENTS

Some abstracts in this book refer to images that were not made available by the publisher

509452 Novel Approaches for the Production of Polymer Powders for Additive Manufacturing

Abstract:

Additive manufacturing (AM), often simply referred to as '3d printing' allows the production of individualized and complex parts without the need of any tools or moulds. Especially powder and beam-based AM processes like selective laser sintering (SLS) of polymers are promising. SLS already is used to manufacture special parts, although, the opening of new fields of application is currently hindered by the limited availability of materials. To further extend the applications of SLS new and cheap materials with good processability must be supplied. Innovative and versatile processes being applicable to a wide variety of thermoplasts are desirable. Currently, the spectrum of commercial SLS powders is very limited, about 95 % of the total market share is made up of polyamide powders that are typically produced by bottom-up approaches like emulsion polymerization or precipitation processes.

Two innovative methods for production of spherical polymer micro particles for SLS will be discussed: The first approach is a process chain consisting of wet grinding of polymers in a stirred media mill followed by rounding of the obtained irregular-shaped particles in a heated downer reactor and subsequent dry coating of the spherical particles for improvement of powder flowability. The process chain is applicable to a various thermoplasts and will be discussed for pure polymers (polystyrene, polybutylene terephthalate) and for filled polymer systems. Bulk powder properties and polymer structures are thoroughly characterized along the process chain. The second approach, melt emulsification with subsequent dry coating, will be exemplified for polypropylene. In this process a polymer melt is emulsified by elongational and shear stresses in a rotor stator device.

Both approaches allow for production of spherical polymer micro particles of good flowability which are promising, novel powder materials for SLS.

Schmidt, J.; Sachs, M.; Fanselow, S.; Zhao, M.; Romeis, S.; Drummer, D.; Wirth, K.-E.; Peukert, W., Chemical Engineering Science 156 (2016) 1–10.

Schmidt, J., Plata, M., Tröger, S., Peukert, W., Powder Technology 228 (2012) 84–90.

Schmidt, J.; Sachs, M.; Blümel, C.; Winzer, B.; Toni, F.; Wirth, K.-E.; Peukert, W., Powder Technology 261 (2014) 78–86.

Fanselow, S.; Emamjomeh, S. E.; Wirth, K.-E.; Schmidt, J.; Peukert, W., Chemical Engineering Science 141 (2016) 282–292.

509465 Characterization of Powders for Additive Manufacturing

Abstract:

Powder-based additive manufacturing (AM) processes like selective laser sintering (SLS) of polymers, selective laser beam melting of metals (SLM) or selective electron beam melting of metals (SEBM) allow for freedom of design and the manufacture of parts with adequate mechanical properties and dimensional accuracy. Within this contribution, powder requirements for AM processes, resulting structure-property relationships and appropriate methods for powder characterization will be discussed.

The effect of particle size, particle shape and surface functionalization on bulk solid properties (flowability, packing density) will be outlined in detail for polymer powder systems including commercial PA12 laser sintering powder and PE powders, and TiAl6V4 powder for SEBM. Possibilities to assess AM processability are exemplarily demonstrated for polymer powder systems to be used in SLS. The capabilities of a Schulze ring shear tester, a powder tensile strength tester to assess powder flowability and a model experiment mimicking the powder deposition process in AM are discussed. The possibilities of dry particle coating as an efficient method to optimize powder flowability and bulk density are demonstrated. Moreover, the effect of structural properties of the polymer (e.g. crystallinity, molar mass) on thermal and rheological properties (melting behavior, coalescence), appropriate analytical methods for characterization and the resulting implications for SLS processability are addressed.

Powder properties, AM processability and part properties are closely connected: powders of good flowability and high bulk density are mandatory to obtain dense devices.

Schmidt, J.; Sachs, M.; Fanselow, S.; Zhao, M.; Romeis, S.; Drummer, D.; Wirth, K.-E.; Peukert, W., Chemical Engineering Science 156 (2016) 1â€“10.

BlÃ¼mel, C.; Sachs, M.; Laumer, T.; Winzer, B.; Schmidt, J.; Schmidt, M.; Peukert, W.; Wirth, K.-E., Rapid Prototyping Journal 21 (2015) 697â€“704.

508105 Ceramic Powders in the Laser Sintering Processes

Abstract:

In selective laser sintering (SLS) a laser beam is used to partially melt particles in a layer of powders [1]. With subsequent increments of powder layers it is possible to create three dimensional structures. This technique can be used in prototyping applications able to produce customized objects of predetermined shape. With SLS it is possible to process powders polymer, ceramics and metal, therefore it appears to be particularly suited to build three dimensional scaffolds made of biocompatible materials. In SLS techniques increasing the energy transferred by laser increases the strength of the structure obtained, but it also produces a volume contraction of the sintered material which reduces the precision of the final object [2]. Since the energy required for sintering depends on the particle size. The use of smaller particles allows the use of smaller sintering energies and therefore the production of complex and more precise structures due to reduced melting [3]

In this paper powders with different particle size distributions of ceramics were obtained by mixing a coarser grained powder, with particles in the range between 10 and 100 microns, and a finer grained powder, with particles in the submicron range. The powder composition and the powder laying procedure were studied and optimized in order to obtain specimens with of the desired mechanical properties. The equipment used is a three dimensional laser sintering equipment using a 40W CO_2 laser beam purposely built. The density and the mechanical resistance of the specimens are studied as a function of the fines content and of the amount of energy released by the laser beam on the unit surface of the lighted area. A simple model approach is used to estimate the strength of the single sintered contact.

References

[1] Tang, Y., Fuh, J.Y.H., Loh, H.T., Wong, Y.S., Lu, L. 2003 Direct laser sintering of a silica sand. Materials and Design 24, 623–629

[2] Tia, X., Günster, J, Melcher, J., Li, D., Heinrich, J.G. 2009 Process parameters analysis of direct laser sintering and post treatment of porcelain components using Taguchi's method. J. Europ. Ceram. Soc. 29, 1903–1915.

[3] Salmoria, G.V., Klauss, P., Modolon Zepon, K., Kanis, L.A. 2013 The effects of laser energy density and particle size in the selective laser sintering of polycaprolactone/progesterone specimens: morphology and drug release. *Int. J. Adv. Manuf. Technol. 66*, 1113–1118.

507086 Powder Characterization Workflow for Powder-Bed Based 3D Printing Processes

Abstract:

The flow properties of a powder are the result of a complex interplay between the characteristics of the grains (size, morphology, water content, crystallinity, etc.) and the interactions between the grains (friction forces, van der Waals interactions, electrostatic forces, capillary forces). In order to control and to optimize processing methods, these powders have to be precisely characterized from different perspectives. In particular, 3D printing applications involving powder require precise, reproducible and appropriate powder characterization methods. In order to obtain thin and homogenous powder layers, a compromise between grain size and flowability has to be found. Unfortunately, when the grain size decreases, the cohesiveness increases and the flowability decreases. Moreover, the powder becomes sensitive to moisture and electrostatic charges. In addition, it is important to reach an optimal density to improve the mechanical properties of the produced part.

We present a methodology [1] based on a set of complementary instruments to perform this characterization: (i) an improved repose angle measurement, (ii) a flowmeter based on the rotating drum geometry, (iii) an automated tapped density measurement, (iv) a laboratory silo and (v) a powder tribo-charger. The classical repose angle measurement method has been revisited to measure powder cohesiveness from the heap irregularities. The rotating drum method is the dynamical version of this cohesiveness measurement to obtain rheological information. The tapped density measurement method has been automated to measure precisely the bulk density, the tapped density and also the compaction dynamics of a powder. The classical laboratory silo handled manually has been automated to obtain Beverloo flow curve. Finally, the tribo-charger measures the ability of a powder to create electrostatic charges (leading to cohesiveness) during a flow in contact with a selected material.

We show how the obtained set of results can be used to solve practical problems encountered in powder-bed based additive manufacturing processes [2].

[1] Measuring the flowing properties of powders and grains, Powder Technology 224, 19â€“27 (2012)

[2] Rheological behavior of Î²-Ti and NiTi powders produced by atomization for SLM production of open porous orthopedic implants, Powder Technology 283, 199â€“209 (2015)

507762 Production of Spherical Polyamide Particles for Additive Manufacturing By Precipitation

Abstract:

The application of polymer-based technologies for additive manufacturing is one the rise. With technologies like fused deposition modeling, stereolithography or binder jetting, a wide variety of solid to flexible parts are becoming available [1]. However, if functional parts with excellent mechanical properties are needed, only few additive manufacturing technologies are feasible. Among these are mostly exclusively powder-based technologies like selective laser sintering (SLS) [2]. While filament-based technologies offer already a wide variety of different materials including blends and composites, powder materials are still mainly polyamides (PA) with only few available commercial products and even less commercial producers [3]. Although the main route of production for these powders, namely thermal induced precipitation processes, is known in principle [4], hardly any information on the process outside of the patent literature [5–7] is available.

In this contribution, we shed light on the formation process of spherical micron-sized polyamide particles via precipitation. Short-chained alcohols for polyamides are used in pressure resistant autoclaves at high temperatures to dissolve the polyamide and subsequently to precipitate spherical particles upon decreasing the temperature. This way, direct transformation of granular bulk material into spherical particles with narrow size distribution can be achieved. The precipitation products are thoroughly characterized regarding morphology and size, structural and thermal properties. Therefore analysis via electron microscopy, static laser scattering, X-ray diffraction and dynamic scanning calorimetry is conducted. Furthermore, the flowability of the powders is investigated and product properties are correlated to process parameters, especially to temperature-time dependencies.

References

[1] I. Gibson, D. Rosen, B. Stucker, Additive manufacturing technologies, 2. ed., Springer, New York, NY, 2015.

[2] B. Wendel, D. Rietzel, F. Kühnlein, R. Feulner, G. Hülder, E. Schmachtenberg, Additive processing of polymers, Macromol. Mater. Eng. 293 (2008) 799–809. doi:10.1002/mame.200800121.

[3] T.T. Wohlers, Wohlers Report 2016, Wohlers Associates, Fort Collins, Col., 2016.

[4] M. Schmid, Selektives Lasersintern (SLS) mit Kunststoffen: Technologie, Prozesse und Werkstoffe, Hanser, 2015.

[5] F.E. Baumann, N. Wilczok, DE19708946

[6] S. Mumcu, H.-J. Panoch, J. Rüter, EP0200852.

[7] K.-R. Meyer, K.-H. Hornung, R. Feldmann, H.-J. Smigerski, EP0014772.

508403 Comparison of Different Flow Assessments for Selective Laser Sintering Powders

Abstract:

Selective Lase Sintering (SLS) is a rapidly expanding field of manufacturing technology. Its raw materials are manifold spanning both high density as well as low density metal powders and extending into polymers and ceramics as well.

While currently strict criteria are applied towards the sphericity of the particles as well as having an extremely narrow size distribution, this puts a considerable strain on the manufacturers of such powders, especially cost wise, and severely limits the availability of exotic materials for high performance engineering purposes.

While ring shear testers are a well-developed and standardized method for the characterization of flow behaviour it is often found to be lacking with especially free flowing powders such as the raw materials for SLS machines.

Several alternative methods (two methods for tensile strength testing, Rotational Cohesion Testing, uniaxial testing, penetration testing) will be compared both from the standpoint of particulate mechanics as well as the perspective of the practical process engineer as well as finally compared to the effectiveness of the proven ring shear tester.

This includes an assessment of the particulate mechanics of such systems as well as extensive experimental work and application of low and high density metals as well as polymer raw materials.

508096 A Demonstration of Bootstrap Method for Particle Diameter Distribution Analysis

Abstract:

Bootstrap method, which is a method of statistics used widely recently, has a very strong feature like non-parametric, simplicity so on. Therefore, it is suitable to apply to the analysis of particle diameter (size) distribution. For a number based sample, of particle diameter, this method can estimate the statistic uncertainty or confidence interval for any of statistical quantities of interest from the distribution, with very simple protocol and without any parametric assumptions. This paper gives a demonstration to introduce the method to the society of particle diameter (size) analysis.

508648 A New Particle Morphology Characterization to Account for Surface Angularity and Experimental Validation through 3D Printing

Abstract:

Granular materials is pervasive in nature, and is known as the second-most manipulated material in industry. It has significant importance in the fields such as construction, agriculture, aerospace and pharmaceutical applications, just to list a few. However, complex behavior of granular material is still poorly understood, e.g., phase transition from solid to liquid-like behavior and vice versa, due to its particulate nature at the grain scale, where the particle morphology plays a critical role in the inter-particular interaction.

Majority of particulate research community have accepted the concept of defining particle morphology at three different scales: global form (at large scale), local angularity (at intermediate scale), and surface texture (at small scale). The global form is the first-order morphological property that characterizes the overall particle shape compared to sphere (or how much it is elongated). A number of morphology descriptors have been developed with some variations including Sphericity, Shape Factor, Flat & Elongated Ratio, etc. Local angularity is the second-order morphological property that characterizes the sharpness of particle corners, for which a set of morphology descriptors also has been developed such as Roundness, Angularity Factor, and Angularity Index. On the other hand, surface texture has been often mechanically characterized in terms of surface roughness (i.e., effect of surface texture) such as inter-particle friction angle.

This approach that characterizes the particle morphology using the three factors at three different scales has been widely perceived as an effective method. However, there still exists a gap in the body of knowledge, in particular describing the local angularity (i.e., morphology at the intermediate scale). The existing local angularity descriptors mainly focus on the "corner" angularity (i.e., sharpness of particle corners), therefore do not robustly capture the "surface" angularity that increases with the presence of a series of local surface concavities. For example, exiting descriptors characterize tetrahedron as highly angular due to its sharp corners, while quantifying the angularity of gear-like particle shapes relatively low. Although it may be reasonable we call the shape with sharp corners "geometrically" angular, it may not be the case when it comes to "mechanical" properties, because the gear-shaped particles would provide a higher shear strength and modulus than those of the tetrahedrons due to a stronger interlocking mechanism if the bulk materials are mechanically tested.

The objective of this study is to develop a new particle morphology descriptor that can robustly characterize the "surface" angularity in addition to the "corner" angularity. A set of 3D numerical particle models are developed and characterized using the existing descriptors and the proposed descriptor for comparison of the quantified angularities. The bulk specimens are then developed for each particle models and mechanically tested for validation of the proposed descriptor. A key feature of the validation approach adopted in this study is employment of 3D printing technique to produce 3D synthetic particles from the 3D numerical particle models. The printed synthetic particles are used for the experimental validation, while bypassing the high computational costs that incur in discrete element simulation of numerical particles with complex geometries. The initial results in the study will be highlighted and discussed.

509472 Determination of Primary Particle Size Distributions of Agglomerates on Transmission Electron Microscopy Images By Artificial Neural Networks

Abstract:

The properties of nanomaterials (e.g. mechanical, optical, catalytic, biological, etc.) are determined by their characteristic length as well as their shape [1]. In case of nanoparticles, the characteristic length is usually some kind of equivalent diameter (e.g. hydrodynamic, volume-based, surface-based, etc.) [2]. There are various techniques available for the determination of the different types of equivalent diameters, one of them being transmission electron microscopy (TEM) [3].

The determination of particle size distributions (PSDs) of non-agglomerated particles on TEM images can already be performed partially or fully automated since several decades [4]. By contrast, the determination of PSDs of agglomerated particles still depends on the recognition of primary particles by a human operator. This practice is not only laborious, expensive and repetitive but also error-prone due to the subjectivity and exhaustion of the operator [5].

To solve this problem, a new approach to imaging particle analysis, with help of artificial neural networks (ANNs), was proposed, implemented and validated. ANNs were trained on the determination of the projected areas of primary particles of agglomerates via regression and the number of primary particles per agglomerate via classification.

A major challenge of the proposed method is the fact that the training of ANNs requires up to several hundreds of thousands of samples with known properties to avoid overfitting [6], i.e. that the ANNs are not able to generalize when being confronted with previously unknown data. Unfortunately, there is no publicly available source of already evaluated samples and a manual evaluation of such a large number of images is hardly feasible, due to the reasons given before. Therefore, TEM images with defined characteristics were synthesized and used for the training of the ANNs.

To speed up the image synthesis as well as the training of the ANNs the necessary calculations were performed on graphics processing units (GPUs), whenever possible.

[1] G. Guisbiers, S. Mejía-Rosales and F. Leonard Deepak. 'Nanomaterial Properties: Size and Shape Dependencies'. In: Journal of Nanomaterials 2012 (2012), pp. 1–2.

[2] J. G. Calvert. 'Glossary of atmospheric chemistry terms (Recommendations 1990)'. In: Pure and Applied Chemistry 62.11 (1990).

[3] E. J. Cho et al. 'Nanoparticle characterization: state of the art, challenges, and emerging technologies'. In: Molecular pharmaceutics 10.6 (2013), pp. 2093–2110.

[4] R. P. King. 'Determination of the distribution of size of irregularly shaped particles from measurements on sections or projected areas'. In: Powder Technology 32.1 (1982), pp. 87–100.

[5] T. Allen. Powder sampling and particle size determination. 1st ed. Amsterdam and Boston: Elsevier, 2003.

[6] M. A. Babyak. 'What you see may not be what you get: a brief, nontechnical introduction to overfitting in regression-type models'. In: Psychosomatic medicine 66.3 (2004), pp. 411–421.

509545 Multi-Wavelength Photo-Centrifugation for Particle Sizing without Reference to Models for the Optical Behaviour

Abstract:

Analytical photo-centrifugation is widely used for the determination of particle size distributions in finely dispersed suspensions and emulsions. It determines extinction-weighted size distributions of the Stokes diameter. Transformation into volume-weighted size distributions (Q_3) requires analytical models for the extinction at single particles, which depend on particle shape and the complex refractive index. These two features are typically unknown in real-life situations.

However, optical models are not required with multi-wavelength detection systems, which provide the optical model as an additional result of the analysis. The fundamental idea behind is that the wavelength dependency of turbidity for defined particle sizes is approx. equivalent to the size dependency of turbidity at fixed wavelengths. Despite this well-understood principle, there are several critical issues that may impede its transfer to practical situations (e.g. wavelength-specific detection limits).

The performance of the described approach was tested with a newly developed photo-centrifuge with a multi-wavelength detection system (LUM GmbH). Various colloidal particle systems were analysed. The approach yields optical models of the particles, which fairly well agree with the relevant analytical models (e.g. Mie theory). Correspondingly, the multi-wavelength analysis yields the same Q_3 as the traditional analysis, which is based on the particleâ€™s refractive index.

507440 Characterization of Particle Size and Shape with a Novel 3D Light Scattering Sensor (3D-LSS) for Aerosols

Abstract:

In the submicron regime particles gain a variety of interesting properties, not only by their size, but also by their structure. Therefore, these particles become nowadays a wide field of applications in pharmaceutical, optical and electronic industry **[1,2]**.

In order to control a production process for this kind of particles properly, the particle size as a one-dimensional quantity is not sufficient. To ensure for example the performance of a catalyst, information about the structure is indispensable. The state of the art methods to get information about the structure of particles in the gas phase in the submicron regime are imaging methods like SEM or TEM. These methods can only be used offline. In order to reduce the analysis time and to control the production processes it is necessary to get information about the particle size and the structure online.

This information can be estimated through analysis of the distribution of the light scattered by an aerosol particle. Common used optical particle sizers for the submicron (i.e. the Mie-) regime measure the intensity of scattered light emitted by a single particle at one particular position. In this case, the effect of elastic light scattering (ELS) is utilized and the result is a one-dimensional equivalent diameter.

The novel developed 3D Light Scattering Sensor (3D-LSS) is also based on the optical technique of ELS, but the setup uses 16 photomultipliers at fixed scattering angles acquiring scattering intensities of single particles for size measurement.

The centerpiece of the system is a hollow, spherical aluminum shell with an inner diameter of 140 mm. It consists of two hemispherical shells which are bolted together. Perpendicularly, two bores are installed, through which the examined aerosol is introduced by aerodynamic focusing and sucked off again. A window for feeding the laser beam and a beam trap are installed horizontally. The intensities of the light scattered by particles can be measured at a variety of positions (up to 16 postions simultaneously in a wide range of scattering and azimuthal angles). For this purpose, bores are provided in the shell. Detectors, which can be introduced into these, pass the scattered light via glass fibers to photomultipliers. The output current of the photomultipliers is converted into a voltage signal by operational amplifiers and low-pass filters. These signals can then be saved on a PC for further investigation. Due to the flexible structure with regard to the positioning possibilities of the detectors and the possibilities to influence the laser beam, the system can be adapted to different measuring tasks.

By the use of multiple detectors in the 3D-LSS at different positions and suitable algorithms for data inversion, information about the 3D structure of the investigated particle can be gained online. E.g. for spherical particles the light scattered in the azimuthal domain at a scattering angle of 90° is homogenous at every position, while the azimuthal distribution of the scattered light emitted by a cylindrical particle is inhomogeneous if the incident light is circularly polarized. Thus, if the intensity of the scattered light is measured at different azimuthal angles and the variance of these values is then determined, conclusions can be drawn on the sphericity of the particles being examined **[3]**.

The design and function of the sensor system in detail as well as measurements on different non-spherical particle systems will be presented.

We gratefully acknowledge for the financial support: Deutsche Forschungsgemeinschaft (DFG, project AN782/6-2) in the framework of the joint research program â€œmulti-parameter characterization of particle-based functional materials by innovative online measurement technologyâ€ (PAK688).

[1] J. A. Champion, Y. K. Katare, S. Mitragotri; Journal of Controlled Release 121 (2007) 3-9

[2] Q. Xue; European Polymer Journal 40 (2004) 323-327

[3] H. Barthel; *PhD Thesis* (1999), TU Kaiserslautern, Kaiserslautern, Germany

507969 Importance of Cohesion Strength Measurements As Flowability Indicator of Powders in a Fluidized Bed

Abstract:

It is well known that the difference in bulk material properties such as particle size, shape and density as well as surface moisture influences the flowability of fine powders [1]. Fluidized beds are commonly treated as i) at "incipient fluidization" mode whilst the packed-bed of particles are about to "break" solid-solid surface contact and ii) in "suspension mode" whilst the particles are fully-suspended in the carrier fluid which supports their weight and is subject to fluid-solid surface interactions [2]. This study links cohesion strength with the flowability of commonly used industrial fine powders (i.e. Cement, Barite, Bentonite, Alumina and Silicon carbide). The solid like structures observed in cohesive powders can be broken by pre-consolidating and initiating shearing using an Anton Paar MCR Rheometer with Powder Cell and a fluidization set with controlled gas flow [3].

Cohesion strength is determined via shearing the sample following a protocol where the sample is compressed at low consolidations with an air-permeable piston and then a spoke-shaped geometry penetrates the sample and shears it. Warren Spring cohesion is calculated as the maximum value.

Furthermore comparisons on shear rate dependent viscosity and Brookfield Powder Flow Tester (using annular shear cell) contribute to the calculations which determine the flowability of the powders and allow the characterization of harsh cohesive powders.

[1] Valverde J.M., Castellanos A. Effect of particle size and interparticle force on the fluidization behaviour of gas fluidized beds. Phys. Rev. E 67, 051305 (2003)

[2] Rietema, K The dynamics of fine powders, Elsevier, London (1991)

[3] Salehi H, Schutz D, Romirer R, Barletta D, Poletto M. On the use of a powder rheometer to characterize the powder flowability at low consolidation with torque resistances. AIChE Journal http://dx.doi.org/10.1002/aic.15934

507965 Investigation into the Attrition Behaviour of FCC Catalysts

Abstract:

Fluidized bed operations have a key importance in fluid catalytic cracking units. Here the intensive mechanical stress causes attrition and consequently important material loss. The distributor region, the jets and the cyclone are the most common contributors to this phenomenon since attrition of friable particles is most likely to occur at impact points where the impact forces are high. Therefore, attrition can be minimized by minimizing the impact angle Î,, and minimising stream velocity.

In this work, the attrition of fluid cracking catalyst is considered in detail with emphasis on the link between single particle properties and bulk attrition behaviour, by studying two different sample types and considering particle-wall impacts. The experiments involve the bench scale centrifugal attrition tester (developed by the Wolfson Centre for Bulk Solids Handling Technology, UK) which simulates a particle-wall impact scenario. This apparatus has more recently been used for the evaluation of particle attrition by impact [1] and comprises a rotating disc surrounded by an array of targets. The particulate samples are fed into the centre of the rotating disc and radially accelerated via the tubes and impacted against the target array. Typically, as the impact velocity and impact angle increase, the level of attrition is also observed to increase. Good correlation was found at high impact velocities/gas velocities considering real conditions found in cyclones.

This paper presents a degradation model for particles in fluidized bed conditions. The model utilises a breakage matrix approach. Where the breakage matrix is empirically-obtained from the centrifugal attrition tester, and transforms the input particle size distribution in fluidizing conditions, to the output particle size distribution after the test. Particle size analysis and particle shape analysis were carried out using Sympatec and Malvern Morphologi G3 respectively.

Regarding the operational capacity of the apparatus, the centrifugal tester allows greater control of particle impact velocity; particle-gas interactions are lower than those of the gas-blast tester and steady state conditions are more readily defined as recently summarised in a review paper [3].

[1] B. Kotzur, M.S. Bradley, R.J. Berry, R.J. Farnish, Influence of Solids Loading Ratio on Particle Attrition within a New Centrifugal Accelerator Impact Tester, in: PARTEC Int. Congr. Part. Technol., 2016.

[2] B.A. Kotzur, M.S.A. Bradley, R.J. Berry, R.J. Farnish, Breakage Characteristics of Granulated Food Products for Prediction of Attrition during Lean-Phase Pneumatic Conveying, Int. J. Food Eng. 12 (2016) 835â€“850

[3] B.A. Kotzur, R.J. Berry, S. Zigan, P. Garcia-Trinanes, M.S.A. Bradley, Particle attrition mechanisms, their characterisation and application to lean phase pneumatic conveying systems: A Review. (submitted to Powder Technology 05/05/2017)

508693 Particle Tracking Velocimetry Measurements for CFD-DEM Validation

Abstract:

The coupled Eulerian-Lagrangian method known as CFD-DEM is a powerful modeling tool that provides particle-scale, detailed insight into gas-solid multiphase flow devices. However, since CFD-DEM tracks the motion of every particle in a given system, many of the validation-grade experiments conducted previously for Eulerian-Eulerian (TFM) methods are too large (particle count too high) for CFD-DEM. By validation-grade, we mean high-quality experimental data where all particle properties and flow boundary conditions relevant to the computational model have either been well-characterized or the uncertainties quantified. Some more recent works, see e.g., Gopalan et al. (2016), have collected validation-grade experimental data specifically targeted at CFD-DEM. Previous works focused on simple systems and simple particle physics, most commonly in rectangular fluidized beds with a single gas inlet, leaving a gap between the existing CFD-DEM validation database and the TFM validation database and more realistic, engineering data. We aim to help bridge this gap by providing validation-grade data of increasingly more complex problems and physics which are still amenable to CFD-DEM--both current capabilities and projected short-term future capabilities.

In this presentation, we summarize recent experiments conducted at Particulate Solid Research, Inc. and the University of Colorado which utilize particle tracking velocimetry (PTV). At one end of the Geldart spectrum, large Group D particles are studied in a semicircular fluidized bubbling bed that is operated slightly above and slightly below minimum fluidization. Additional fluidization is provided near the flat, front face of the unit via four horizontal air jets. The two pairs of jets are submerged, subsonic but high-velocity and set at the right and left hand sides of the bed directed inwards towards each other at elevations of approximately 5 and 15 cm above the distributor. High speed video (HSV) imaging is taken of the face of the semicircular bed to capture the dynamics of the opposing jets. PTV is used to collect data from the HSV images which is then averaged into bins (grid) to create 2-D time averaged solids velocity profiles and ultimately used to calculate momentum-based jet penetration depth measurements. At the other end of the Geldart spectrum, experiments have also been performed for mildly cohesive Group A and A/C particles in the dilute riser section of a circulating fluidized bed. The entrained fines tend to form agglomerates: groupings of a few particles in size held together by interparticle forces. Local regions of the riser flow can be imaged with HSV and a new PTV method has been developed that can distinguish between primary particles and agglomerates. With the new PTV method, the hydrodynamic behavior of particles and agglomerates are extracted, such as the size, concentration, mean velocity, solids volume fraction and kinetic energy associated with relative particle and agglomerate motion (e.g. granular temperature).

CFD-DEM simulations using the open source MFiX code have also been conducted for some of the experimental conditions and we will briefly provide preliminary results, both successes and challenges. In general, the numerical simulations of the air jets in the semicircular bed compare quite favorably to the experimental data. The most noticeable bias that exists is that the simulated jets tend to penetrate further into the bed than observed experimentally, perhaps a consequence of the (momentum) point source modeling of the jets. For simulations of the cohesive particles (Geldart Group A), the CFD-DEM predictions are complicated by the fact that interparticle forces need to be modeled. We briefly report on simulations which utilize a recently developed adhesion model (LaMarche et al. 2017). Corresponding with the increase in complexity of the physics of the problem, the numerical results are degraded compared to the more idealistic (spherical, monodisperese and non-cohesive) Group D particle results, although some good agreement is still observed. Notably, when the uncertainty in local superficial gas velocity is considered, the agreement in the entrainment flux improves substantially.

Gopalan, B., M. Shahnam, R. Panday, J. Tucker, F. Shaffer, L. Shadle, J. Mei, W. Rogers, C. Guenther, M. Syamlal, Measurements of pressure drop and particle velocity in a pseudo 2-D rectangular bed with Geldart Group D particles, Powder Technology, 291, 299-310 (2016).

LaMarche, C.Q., S. Leadley, P. Liu, K. M. Kellogg, C. M. Hrenya, "Method of quantifying surface roughness for accurate adhesive force predictions," Chemical Engineering Science, 158, 140-153 (2017).

508034 Experimental Verification of Solid-like and Fluid-like States in Homogeneous Fluidization Regime of Geldart a Particles

Abstract:

Experimental Verification of Solid-like and Fluid-like States in Homogeneous Fluidization Regime of Geldart A Particles

Although homogeneous fluidization for Geldart A particles has received considerable attention for a long time, the underlying mechanism has not yet been fully understood.[1] Some ascribed the stability to inter-particle forces,[2, 3] while others insisted a purely hydrodynamic explanation.[4] Valverde et al.[5] fluidized fine particles with the average size of 8.53 ¦Ìm (i.e., Geldart C particles) by the addition of fumed silica nanoparticles and found that even during homogeneous fluidization both solid-like and fluid-like behaviors can be distinguished, indicating that the stability of homogeneous fluidization may have two distinct origins: in solid-like state, the particle contact forces dominate; while in fluid-like state, the hydrodynamic factors dominate. However, it is evident that the above conclusions were drawn based on the modified Geldart C particles. To our best knowledge, there is no experimental report on whether both solid-like and fluid-like states exist for true Geldart A particles.

It should be stressed that, however, fluidization behaviors of the modified Geldart C particles and true Geldart A particles should be different from each other: besides some apparent fluidization characteristics such as the expansion ratio, maximum free volume and the ratio of the minimum bubbling velocity to the minimum fluidization velocity, the modified Geldart C particles undergo a dynamical aggregation process,[6] while for Geldart A particles, it is still questioned whether or not aggregation presents in homogeneous expansion region.[7] Homogeneous fluidization of Geldart A particles is of great relevance: for fundamental research, it is closely associated with the instability and origin of meso-scale structures in fluidized beds;[8] for industrial application, the optimal operation for particles circulation in fluidized bed reactor-regenerator systems, such as in fluid catalytic cracking (FCC) and methanol to olefins (MTO) process, is achieved when particles flow downwards a standpipe in a bubble-free condition,[9] which is actually homogeneous fluidization. Therefore, it is significant to investigate if both solid-like and fluid-like states exist in homogeneous expansion regime for true Geldart A particles as well.

In this work, incipient fluidization characteristics of FCC catalyst, which is a typical Geldart A powder, were carefully investigated by the use of electrical capacitance tomography (ECT), pressure fluctuation and camera recording. It was found that the dependency of the standard deviation of the overall solid concentration measured by ECT on superficial gas velocity shows three platforms, which were further verified by pressure fluctuation and camera recording to correspond to fixed bed, solid-like homogeneous and fluid-like homogeneous regimes, respectively, and the gas velocity in the terminal point of each platform is the minimum fluidization velocity, the minimum bubbling velocity and the critical velocity between the solid-like and fluid-like homogeneous fluidization states, respectively. Once the existence of both states is verified, it is reasonable to speculate that the mechanism of homogeneous fluidization for Geldart A particles have two distinct factors: one dominated by inter-particle forces in solid-like state and the other dominated by fluid dynamics in fluid-like state.

Figure 1. Average and standard deviation of the overall solid concentration against superficial gas velocity.

Literature Cited

1. Oke O, Lettieri P, Mazzei L. An investigation on the mechanics of homogeneous expansion in gas-fluidized beds. *Chem Eng Sci*. 2015;127:95-105.

2. Rietema K. The effect of interparticle forces on the expansion of a homogeneous gas-fluidised bed. *Chem Eng Sci*. 1973;28:1493-1497.

3. Menon N, Durian DJ. Particle motions in a gas-fluidized bed of sand. *Phys Rev Lett*. 1997;79:3407-3410.

4. Cody GD, Goldfarb DJ, Storch GV, Norris AN. Particle granular temperature in gas fluidized beds. *Powder Technol*. 1996;87:211-232.

5. Valverde JM, Castellanos A, Quintanilla MAS. Self-diffusion in a gas-fluidized bed of fine powder. *Phys Rev Lett*. 2001;86:3020-3023.

6. Valverde JM, Castellanos A. Effect of vibration on agglomerate particulate fluidization. *AIChE J*. 2006;52:1705-1714.

7. Sande PC, Ray S. Fine mesh Computational Fluid Dynamics study on gas-fluidization of Geldart A particles: Homogeneous to bubbling bed. *Ind Eng Chem Res*. 2016;55:2623-2633.

8. Sundaresan S. Instabilities in Fluidized Beds. *Annu Rev Fluid Mech*. 2003;35:63-88.

9. Xie H-Y, Geldart D. Fluidization of FCC powders in the bubble-free regime: effect of types of gases and temperature. *Powder Technol*. 1995;82:269-277.

508190 Magnetic Resonance Imaging and Modeling of Liquid-Solid Fluidization

Abstract:

Magnetic Resonance Imaging and Modeling of Liquid-Solid Fluidization

Christopher M. Boyce[1,2], Alexander Penn[2,3], Klaas P. Pruessmann[3], Christoph R. Müller[2]

[1]Department of Chemical Engineering, Columbia University

[2]Department of Mechanical and Process Engineering, ETH Zürich

[3]Institute for Biomedical Engineering, University of Zürich and ETH Zürich

Although less studied than gas-solid fluidized beds, liquid-solid fluidized beds have a variety of applications ranging from particle separation to thermal energy storage to bioreactors[1]. On a fundamental level, liquid-solid fluidized beds typically exhibit homogeneous fluidization, making them an ideal candidate for studying particle motion in a granular gas[2]. Studies of liquid-solid fluidized beds have been somewhat limited by difficulties in studying particle and liquid motion in the interior of 3D opaque systems.

Here, we use rapid magnetic resonance imaging (MRI) to study both liquid and particle dynamics inside of a 3D liquid-solid fluidized bed. MRI techniques are used to create instantaneous maps of particle and liquid velocity over time[3], as well as the distribution of particle and liquid velocities[4] in the system at the molecular level. The findings are compared with various theories for fluid and particle motions in a granular gas. Additionally, the results are compared with computational fluid dynamics – discrete element method (CFD-DEM)[5] simulations to understand the abilities and deficiencies of such a model in capturing liquid and solid dynamics in a liquid-solid fluidized bed.

References:

(1) Epstein, N. Applications of Liquid-Solid Fluidization. *Int. J. Chem. React. Eng.* **2003**, *1*, 1–16.

(2) Rouyer, F.; Menon, N. Velocity Fluctuations in a Homogeneous 2D Granular Gas in Steady State. *Phys. Rev. Lett.* **2000**, *85* (17), 3676–3679.

(3) Penn, A.; Tsuji, T.; Brunner, D. O.; Boyce, C. M.; Pruessmann, K. P.; Müller, C. R. Real-Time Probing of Granular Dynamics with Magnetic Resonance. *Sci. Adv.* **2017**, *Accepted*.

(4) Boyce, C. M.; Rice, N. P.; Ozel, A.; Davidson, J. F.; Sederman, A. J.; Gladden, L. F.; Sundaresan, S.; Dennis, J. S.; Holland, D. J. Magnetic Resonance Characterization of Coupled Gas and Particle Dynamics in a Bubbling Fluidized Bed. *Phys. Rev. Fluids* **2016**, *1* (7), 074201.

(5) Tsuji, Y.; Kawaguchi, T.; Tanaka, T. Discrete Particle Simulation of Two-Dimensional Fluidized Bed. *Powder Technol.* **1993**, *77* (1), 79–87.

508810 Quartz Crystal Microbalance As a Technique to Probe the Rheological Properties of Particulate Suspensions

Abstract:

Quartz crystal microbalance (QCM) is an ultra-sensitive measurement device which has been applied to measure the physical properties of deposited layers. These layers are often thin-film, mechanically rigid, with the deposited mass and layer thickness determined using the conventional Sauerbrey relationship [1]. For soft materials, viscous losses introduce added complexity where both the resonance frequency and dissipation/resistance are required to model the physical properties of the deposited film [1-5]. In the current research we move away from the conventional applications of QCM and assess the capability of this simple resonance technique to provide useful rheological properties of particulate suspensions.

QCM is a simple mechanical device where an AT-cut quartz sensor is driven into oscillation in the MHz range (5 MHz in the current study). The oscillatory motion of the sensor and the dampening of the sensor following excitation can provide great insight to the physical properties of deposited and/or adsorbed layers, with the technique providing nano-gram resolution [6]. Our previous work (to be discussed briefly in the current presentation) showed that the oscillatory resistance correlated to the shear yield stress of colloidal sediments. This was the first study to demonstrate that QCM can provide meaningful data when interacting with concentrated particle suspensions. The interaction potential between the sensor and suspension remained constant, and the yield stress of the suspension varied by increasing the particle concentration. The simple approach of submerging the resonating sensor into the sediment and measuring the air-to-sample resistance shift, showed that the magnitude of the resistance shift increased exponentially with increasing sediment yield strength â€" i.e. the motion of the sensor becomes increasingly constrained by the overlying sediment.

The current research explores the frequency and resistance responses of the QCM when the strength of interaction between the sensor and particle suspension is varied. The typical sensor used in the study is gold-coated and the particle type was TiO_2 (titania). Titania was chosen because it is well defined in literature, and its anatase phase exhibits an isoelectric point at pH ~6.5, hence the particles were positively charged in acidic conditions and negatively charged in basic conditions. Across the pH range the gold sensor remained negatively charged with slight dependence on the pH [7]. At a fixed solids concentration (16.2 vol%) the pH of the titania suspension was varied between pH 3 and pH 10 and the suspension shear yield stress varied between 0 Pa and ~64 Pa, with the maximum yield stress measured at the suspension isoelectric point. Submerging the QCM into the titania samples and measuring the frequency and resistance responses across the pH range, the trending responses were complex with the magnitude of the resistance peaking (+ve) at pH 4, decreasing to pH 6.0, and remaining independent of pH in basic conditions. While the suspension yield stress is changing, the response of the QCM appears to be more sensitive to the interaction potential between the sensor and the overlaying suspension. Atomic force microscopy showed that the attractive force between a titania sphere and QCM gold sensor decreased as the pH was increased from pH 4 to 6.5. Beyond this pH both surfaces were negatively charged and the overall interaction potential was purely repulsive, with the colloidal system stabilized by electrical double layer forces. As such, in basic conditions both the frequency and resistance of the QCM sensor were independent of the suspension pH, thus the response did not track the changes in suspension yield stress. This series of experiments highlights the importance of a strong (positive) interaction between the sensor and particle suspension to measure suspension rheological properties. However, it does also highlight the ability of QCM to measure interaction forces between colloidal particles and the sensor.

Finally, the gold-coated sensors were coated with a 200 nm thick layer of titania. The sensor resistance showed a peak value at the suspension isoelectric point (pH ~6.5), indicating strong interaction between the suspension and sensor, with the resistance reducing as the pH became slightly more acidic and basic. Such response remains qualitatively consistent with the change in sediment yield stress, although the work remains ongoing to fully understand the QCM response.

The current research will for the first time present the application of QCM to probe the colloidal forces and rheology of concentrated particulate suspensions. Through further development and fundamental understanding of the QCM signal, the technique can be potentially applied to act as a simple QC rheometer.

References

1. Sauerbrey, G., *Verwendung Von Schwingquarzen Zur Wagung Dunner Schichten Und Zur Mikrowagung.* Zeitschrift Fur Physik, 1959. **155**(2): p. 206-222.

2. Kanazawa, K.K. and J.G. Gordon, II, *Frequency of a quartz microbalance in contact with liquid.* Analytical Chemistry, 1985. **57**(8): p. 1770-1.

3. Denolf, G.C., et al., *High frequency rheometry of viscoelastic coatings with the quartz crystal microbalance.* Langmuir : the ACS journal of surfaces and colloids. **27**(16): p. 9873-9.

4. Martin, S.J., V.E. Granstaff, and G.C. Frye, *Characterization of a Quartz Crystal Microbalance with Simultaneous Mass and Liquid Loading.* Analytical Chemistry, 1991. **63**(20): p. 2272-2281.

5. Johannsmann, D., *Viscoelastic, mechanical, and dielectric measurements on complex samples with the quartz crystal microbalance.* Physical Chemistry Chemical Physics, 2008. **10**(31): p. 4516-4534.

6. O'Sullivan, C.K. and G.G. Guilbault, *Commercial quartz crystal microbalances - theory and applications.* Biosensors & Bioelectronics, 1999. **14**(8-9): p. 663-670.

7. Sylvestre, J.P., et al., *Surface chemistry of gold nanoparticles produced by laser ablation in aqueous media.* Journal of Physical Chemistry B, 2004. **108**(43): p. 16864-16869.

507772 Use of Acoustic Backscatter Systems to Characterise Concentrated Dispersions

Abstract:

The chemicals industry is increasingly looking for modern instrumentation for process monitoring and control. For the production of particulate suspensions and colloids (from high value products such as paint, to minerals processing and water treatment) the complexity of these multiphase systems significantly limits the applicability of many types of online measurement equipment. One technique with wide potential application however is pulse-echo ultrasonics. This presentation is focused in particular, on understanding the use of acoustic backscatter systems (ABS) as a cost effective *in situ*/online technique for measuring dispersion dynamics in real time, including solid-liquid separation and mixing operations. Critically, backscatter transducers offer a flexible solution capable of remote operation, and can measure suspensions at high concentrations. They operate by recieving the backscatter voltage response of a high frequency (1 â€“ 5 MHz) echo pulse, where the scattering and attenuation of particles can quantify important system parameters such as concentration and particle size. Here we develop a new calibration method to gain attenuation and scattering constants for any arbitrary particle type from large micron size to just within the colloidal range. By enabling the measurement of these constants, we are able to use a range of fully resolved and semi-empirical theories of acoustic scatter to characterize different systems. Additionally, the effect of particle aggregation, especially in the measurement of mineral wastes, has been analyzed in detail, with comparison to the fractal dimension of aggregates. This technique enables real time measurement of a range of important suspension process systems, including wet-milling, batch-scale solid-liquid separation, large-scale clarifier trials and liquid jet mixers, as part of nuclear waste processing operations (e.g. Bux *et al.* AIChE Journal, 63(7), p2618â€“2629 (2017)).

507805 Classification and Regression Modelling for Investigating the Effect of Particle Size and Morphology on the Functionality of Industrial Spray Dried Milk Powder

Abstract:

The rehydration of instant whole milk powder (IWMP) is a complex process and is quantified using a a variety of offline tests, all of which have poor repeatability, are labour intensive, and can only be carried out post production. Consequently the ability to infer the rehydration properties of the powder whilst manufacturing would be an invaluable tool, potentially enabling corrective measures to be taken during production. However, the evaluation of alternative measurement techniques, and their subsequent correlation with subjective rehydration tests forms a considerable challenge. This work compares partial least squares regression (PLS) and classification trees, a classification technique, for determining if particle size and morphology can predict two IWMP functional rehydration tests; namely dispersibility and slowly dissolving particles (SDPs).

Dispersibility measures how quickly IWMP breaks up into individual particles and is measured by dissolving a specific amount of powder in water for a short amount of time, before being passed through a mesh, with more residue (or total solids of the reconstituted milk) indicating poor performance. SDPs on the other hand are measured by rehydrating milk, pouring it in and out of a test tube and monitoring the amount of undissolved white particles on the surface of the tube. Although the particle size is implicated in both, both tests are subjective and are sensitive to the measurement conditions, such as the stirring speed, time and temperature. Both tests infer important characteristics of bulk powder behaviour.

Particle morphology on the other hand measures individual particle characteristics, and it is proposed that morphology can be used as a proxy variable for the dispersibility and SDPs. Particle morphology includes circularity, convexity, solidity, minimum and maximum Feret diameters and elongation. Different particle morphology distributions characteristics assumed to be determined by the drying conditions and agglomeration in the dryer. For example, it is hypothesised that convexity and solidity would be an indicator for the â€˜loosenessâ€™ of the agglomerate shape, and hence how quickly it can disperse in the water.

In order to investigate the effect of morphology on dispersibility and SDPs, 35 milk powder samples were collected in-situ from a large scale industrial spray dryer. The dispersibility and SDPs were measured at the industry laboratory, and the particle morphology was measured using light microscopy and image processing. The morphology characteristics were extracted for all particles larger than 100 Î¼m, as these particles were considered to be best indicator of the agglomeration state of the powder. The distribution for each parameter was characterised using the mean, median, standard deviation, spread, and the 10th and 90th percentiles.

As every particle is slightly different, the result of each morphological characteristic measurement is a distribution, as opposed to a single value. This makes the elucidation of any relationship with dispersibility or SDPs more complex. The first method treated the problem by assuming that the distribution data can be treated similarly to spectral data, such as from a near infra-red (NIR) instrument, and then correlating it against the quality parameters using PLS. However, given the nature of the tests the data for SDPs and dispersibility are highly discretised. Therefore, random forests, a classification technique was also investigated for modelling the data, to determine if this is a more appropriate method. Because there were 35 total samples compared with 36 morphological predictor variables, bootstrapping was chosen to validate whether the models had predictive ability. The models were re-run multiple times with a random re-selection of testing data to determine their robustness.

For SDPs, the mean square error (MSE) varied significantly for PLS, between 15-50 % using the validation data set and 10 principal components. On the other hand, random forest classification modelling was more consistent, with re-runs reducing the root mean square error going down to 15 % within 10 trees. Dispersibility models showed slightly worse performance, with higher MSE for both PLS and random forests. However, random forest again produced more consistent models with multiple model re-runs.

PLS is a linear regression technique, often applied for spectral absorbance data. For quality parameters such as concentration of a substance, an increase in absorbance at particular wavelengths with an increase in concentration is often reasonable. However, there is no clear reason to assume that a worsening in dispersibility would change linearly with particular morphology characteristics. The physical parameters that influence the functional tests are not fully understood, and it is possible that the mechanisms that give good, moderate and poor dispersibility or SDP results are entirely different in their nature. Given that classification trees does not require this assumptions, they were chosen for interpretation. They also gave more consistent model results.

The size and morphological distribution characteristics that came up with the most importance to the model for dispersibility were mass of particles smaller than 125 μm, the solidity distribution standard deviation and 90th percentile, and the 90th percentiles of circularity. It is known that the amount of fine particles affects the dispersibility of the powder, as they aggregate and prevent further water from penetrating into the solid. It was hypothesised that the solidity and circularity would be representative of the agglomeration state of the dryer, and would thus be predictive variables. They are a measure of how open the agglomerate is, and the distribution describes how many open agglomerates there are. Therefore, it may be possible to use this to assess where in the spray dryer the agglomeration is occurring.

The morphological distribution characteristics that were found to be the most important for SDPs were similar to those of dispersibility, with fraction of fine particles being important, but also coarse ones, larger than 450 μm. The fraction of coarse particles may be important as they dissolve more slowly simply due to size. Both solidity and circularity were important, similar to for dispersibility, the solidity and circularity could be indicative of drying history and agglomeration conditions.

The standard functional tests being studied are complex, thus it is important to select the appropriate modelling technique to extract useful information, as certain assumptions regarding linearity may not be valid. This is especially true when the range of functional performance narrow, with only a small number of the samples being off-specification. Classification trees were investigated for their ability to model discretised outputs where linearity cannot be assumed, as required for standard PLS regression. Solidity, circularity and size were found to be correlated with SDPs and dispersibility, and are expected to indicate the agglomeration happening in the spray dryer. This data shows that morphology and size distributions have strong potential for inferring powder functionality and spray dryer operation.

508182 Electrochemical Characteristic Evaluation of Lithium-Ion Battery Slurry By a Multi-Sensing Electrochemical Impedance Spectroscopy System

Abstract:

Lithium-ion battery (LIB) is widely used in electronic equipment on account of their high energy density, no memory effect, relatively good safety and cycling-life performance. The performance of LIB is affected mainly by the electrochemical characteristic of the LIB slurry
. However, in the previous research, most of the researchers focused on the performance of the anode electrodes instead of the slurry. Therefore, the main influenced factors for the performance of LIB need to be more considered during the slurry stage.
In the present study, we aimed to evaluate the electrochemical characteristic of Lithium-ion battery slurry by using of a Multi-sensing Electrochemical impedance spectroscopy system. The Multi-sensing EIS system is composed of a main board with a field programmable gate array (FPGA), a switching module and an eight electrode EIS sensor. Compare to the conventional used impedance analyzer for EIS measurement, the developed system achieved agile and compact feature for the real-time measurement of impedance spectroscopy. Instead of conventional two electrodes method, an eight electrode EIS sensor was used. The eight electrodes are attached at equal intervals on the outside of a circular glass beaker. Measurement is performed between various combinations of electrodes including adjacent and opposing electrodes or between opposing electrodes to obtain more comprehensive measurement information of the measurement domain.
By using the Multi-sensing EIS system, a Nyquist diagram between the respective electrodes is obtained. The electrode combination dependence on the measured Nyquist diagram is discussed firstly. The electrodes independent EIS data was analyzed deeply to evaluate the electrochemical characteristic of Lithium-ion battery slurry. Eventually, validation of the proposed multi-sensing EIS system was also conducted to clarify its capability to realize the online measurement of electrochemical characterizations for LIB slurry.

508262 Influence of Shearing As Pretreatment on the Separation Properties of Exopolysaccharides (EPS) Producing Bacterial Starter Cultures

Abstract:

BACKGROUND: Bacterial starter cultures such as lactic acid bacteria (LAB) are essential for the manufacture of fermented foods. The industrial production of these starter cultures takes place at specific conditions regarding, for example, temperature, pH and nutrient composition. Due to the diversity of highly specialized starter cultures, an individual adaption of the fermentation process is necessary. However, changing fermentation parameters also leads to altered separation properties that show a large influence on the subsequent separation step.

OBJECTIVE: Aim of this project is to investigate the influence of shear energy input on the sedimentation properties of different EPS producing lactic acid bacteria. Furthermore, a guideline for the determination of the optimal pretreatment and centrifugal properties of different EPS-producing starter cultures should be developed.

METHODS: Shear forces are generated either with a rotor-stator system (Ultra-Turrax) or with a high pressure homogenizer (Emulsiflex C5). Shear energy is applied to the samples by a fast stirring rotor or by pressing the samples through a cross-section constriction, respectively. Sedimentation velocity and sediment compression are measured with an analytical centrifuge (LUMisizer), and particle size is measured with a disc centrifuge.

RESULTS: In case of bacteria with cell-bound (capsulare) exopolysaccharides (EPS) that could be sheared off the cells, shearing of the cultures leads to a higher sedimentation velocity. Furthermore, a decrease of the sedimentation velocity is induced when cell chains were disrupted because of high shear forces. Hence, there exists an optimum shear treatment for each culture that depends on the type and amount of produced EPS. Shearing in a high pressure homogenizer affected sedimentation velocity in a similar manner as did shearing with the Ultra-Turrax, but at a significantly lower energy input. Shearing with both shear devices leads to decreasing sediment heights of the examined cultures.

CONCLUSIONS: It was shown that, depending on the type and amount of the strain specific EPS, shearing EPS producing bacterial starter cultures may result in an increase or in a decrease of the sedimentation velocity. An increase in sedimentation velocity is mainly the result of shearing off capsular EPS from cell surface, whereas a decreased sedimentation velocity indicates a destruction of cell-chains due to shearing. Furthermore, sediment compression could be detected after shear treatment of EPS producing (capsular and free in the medium) LAB, which is caused by shearing of capsular EPS respectively disrupting cell-chains in case of free EPS producing LAB.

This IGF Project 18605 BG of the FEI is supported via AiF within the programme for promoting the Industrial Collective Research (IGF) of the German Ministry of Economic Affairs and Energy (BMWi), based on a resolution of the German Parliament.

508707 The Effect of Surface Chemistry on the Surface Energetics of Homogenous and Heterogeneous Physical Mixtures of Mannitol

Abstract:

The surface energy plays an important role in guiding number of phenomenon like wettability, cohesion/adhesion, flowability etc. Inverse gas chromatography has emerged as a useful tool for surface energy determination, and more recently, the surface energy heterogeneity of the solids using finite dilution inverse gas chromatography (FD-IGC) measurements. Notionally for mixtures of solids, this technique should probe the high energy domains preferentially while ignoring lower energy domains, however practically it appears that the problems facing the relative heterogeneities of materials or mixtures are not simple. The first part of this work involves the investigations on the mixtures produced by taking the weighted fractions of the homogenous (surface modified) and heterogeneous components. In the second part, investigations on the blends of homogenous components with different chemical surface functionalities was carried out. Different surface chemistries for the mannitol samples was obtained by the process of silanisation of surfaces with alkyl- and haloalkyl- groups to contain specific end functional groups. Mixtures of homogeneous materials was found to yield similarly homogeneous distributions, however the value was found to be an average of the constituent components. Homogenous binary mixtures containing components with different chemical surface functionalities were found to restore a heterogeneous effect in surface energy. A 50:50 weighting of the mixture was found to vary in the range of 28-32 mJ/m^2. Increasing haloalkyl fraction yielded decreasing energetics which was in the range of 26-28 mJ/m^2.

Furthermore, a novel method for measurement of energetic heterogeneity via modelling finite dilution IGC measurements using a Boltzmann distribution model and simulation fitting approach was developed. The surface energy distributions was determined, using a multi-solvent system site filling. The modelling was able to predict the surface energy distribution of the homogeneous as well as heterogeneous binary mixtures.

507201 Flowability Assessment of Weakly Consolidated Powders

Abstract:

Flowability Assessment of Weakly Consolidated Powders

Alexandros Georgios Stavrou[1], Colin Hare[1], Ali Hassanpour[2], and Chuan-Yu Wu[1]

[1] Department of Chemical and Process Engineering, University of Surrey, Guildford, GU2 7XH, UK.
a.stavrou@surrey.ac.uk

[2] School of Chemical and Process Engineering, University of Leeds, Leeds, LS2 9JT, UK

The inability of cohesive powders to flow consistently and reliably is one of the most dominant problems across the powder industries. Numerous techniques have been established to study granular flow, mainly measuring the onset of flow at high consolidation pressures to simulate large scale processes, which is relevant to hopper design. In contrast, many industries, e.g. pharmaceutical, involve processes that expose powders to low stress levels (< 1 kPa), such as dry powder inhalers, feeders for packing and tabletting machines, and filling and dosing of powders in capsules. In addition it is important to understand powder flowability at high strain rates to optimise dynamic processes, such as packaging, conveying and mixing. Ball indentation is a new technique that can be used to study these difficult facets of bulk solids flow [1]. In this technique a consolidated powder bed is penetrated by a spherical indenter. The flowability is determined by calculating hardness from the force-displacement response of the indenter. Hardness can be linked to the unconfined yield strength, which is derived by the commonly used shear testers, via the constraint factor [2].

Indentation tests were carried out at low (0.1-1 kPa) and high stresses, and shear cell measurements at high stresses, using a series of materials. The technique exhibits very good reliability. Some of the materials exhibited a linear increase of hardness with the increase of pre-consolidation stress, whereas others showed a more pronounced increase in the region of low stresses. The constraint factor of all the materials was found to be almost constant at high stresses. By inferring the unconfined yield strength at low stresses from ball indentation measurements, for some materials it was found that shear cell extrapolations led to an overestimation of cohesion. By broadening the range of experimentally tested materials it will be evaluated how single particle shape, agglomerate shape, and surface energy influence the constraint factor, while the constraint factor behaviour at low stresses will be studied using DEM.

References

[1] Hassanpour, A. & Ghadiri, M., 2007, *Particle and Particle Systems Characterization*, **24(2)**, 117â€"123.

[2] Zafar, U., Hare, C., Hassanpour, A., Ghadiri, M., 2017, *Powder Technology*, **310**, 300-306.

FLOWABILITY OF DRY SPENT COFFEE GROUND (SCG) POWDERS

Lucas Massaro Sousa and Maria do Carmo Ferreira
Chemical Engineering Department, Federal University of São Carlos, São Carlos, Brazil
E-mail addresses: lucas.massarosousa@gmail.com AND mariaf@ufscar.br

Abstract

Although the spent coffee grounds are used daily to produce renewable energy in the soluble coffee industry, there is a lack of information concerning their bulk properties and flowability behavior in the literature. The knowledge of these properties can assist the handling of SCGs in several unit operations. Here we measured the bulk densities and the Hausner ratio (HR) of several SCG mixtures with mean sizes from 225 to 550 μm and verified whether the Densification Model and the Modified Linear-Mixture Packing Model can be applied to predict the bulk properties of the mixtures. The HR of the SCGs is found to be overwhelmingly driven by the mass fraction of fines in the mixture, and the flowability classification varied from *fair* to *very poor*. The Densification Model predicted well the experimental packing behavior, with physically consistent fitting parameters. The Modified Linear-Mixture Packing Model predicted well the loose and tapped bed porosities, with mean deviations less than 2%. Therefore, the knowledge of the experimental data reported here, as well as using the models industrially can assist handling of SCG powders.

Introduction

In 2016, more than 9 million tons of coffee beans were produced worldwide [1], generating a revenue around US$ 25 billion [2]. Fruit and beans are processed to produce coffee powders that can be forwarded to soluble coffee industries (SCI) or for direct use in beverage preparation contributing to produce a significant amount of Spent Coffee Ground (SCG) powders. The SCGs are the most abundant coffee by-product with proven applications in composting, animal feeding, recovering of antioxidant components and as a biomass source for production of renewable energy by gasification, combustion or pyrolysis. In 2016, the SCI alone generated around 2.5 million ton of SCG readily available for industrial processing [3-4].

In most processes, SCG is submitted to feeding, storage, compaction and transporting operations whose success is highly dependent on the powders' bulk properties and flowability. Because the SCG powders come from quite different and uncontrolled sources, they usually present variable mean sizes, multimodal particle size distributions (PSDs) and wide size span, making the prediction of their flowability behavior a challenging task. Using non-mechanical valves to feed dry SCG powders in a pilot scale circulating fluidized bed reactor is currently under investigation in our research group, motivated by its appealing potential for renewable energy production. During cold flow tests in a non-mechanical L-valve, it was observed that feeding stability and mass fluxes depend significantly on the mean size and flowability features of the SCG samples.

This study is aimed at evaluating the bulk densities and the Hausner ratio of SCG powders of different PSDs and in a broad range of sizes, as well as testing the Densification Model and the Linear-Mixture Packing Model to predict the bulk properties of the mixtures.

Methodology

The SCG was obtained after brewing Brazilian ground coffee (brand Três Corações), acquired in São Carlos-SP, Brazil. To reduce the powder's moisture content, 5 mm thick layer samples were oven-dried at 105±2°C for 24h and stored under ambient temperature until further use. The particle size analysis of the original dried sample (X=0.06 dry basis) is presented in Figure 1 and was performed by sieving according to the ASTM standards [5], in triplicate assays.

The powder mass fraction retained in the sieves of size aperture above 600 μm, which was only 20% of the total, was discarded and the remaining portion was sieved again from 600 to 150 μm. This portion was separated into three base samples denominated as A, B and C. Sample A is made up of particles with sizes between 600 and 500 μm (d_{svA}=550 μm), sample B contains the particles retained between 500 and 300 μm (d_{svB}=400 μm), and sample C those retained between 300 and 150 μm (d_{svC}=225 μm). Additional mixtures were produced by combining mass fractions (y) of 20, 40, 60 and 80% of the base samples, resulting in 12 binary and 6 ternary mixtures. Tests were carried out in triplicate for the base powders, binary and ternary mixtures, therefore 63 dry samples were prepared and analyzed.

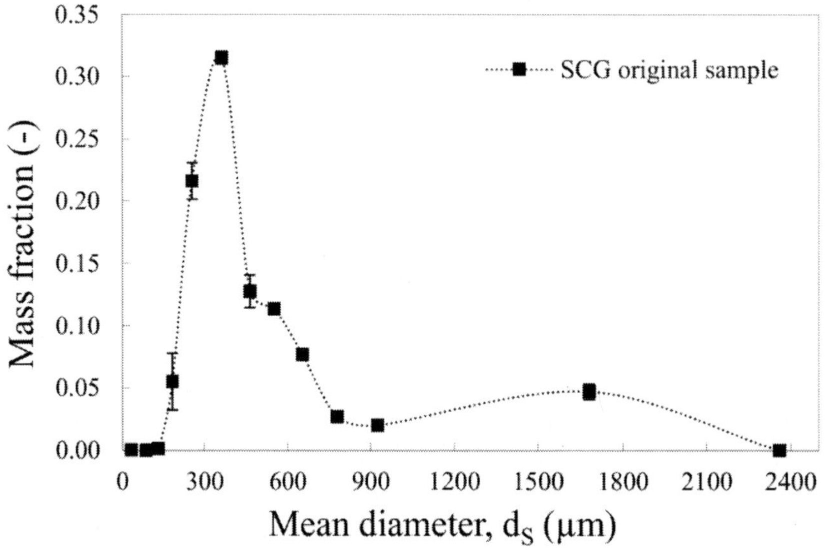

Figure 1 – Particle-size distribution of the SCG original sample

The bulk densities of the SCGs were assessed through the QAS/11450 standard procedure [6], in which 50-70g of powder samples were poured through a funnel into a graduated glass cylinder. The cylinder was placed into a proper tapping device [7] that allowed the vessel to be successively lifted at a height of 3 cm and released. The powder volume was periodically recorded and the loose (ρ_{lb}) and tapped bulk densities (ρ_{tb}) were measured, respectively, for 0 and 1250 taps. The Hausner ratio (HR) was calculated for each sample through Equation 1. Loose and tapped bed porosities (ε_{lt} and ε_{bt}) were calculated through Equation 2, using the true density of the SCG (ρ_p) measured using helium. The mean diameter of the samples was calculated through Equation 3, in which y_i is the mass fraction of the base sample (A, B or C) in the mixture and d_{svi} is the sieve diameter of the base sample.

$$HR = {}^{\rho_{tb}}/_{\rho_{lb}} \quad (1)$$

$$\varepsilon = 1 - \rho_b / \rho_p \quad (2)$$

$$d_S = \frac{1}{\Sigma(y_i/d_{svi})} \quad (3)$$

The Densification Model proposed by [8] was applied to predict the packing behavior by tapping of the SCG mixtures. The model is described by Equations 4 and 5. According to this model, N_1 represents the number of taps at which 63% of the large pores or voids have been eliminated from the packed-bed, while N_2 represents the number of taps at which 63% of the small pores were filled by the particles in the bed. Finally, the parameter f represents the densification fraction that is achieved by the elimination of the large pores from the bed.

$$\frac{\xi}{\xi_{max}} = \left[1 - f.e^{\left(-\frac{N}{N_1}\right)} + (1-f).e^{\left(-\frac{N}{N_2}\right)}\right] \quad (4)$$

$$\frac{\xi}{\xi_{max}} = \frac{V_{p0}-V_p}{V_{p0}-V_{p\infty}} \quad (5)$$

The Modified Linear-Packing Model proposed by [9] was applied to predict the loose and tapped bed porosities of the SCGs mixtures. The bed porosity (ε) was calculated by Equation 6, in which V is the specific volume of the powder mixture. The variable V was obtained by Equations 7 to 10, in which V_A, V_B, and V_C were calculated by Equation 6 using the experimental values of loose or tapped porosities (Table 1), according to the packing state to be determined.

$$\varepsilon = (V-1)/V \quad (6)$$

$$V = \max\{V_A^T, V_B^T, V_C^T\} \quad (7)$$

$$V_A^T = V_A.y_A + V_B.(1 - f(R_{BA})).y_B + V_C.[1 - f(R_{CA})].y_c \quad (8)$$

$$V_B^T = [V_A - (V_A - 1).g(R_{BA})].y_A + V_B.y_B + V_C.[1 - f(R_{CB})].y_C \quad (9)$$

$$V_C^T = [V_A - (V_A - 1).g(R_{CA})].y_A + [V_B - (V_B - 1).g(R_{CB})].y_B + V_C.y_C \quad (10)$$

The small-to-large ratios of the mixtures (R_{ij}) were calculated from the mean sizes of A, B and C ($d_{svA}>d_{svB}>d_{svC}$), and yielded the values R_{BA}=0.73, R_{CB}=0.56, and R_{CA}=0.41. The interaction functions f(Rij) and g(Rij), that depend only on Rij, were determined by Equations 11 and 12, recommended for spherical and non-spherical particles [11].

$$f(R_{ij}) = (1 - R_{ij})^{3.33} + 2.81.R_{ij}.(1 - R_{ij})^{2.77} \quad (11)$$

$$g(R_{ij}) = (1 - R_{ij})^{1.97} + 0.36.R_{ij}.(1 - R_{ij})^{3.67} \quad (12)$$

Results and Discussion

The main physical properties of the base samples are summarized in Table 1. The true density of particles is 1315±3 kg/m^3. Based on their mean size and density, the base powders A and B are classified as Geldart B (*coarse powder*) and powder C is classified as Geldart A (*fine powder*) [11]. The loose and tapped bulk densities of the A and B powders were not significantly different considering the measurements' uncertainties, while the bulk densities of C were lower than those of A and B powders, which is probably due to the high intensity of van der Waals forces and aggregates formation, which is typical of fine powders [12-13].

Table 1 - Sieve mean diameter, moisture content, true density, loose and tapped bulk densities and porosities for the base powders.

SCG sample	d_{sv} (μm)	Moisture content (% w.b.)	ρ_{lb} (kg/m³)	ρ_{tb} (kg/m³)	ε_{lb} (-)	ε_{tb} (-)
A	550	5.4 ± 0.1^a	390 ± 20^a	471 ± 7^a	0.700 ± 0.010^a	0.642 ± 0.007^a
B	400	2.8 ± 0.1^b	380 ± 20^a	460 ± 10^a	0.710 ± 0.010^a	0.650 ± 0.010^a
C	225	3.2 ± 0.1^c	262 ± 2^b	393 ± 7^b	0.801 ± 0.001^b	0.701 ± 0.006^b

Values with different letters in the same column are significantly different at a 0.05 significance level.

The packing behavior of the AB, AC and BC binary mixtures is presented in Figure 2. For all the samples, a marked change in the bulk density is observed after a low number of taps (N<10) and a density practically constant is achieved for a number of taps between 100 and 300, depending on the SCG sample. These results are in accordance with the literature, as the packing by tapping is governed by two main mechanisms: the occupation of large pores (voids much larger than the particle size) and the rearrangement of particles to fill the small pores (of a size similar to the particle size) [8]. At the beginning of the packing process, the occupation mechanism is predominant, increasing the bulk density of the powder after a small number of taps. After elimination of the large pores, the particles are rearranged to fill the small voids, consequently the bulk density increases at a slower rate.

a)

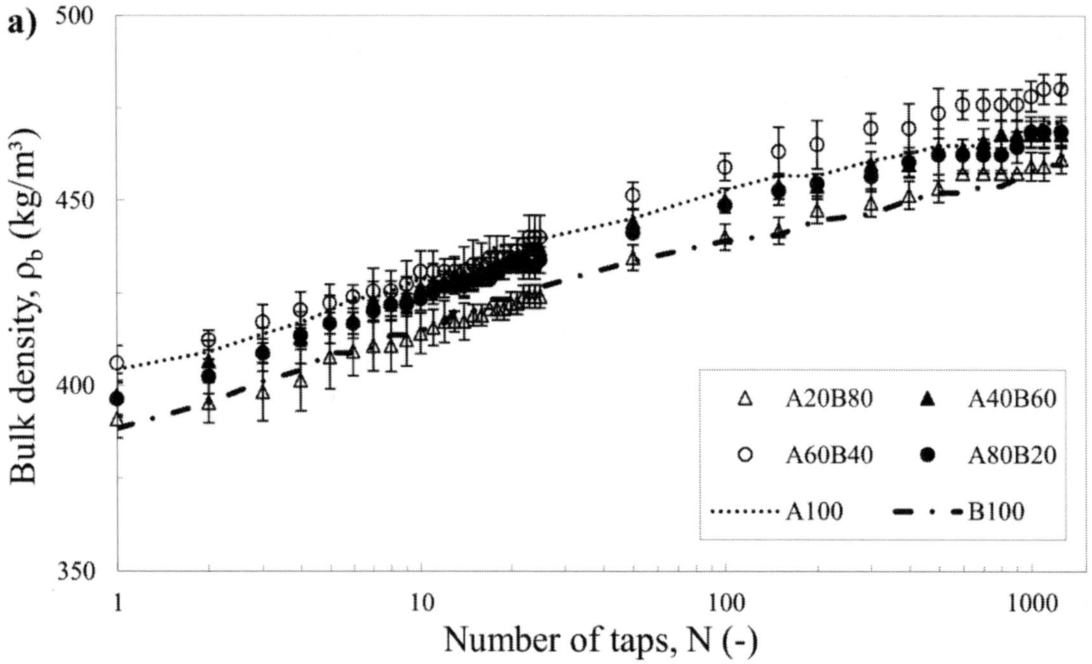

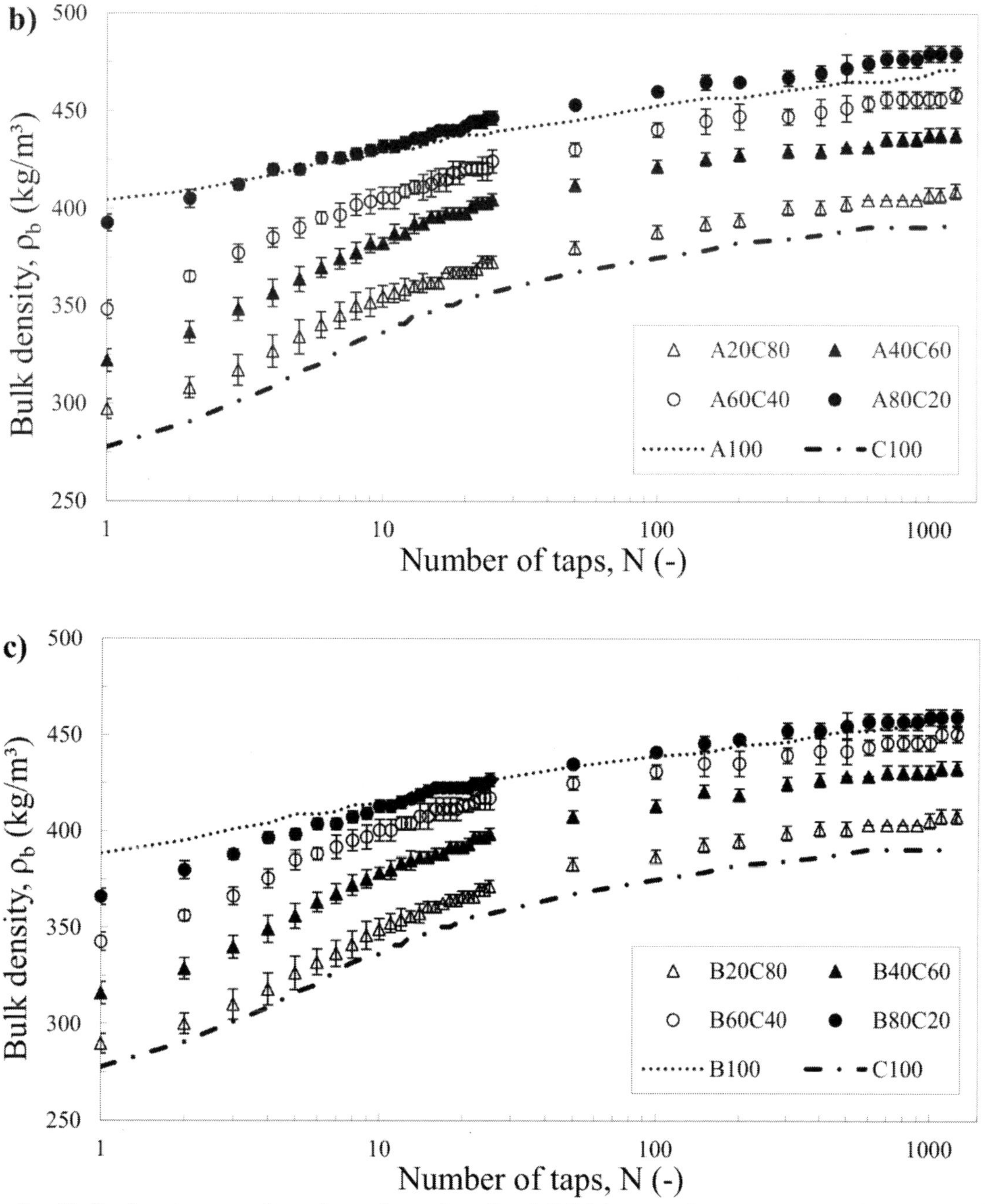

Figure 2 – Bulk density as a function of tapping for SCG binary mixtures: a) AB, b) AC, and c) BC.

It is worth noting that all mixture curves are limited by the values of densities observed for the base samples. Moreover, the addition of 20, 40, 60 and 80% of finer particles to samples A and B led to a consistent differentiation of the packing curves, while the AB curves were not significantly different considering the measurements' uncertainties. According to the literature [9], for size ratios >0.154, as in this case (R_{BA}=0.73, R_{CB}=0.56, and R_{CA}=0.41), the packing of the binary and ternary powders occurs mainly by the occupation mechanism. Although under this mechanism the packing is controlled by all the components in the mixture, it is noticeable that it is most influenced by the presence of sample C.

This is due to the similar bulk densities of samples A and B. Besides, the higher is the amount of C in the mixture, the closer is the mixture bulk density to that of the sample C.

The fitting parameters of the Densification Model are presented in Table 2 for each SCG sample. The model predicted the packing curves behavior for all the SCG samples and the values of parameters N_1 and N_2 are consistent to those experimentally reported in the second paragraph of this section. A correlation between the model parameter f and the maximum degree of densification (ξ_{max}) with the mean diameter of the mixtures is presented in Figure 3. A decrease in the mean diameter leads to larger differences between the aerated and compacted bulk volumes (i.e. to higher values of ξ_{max}), and to a larger contribution of the rearrangement mechanism to compaction (i.e., to lower values of f). This behavior is physically consistent, as the reduction of d_S allow particles to fill voids with smaller sizes throughout tapping.

Table 2 – Fitting parameters of the Densification Model, mean size and ξ_{max} for the SCG samples

SCG sample	**d_S (μm)**	**f (-)**	**N_1 (-)**	**N_2 (-)**	**ξ_{max} (-)**
A_{100}	550.0	0.59	4.4	201.6	0.17
B_{100}	400.0	0.59	4.1	237.3	0.18
C_{100}	225.0	0.74	4.5	113.1	0.33
$A_{80}B_{20}$	511.6	0.57	3.4	175.8	0.19
$A_{60}B_{40}$	478.3	0.48	2.4	119.2	0.19
$A_{40}B_{60}$	449.0	0.62	2.8	141.4	0.20
$A_{20}B_{80}$	423.1	0.55	3.4	172.9	0.19
$B_{80}C_{20}$	346.2	0.67	3.6	123.2	0.24
$B_{60}C_{40}$	305.1	0.75	3.3	155.1	0.30
$B_{40}C_{60}$	272.7	0.74	4.6	129.3	0.30
$B_{20}C_{80}$	246.6	0.74	4.6	126.6	0.34
$A_{80}C_{20}$	426.7	0.65	3.1	150.4	0.22
$A_{60}C_{40}$	348.6	0.70	3.6	112.5	0.28
$A_{40}C_{60}$	294.6	0.74	3.9	104.1	0.31
$A_{20}C_{80}$	255.2	0.73	5.0	172.9	0.30
$A_{20}B_{60}C_{20}$	363.3	0.57	3.4	150.9	0.23
$A_{40}B_{40}C_{20}$	382.2	0.64	3.0	137.8	0.25
$A_{60}B_{20}C_{20}$	403.3	0.64	3.1	120.2	0.23
$A_{20}B_{40}C_{40}$	318.3	0.69	3.8	163.5	0.28
$A_{40}B_{20}C_{40}$	332.8	0.69	3.5	165.8	0.29
$A_{20}B_{20}C_{60}$	283.3	0.74	4.1	132.0	0.31

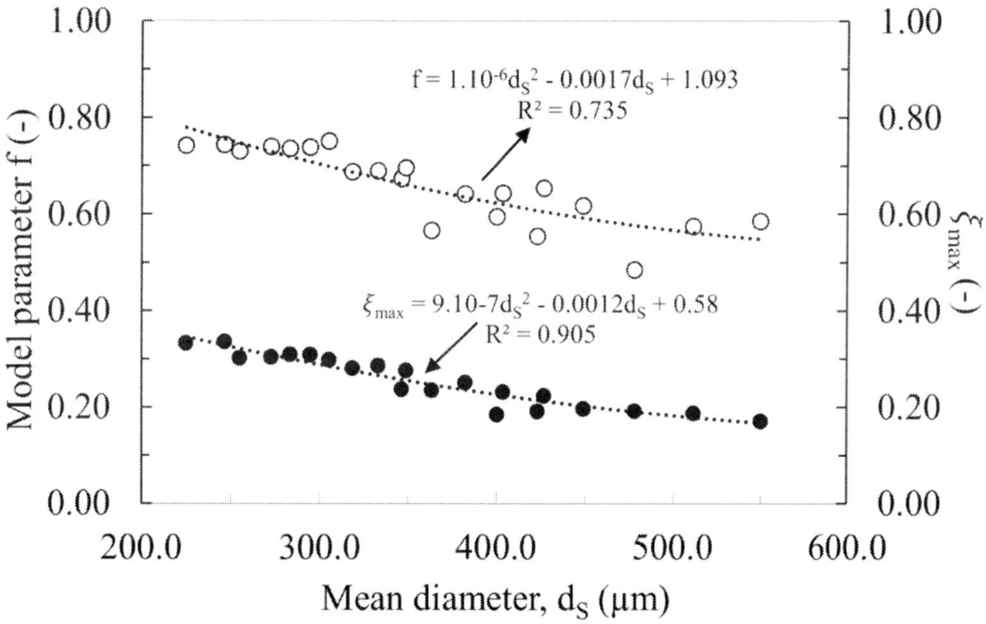

Figure 3 – Model parameter f and ξ_{max} as a function of the SCG mixture mean diameter

According to Figure 2, the bulk densities of the binary and ternary mixtures are in between the values for the largest (A) and finest (C) samples. Therefore, the Modified Linear-Packing Model [9] was applied to predict the porosity of the packed-beds of variable composition. Figure 4 shows a good agreement between the experimental loose and tapped porosities with that predicted by the model, with mean deviations less than 2%.

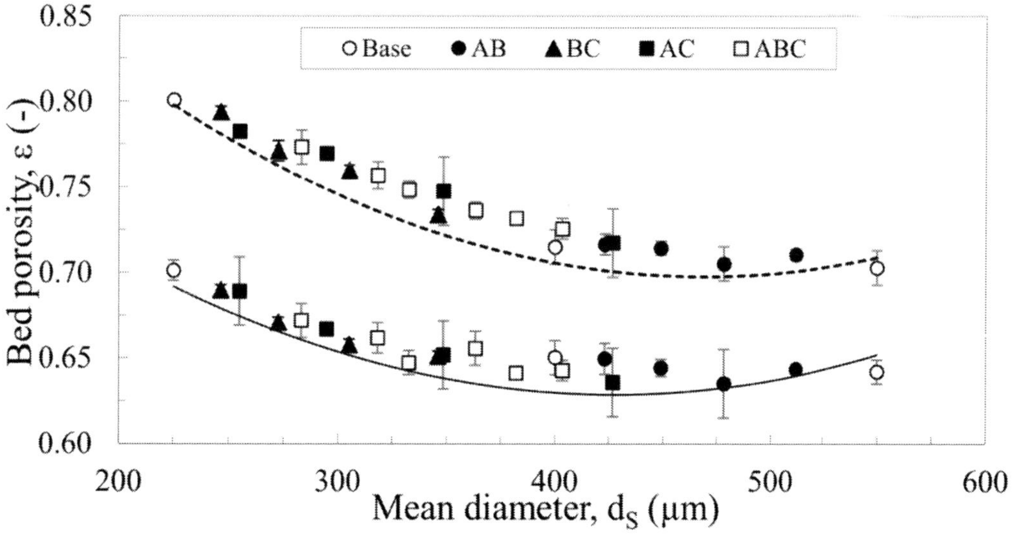

Figure 4 - Porosities as a function of mean diameter for the SCG base samples, binary and ternary mixtures: —, tapped and --- loose porosities predicted by the Modified Linear-Packing model.

The loose and tapped predicted porosities were used in Equation 2 to calculate the bulk densities of powders and HR was obtained for the binary and ternary mixtures from Equation 1. The predicted (solid lines) and measured (symbols) values of HR are presented in Figure 5 as a ternary diagram, where each vertex corresponds to a pure base sample (A, B or C) and the edges to mixtures with mass fractions varying from 0 to 100%.

Based on the HR results shown in Figure 5, it can be concluded that powder flowability worsens as the mass fraction of powder C is raised. Therefore, powders with $y_c=0$ have flowability categorized as fair; with $y_c=0.2$ the flowability as passable; in the interval $0.4\leq y_c\leq 0.6$ as poor and with $y_c\geq 0.8$ as very poor. For $y_c>0.45$ HR increases linearly with the increase in y_c, indicating that powder flowability is governed by the finest particles and the presence of coarse particles did not alter the flowability.

The influence of the moisture content on the packing behavior and Hausner ratio, and other flowability indexes for the SCGs (repose angle and flow function) are under investigation and will be approached in a future paper.

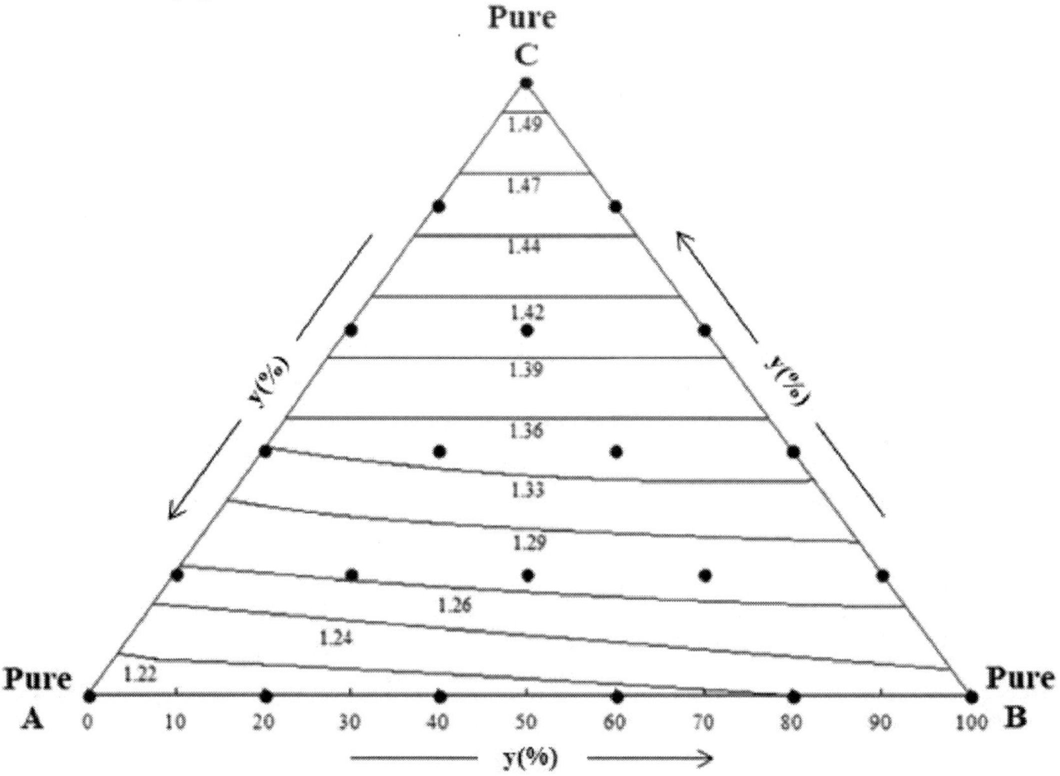

Figure 5 - HR values for SCG powders as a function of base samples composition (y): •, HR measured points; —, predicted by the Modified Linear-Packing model.

Conclusions

This study presented experimental data of bulk densities and HR of non-consolidated SCG powder mixtures prepared from combinations of three base samples with mean sizes from 225 to 550 μm. The measured HR showed that the flowability of a sample worsens as the mass fraction of powder C increases and that the flowability classification of the mixtures tested here varied from *fair* to *very poor*. We verified that the Modified Linear-Mixture Packing Model predicts well the bed porosities of SCGs, and that the Densification Model describes with good accuracy the behavior of packing curves, with physically consistent fitting parameters. The experimental data, the ternary diagram and the fitting parameters presented here can be used industrially to assist in handling of SCGs.

Acknowledgements

The authors would like to thank the São Paulo Research Foundation (FAPESP), grant 2016/25946-2, for the financial support.

References

1. International Coffee Organization. (2017). "Total coffee production by all exporting countries". Retrieved April 16, 2018, from http://www.ico.org/prices/po-production.pdf
2. Brazilian Coffee Exporters Council. (2017). "Monthly coffee report". Retrieved April 16, 2018, from http://www.cecafe.com.br/en/publications/monthly-exports-report/
3. Instant Coffee Market. (2017). "Instant coffee market: global industry trends, share, size, growth, opportunity and forecast 2017-2022". Retrieved March 2, 2018, from http://www.imarcgroup.com/instant-coffee-processing-plant
4. Mussatto, S. I., Machado, E. M. S., Martins, S., and Teixeira, J. A. (2011). "Production, composition, and application of coffee and its industrial residues". *Food and Bioprocess Technology*, 4, pp. 661–672.
5. ASTM International. (2010). "Standard test method for sieve analysis of fine and coarse aggregates". *ASTM C136-01*, pp. 1–5.
6. Organization, W. H. (2012). "Bulk density and tapped density of powders". *QAS/11.450 - The International Pharmacopeia*, pp. 1–6.
7. Campos, M. M., and Ferreira, M. C. (2013). "A comparative analysis of the flow properties between two alumina-based dry powders". *Advances in Materials Science and Engineering*, 2013, pp. 1–7.
8. Mallol, G., Amorós, J. L., Orts, M. J., and Llorens, D. (2008). "Densification of monomodal quartz particle beds by tapping". *Chemical Engineering Science*, 63, pp. 5447–5456.
9. Yu, A. B., & Standish, N. (1991). "Estimation of the porosity of particle mixtures by a linear-mixture packing model". *Industrial and Engineering Chemistry Research*, 30, pp. 1372–1385.
10. Yu, A. B., Zou, R. P., & Standish, N. (1996). "Modifying of the linear packing model for predicting the porosity of nonspherical particle mixtures". *Industrial & Engineering Chemistry Research*, 35, pp. 3730–3741.
11. Geldart, D. (1973). "Types of gass fluidization. Powder Technology", 7, pp. 285–292.
12. Castellanos, A. (2005). "The relationship between attractive interparticle forces and bulk behaviour in dry and uncharged fine powders". *Advances in Physics*, 54, pp. 263–376.
13. Li, Q., Rudolph, V., Weigl, B., and Earl, A. (2004). "Interparticle van der Waals force in powder flowability and compactibility". *International Journal of Pharmaceutics*, 280, pp. 77–93.

507744 Effect of Particle Size and Cohesion on Powder Yielding and Flow

Abstract:

The bulk properties of powders depend on material characteristics and size of the primary particles. During storage and transportation processes in the powder processing industry, the material undergoes various modes of deformation and stress conditions, e.g., due to compression or shear. In many applications, it is important to know when powders are yielding, i.e. when they start to flow under shear; in other cases it is necessary to know how much stress is needed to keep them flowing.

The measurement of powder yield and flow properties is still a challenge and will be addressed in this study.

In the framework of the collaborative project T-MAPPP, a large set of shear experiments using different shear devices, namely the Jenike shear tester, the ELE direct shear tester, the Schulze ring shear tester and the FT4 powder rheometer, have been carried out on eight chemically-identical limestone powders of different particle sizes in a wide range of confining stresses. These experiments serve two goals: i) to test the reproducibility/consistency among different shear devices and testing protocols; ii) to relate the bulk behaviour to microscopic particle properties, focusing on the effect of particle size and thus inter-particle cohesion.

The experiments show high repeatability for all shear devices, though some of them show more fluctuations than others. All devices provide consistent results, with the FT4 powder rheometer gives lower yield/steady state stress values, due to a different pre-shearing protocol. As expected, the bulk cohesion decreases with increasing particle size (up to 150 μm), due to the decrease of inter-particle cohesion. The bulk friction, characterized in different ways, is following a similar decreasing trend, whereas the bulk density increases with particle size in this range.
Interestingly, for samples with particle sizes larger than 150 μm, the bulk friction increases -- presumably due to particle interlocking effects - up to magnitudes comparable to those of the finest powders. Furthermore, removing the fines from the coarse powder samples reduces the bulk cohesion and bulk density, but has a negligible effect on the bulk friction.

In addition to providing useful insights into the role of microscopically attractive, van der Waals, gravitational and/or compressive forces for the macroscopic bulk powder flow behaviour, the experimental data provide a robust database of cohesive and frictional fine powders for industrially relevant designs such as silos, as well as for calibration and validation of models and computer simulations.

507750 Industrially Relevant Powder Characterisation Using a Uniaxial Powder Tester

Abstract:

To optimise productivity whilst ensuring high final product quality, it is necessary to evaluate raw materials, intermediates and end-products rapidly and reliably. The recent development of an Advanced Uniaxial Powder Tester (AUPT) (Freeman Technology Ltd, UK) provides a versatile and automated powder characterisation technique with a well-defined operating procedure. This simple but sensitive tool has the ease and speed of use associated with more traditional powder testing techniques whilst maintaining high repeatability and low operator dependency.

The AUPT measures the load required to break/fail a free-standing powder column which has been consolidated at a known Major Principal Stress (MPS). In addition to these measurements, the AUPT can also determine Bulk Density and Compressibility Percentage data. Powder flow properties can also be evaluated for a range of storage conditions by adjusting MPS and/or consolidation time.

Twelve diverse powders were evaluated using the AUPT, with each tested at three different MPS values. Highly repeatable values for uniaxial Unconfined Yield Stress (uUYS), Compressibility and Bulk Density were generated for each of the materials at each MPS, demonstrating how uniaxial powder testing can be used to quantify and rank the flow properties of diverse materials representing a wide range of industries. Comparable results were also generated between different users demonstrating low operator variability, unlike many other simple techniques.

All twelve samples displayed high sensitivity to the different consolidation loads utilised. This is most prominently observed for a sample of detergent which generated lower uUYS values at lower MPS values but was significantly more sensitive to applied stress than many of the other materials, for example limestone, generating higher uUYS values at 100 kPa. As such, this material may display more problematic process performance in higher stress operations such as hopper discharge.

The highly repeatable Compressibility values generated for each of the materials indicates consistent consolidation across the powder column over repeat runs. The data also illustrates that bulk properties such as Compressibility should not be relied on alone to assess flowability. For example, despite MCC and a sample of carboxymethyl cellulose (CMC) generating comparable Compressibility profiles, clear differences in uUYS were observed.

Powders were also consolidated for extended periods of time in controlled environments using off-line consolidation stations to study the impact of long-term storage as well as elevated temperatures and humidity levels. Clear and repeatable differences were observed between samples after being consolidated for 1 or 16 hours over a range of stress levels.

This study has demonstrated that the AUPT can be used to rapidly quantify a defined set of parameters to improve process efficiency and obtain repeatable data in long term studies. The technique displays the speed and ease of use associated with more traditional characterisation methods with the additional benefits of having a well-defined operating procedure, automation and low operator dependency.

507796 Effect of Particle Properties on Evaluation of Flowability Using a Test of Powder Discharge By Air Pressure

Abstract:

Cohesive powders are widely used in various industries such as chemical reaction, material production, food processing and pharmaceutical development. Recently the cohesive powder used in these processing is increased because it has the advantage, i.e. having superior reaction based on a large specific surface area. However, the cohesive powder handling in the surrounding air is extremely difficult by the influence of the cohesive force which is relatively strong to the gravity force in case of the reducing particle diameter. For example, the bridge and the flooding of cohesive powders are encountered in the practical situation and these troubles need to be avoided. As this issues, it is well known that these phenomena are related to the evaluation of powder flowability. Therefore, it is very important subject to the evaluation of flowability of cohesive powder.

In this study, we experimentally investigated to the flowability and cohesiveness of the cohesive powder using a test of the powder discharge by the air pressure. The powder used was three kinds of Calcium hydroxide having the difference of particle diameter and moisture contest. These powders were belonging to the Geldart group C particle. We were analysis the mass flow rate of cohesive powder and the average air pressure acting on powder bed where these parameters were calculated by the discharge mass of the powder and the air pressure as a function of the elapsed time, respectively. Furthermore, the separation force acting on the powder was estimated by the based on Rumpf equation which the air pressure at the beginning of the powder discharge, the initial void fraction, the particle diameter and the particle density were substituted in. Cohesiveness was expressed as the ratio of separation force and gravity force. Additionally, van der Waals force, electrostatic force and liquid bridge force on the powder were also calculated. Finally, we performed the evaluation of flowability to three kinds of Calcium hydroxide powder and examined the influence of cohesiveness of the powder against flowability of the powder.

Experimental device was consists of the filling part of the cohesive powder and the air supplying section at the top and the bottom of a vessel. The powder filling vessel was made by the cylindrical acrylic pipe which has the inside diameter of 50mm and the height of 150mm. The air distributors to the uniform air supplying were arranged at the top and the bottom of a cylindrical vessel. The supplying air which produced by the compressor was introduced through the three way valve and the air was supplied simultaneously from distributors to the powder bed in a vessel. By using this operation, the powder in a vessel can be discharge from a circular orifice at the central of bed bottom. In this experiment, a circular orifice diameter was fixed 4mm. The flow rate of air to the top and the bottom of the powder bed was adjusted by the mass flow controller. The discharge mass of the powder through an orifice was measured by the load cell which is installed at the bottom of a receiving vessel. The air pressure acting on the top of the powder bed was measured by the pressure sensor which is located on 150mm from the bottom of vessel.

Powder used is as following. Three kinds of Calcium hydroxide which has the different particle diameter, the same density and the different material moisture was used in this study, and these powders called as $Ca(OH)_2$ -A, $Ca(OH)_2$ -B and $Ca(OH)_2$ -C. As for the particle properties, the average particle diameter of $Ca(OH)_2$ -A and B were 7μm and $Ca(OH)_2$ -C was 13μm, the density was 2240kg/m^3, and the moisture content of $Ca(OH)_2$ -A,

B and C were 0.56, 0.1 and 5.5%, respectively. Furthermore, the Carr's flowability index of three powders which measured by the powder tester were 25, 28 and 23 with $Ca(OH)_2$ -A, B and C respectively. According to the evaluation of Carr's index to these powders were classified as "Bad" flowing of the powder. Additionally, all the powders belongs to the group C particle in the Geldart classification. This means that the powder to handle in this study is very cohesive.

The experimental conditions in the present study is as following. The initial filling mass of powder to a cylindrical vessel was fixed 70g. The flow rate of supplying air at the top and the bottom of the powder bed in a vessel were varied from 0.067×10^{-5} to 1.5×10^{-5} m^3/s to give the difference air pressure acting on the powder bed. Three kinds of powder was filling naturally in a cylindrical vessel and then the initial height of the powder bed was recorded because the initial void fraction was estimated to all the experimental conditions, where the average initial void fraction of $Ca(OH)_2$ -A, B and C were 0.840, 0.830 and 0.804.

As the result, we obtained the following results.

(1) The cohesive powder flow by using our method was classified by three phase such as "Dischargeable flow", "Transient flow" and "Non dischargeable flow" as well as previous work. Furthermore, this method could be evaluated to the flowability and the cohesiveness of the powder as following.

(2) Powder flowability to three kinds of Calcium hydroxide was evaluated by the relationship between the mass flow rate and the average air pressure acting on the top of the powder bed. The mass flow rate of $Ca(OH)_2$ -B was increased gradually with increasing the average air pressure. On the other hand, the mass flow rate of other powders clearly appeared the fluctuating trend, especially, the mass flow rate of $Ca(OH)_2$ -C against the average air pressure was confirmed a large change. It is therefore concluded with the powder flowability that $Ca(OH)_2$ -B has a highest flowability and $Ca(OH)_2$ -C was lowest flowability.

(3) Cohesiveness of three kinds of Calcium hydroxide was estimated as the ratio of the separation force and the gravity force. The results showed that the ratio of the separation force and the gravity force on $Ca(OH)_2$ -B was lower than that of other particles. Meanwhile, the ratio of the separation force and the gravity force on $Ca(OH)_2$ -C indicated the highest value. This result suggested that the cohesiveness of $Ca(OH)_2$ -C was the most of strong.

(4) As the result that the flowability of $Ca(OH)_2$ -C was low, it could be infer that the moisture content and the cohesiveness were influenced to this phenomenon.

507880 Characterising Flow Behavior of Wet Powder in Horizontal Rotating Cylinder

Abstract:

Rotating cylinders are widely used equipmentâ€™s in industry for processing and mixing of particulate materials. The quality of mixing and processing of the particulate material is thereby governed by flow behaviour and the amount of time spent by the material in this equipment.

The flow behaviour of dry particulate solids in such equipmentâ€™s is widely studied but little literature is available on the flow of wet particulate matter. In the present work tracer studies were conducted to measure the residence time distribution. The flow behaviour of the wet powder is characterised and compared with the dry powder using axial dispersion model. The study provides interesting insight into the difference in mixing behaviour of a dry and wet powder.

507721 Dry Fine Grinding: Aspects of Particle Stabilization in Mill Classifier Circuit

Abstract:

In many industrial applications the demand on minerals with increasing product fineness rises. The production of these materials is often performed in closed mill-classifier circuits which enable the production of materials with high product fineness and constant product quality. Thereby, the use of dry operated stirred media mills is a promising and emerging option for the energy efficient production of fine particles due to their high stress intensity as well as high stress frequency compared to conventional ball mills. Generally, in dry grinding processes fine powders are not easy to produce and to handle since attractive particle-particle forces become more decisive with decreasing particle size. These forces are leading to a variety of challenges, in particular caused by a high state of agglomeration, a decrease of the powder flowability as well as material adherences on the machine and plant equipment. Obviously, these aspects may complicate the comminution within a mill-classifier circuit on various levels, simultaneously: As the attractive particle-particle forces determine the particle and bulk properties, they have a strong impact on the stress mechanism inside the mill, the particle classification and final product filtration as well as on the material transport and retention time inside the single devices. In order to control these forces, so-called liquid grinding aids which reduce the particle-particle forces by adsorbing on the surface of the product particles are often used as stabilizing additives. As the name suggests, these additives are especially known for improving the grinding performance inside the mill. However, there is only very little information available how far liquid grinding aids affect the process in mill-classifier circuits. Therefore, this study examines the application of grinding aids in mill-classifier-circuits with stirred media mills on different levels:

In a first step, the stressing behavior inside a dry operated stirred media mill was investigated in dependence of the grinding aid as well as the most important process parameters like stirred tip speed, bead size and bead density. Therefore, limestone was ground in a batch operated 1.6 liter laboratory stirred media mill. It became clear, that those mill parameters which determine the stress energy of the grinding beads have a strong impact on the grinding performance inside the mill. Beyond that, an enhancement of the grinding process as well as its control is effectively achieved by an appropriate application of grinding aids. Thereby, a crucial role of the powder flowability was identified: As the grinding aid determines the flow behavior of the product powder and thus the bulk flow as well as particle capturing between the grinding beads, it influences different important values like the number of stressed particles at each grinding bead collision as well as the resulting stress intensity. Hence, the grinding result strongly depends on the applied grinding aid. Additives that show different impacts on the bulk properties require different optimal process settings. Hence, those parameters need to be adjusted to the applied grinding aid very carefully in order to ensure effective fine grinding. As a rule of thumb, lower stress energies and higher collision frequencies are more favorable when adding powder flow assisting additives.

In the second step, similar experiments were carried out in a continuously operated 6.6 liter horizontal stirred media mill. Thereby it was examined, how far findings from the batch grinding experiments may be transferred to continuous grinding. It is shown that the batch results may give good indications for optimizing continuous stirred media milling. However, an overlapping impact caused by the different powder flowabilities was identified: Since the flow behavior of the product powder determines the axial material transport through the mill, grinding aids may imply a strong impact on the material hold-up as well as retention time within the mill. Thereby, a high powder flowability is not necessarily leading to a lower mill retention time. This circumstance is mainly caused by a deflector wheel in the end of the mill which does not only hold back the grinding beads but also big particles that have not been comminuted satisfactory. Thus, the particle stabilization achieved by grinding aids is affecting the grinding result on different levels simultaneously. All of these issues need to be taken into account when designing dry fine grinding processes in stirred media mills.

In addition to the grinding experiments, the impact of different grinding aids on the classifying performance in a dynamic air classifier was investigated independent of the mill-classifier circuit. Within a systematic investigation, the process parameters of the air classifier were examined in dependence of the grinding aid species, grinding aid concentration as well as fineness of the limestone fraction. As the grinding aids strongly affect the attractive particle-particle forces and thus the agglomeration behavior of the product powder, they also influence the classification performance in dynamic air classifier. Consequently, grinding aids also work as dispersion additives or classifying aids. It is shown how several classification characteristics like cut size, separation sharpness and incorrect discharge depend on the grinding aid application. Thereby, an ideal operating range with regard to the particle stabilization was identified.

Finally, grinding results from a mill-classifier circuit consisting of a dry operated horizontal stirred media mill and a deflector wheel air classifier are presented. It is shown how far the grinding aids affect the product quality, product throughput and specific energy consumption. Thereby, the results are correlated with the findings from the investigation of the single units shown above. For an even more comprehensive understanding of the pilot plant, a material circulation factor is considered which describes the mean number of material circulations inside the plant. It enables a more realistic description of the residence of the particles within the pilot plant. Again, effects of the different grinding aids are identified and are correlated with the grinding results.

508125 Influence of Filtration on Particle Size and Shape of Crystalline Material

Abstract:

The production process of pharmaceuticals and fine chemicals often includes a biotechnological synthesis. Subsequent, the target molecule has to be separated from the surrounding liquor selectively and with high purity. Therefore the crystallization is a common step in downstream processing. After the formation of the solid crystalline form, the material has to be separated from the broth using a solid-liquid-separation-step. During the separation the product is exposed to mechanical stress, which can lead to a modification of the outer appearance of the crystals.

In literature the impact of crystallization conditions, for example agitator speed or additional compounds, on the shape and agglomeration of crystalline particle systems is described [1,2]. Besides the formation of the solid phase, also the subsequent processing is influencing the particle form. Events which influence the shape of the particles can be either breakage of crystals or the formation of solid bridges due to solved compounds in the remaining liquid after mechanical dewatering. These changes in particle structure can have a high impact on solubility, bioavailability and further formulation properties of the final product [3,4]. By choosing the right separation device as well as the appropriate machine parameters, unfavorable changes in product shape can be decimated or in best case avoided. Therefore, the intensity of the following formulation steps can be reduced by adjusting the required product characteristics during the separation step.

In the following work, the influence of the separation on the crystal appearance is investigated. As criterion of the product quality the grade of agglomeration is evaluated. As model system for pharmaceutical and fine chemicals amino acids are used. For the quantification of the agglomeration grade of the crystalline systems an automated image analysis and classification was used based on literature [5,6]. The different classification methods of agglomerated and single crystals is applied to different system of crystalline amino acids.

Since this work focuses on the filtration as separation method, different types of filtration, for example pressure filtration and vacuum filtration are compared regarding the impact on the agglomeration degree as well as the particle size distribution. Automated image analysis was used to classify the crystals regarding their appearance. Therewith the grade of agglomeration before and after filtration was analysed and set into relation.

With this information the choice of the appropriate separation device can be complemented by the knowledge of the influence of the separation on the particle shape.

References

[1] D. Wieckhusen, *CHIMIA* 2006 *60 (9)*, 598. DOI: 10.2533/chimia.2006.598.

[2] E.M. Ålander, M.S. Uusi-Penttilä, Å.C. Rasmuson, *Ind. Eng. Chem. Res.* 2004 *43 (2)*, 629. DOI: 10.1021/ie0301280.

[3] S. Schmidt, D. Havekost, K. Kaiser, J. Kauling, H.-J. Henzler, *Eng. Life Sci.* 2005 *5 (3)*, 273. DOI: 10.1002/elsc.200500116.

[4] N. Blagden, M. de Matas, P.T. Gavan, P. York, *Advanced drug delivery reviews* 2007 *59 (7)*, 617. DOI: 10.1016/j.addr.2007.05.011.

[5] L.-M. Terdenge, S. Heisel, G. Schembecker, K. Wohlgemuth, *Chemical Engineering Science* 2015 *133*, 157. DOI: 10.1016/j.ces.2015.03.002.

[6] D.R. Ochsenbein, T. Vetter, S. Schorsch, M. Morari, M. Mazzotti, *Crystal Growth & Design* 2015 *15 (4)*, 1923. DOI: 10.1021/acs.cgd.5b00094.

506374 Prediction of Granule Critical Quality Attributes Using Twin Screw Granulation

Abstract:

Wet Granulation is a commonly used unit operation for particle size enlargement of fine powders to mitigate issues such as blend segregation, dust formation, and poor flowability. Twin Screw Granulation is a robust and continuous approach towards performing wet granulation at various scales of operation. A predictive understanding of the granule properties in Twin Screw Granulation is becoming increasingly important in order to ensure a Quality-by-Design approach in the wet granulation and downstream processes.

Granule breakage is found to be the dominating rate process in Twin Screw Granulation. We have performed breakage isolating experiments to understand the influence of formulation properties, process parameters, and screw design on granule breakage characteristics, specifically, granule breakage probability and daughter size distributions. A model has been proposed for the breakage rate process in Twin Screw Granulation, based on the results from the breakage isolating experiments. We have used screw pullout experiments to understand the development of the critical quality attributes of the granules as a function of the axial distance from the end of the granulator. These experiments provide further insight into the granulation mechanisms in different screw element designs. The results from the screw pullout experiments are consistent with the granule breakage model deduced from the breakage isolating experiments.

PARALLEL GPU-BASED MONTE CARLO TECHNIQUES FOR THE FLOW-SHEET SIMULATION OF SOLID PROCESSES

Gregor Kotalczyk, Krishan Lambach and Frank Einar Kruis, Faculty of Engineering, University Duisburg-Essen, Duisburg, Germany

Abstract

We describe two parallel algorithms (the deterministic 'merging' algorithm and the stochastic 'random removal' method) for the simulation of particulate flow between single units in the context of flow-sheet simulations. These algorithms are tested on an exemplary flow-sheet system modelling typical particle production processes. The mixing of several particulate instreams into single units as well as the separation of particles in screen units according to their size are included into this test system. The proposed methods are validated by comparison with numerical solutions of the corresponding sets of differential equations. Although the 'random removal' technique can be used in the context of coagulation or nucleation, we find that it leads to very high levels of statistical noise in the here presented transport implementation. These high noise levels make the method nearly inapplicable for problems of the same complexity as the here presented one. The presented introduced parallel merging algorithm shows less statistical noise and a very good reproduction of the benchmark results at the cost of greater computational complexity and thus computing times.

Introduction

The description of particle production processes by means of a flow-sheet simulation allows a simple framework for performance analysis and optimization. The modularization of a whole process into single unit operations allows the investigation of specific settings of single units and consequences of these settings for the overall process. The solution of the corresponding equations, however, poses a subject of great difficulty, hence mostly several particle properties have to be taken into account for the complete characterization of the particulate material. The corresponding particle dynamics may exhibit strong non-linear dependencies on these multi-variate properties and are mostly described by the population balance equation (PBE) (see e.g. [1,2]). The Monte Carlo (MC) technique poses a solution method for the PBE, which allows to take several particle properties into account. Such multivariate modelling requires normally a large amount of computational resources, if a sectional method is used for the solution of the PBE (see e.g. [3]).

The modelled particle size distributions (PSD) can describe differences in particle concentrations which can span several orders of magnitudes. The computational efficient description of such large differences of particle concentrations becomes possible by the application of weighted MC simulation particles [4,5]. Recent algorithms for the nucleation, coagulation [6], condensational growth/evaporation [7] and breakage [8] of particles by means of weighted MC simulation particles have been formulated specifically for graphic processing units (GPU). It has been shown that the parallel computation of the GPU architecture could be exploited for the accelerated solution of the PBE, making thus the incorporation of this stochastic technique for the simulation of flow-sheet processes suitable.

Production processes on plant scale can be mostly formulated as a sequence of coupled unit operations (with possible tear-streams), which in turn can be described by the PBEs. There exist several approaches, to solve such a system, next to the sequential modular approach, a (simultaneous) equation-oriented approach can be chosen to describe such a coupled system (see e.g. [9,10]). The parallel formulation of the solution of the PBEs by means of the MC method is best suited for the equation-oriented approach. We present in the following a parallel simulation algorithm for particulate transport, which allows the incorporation of these several mechanisms in an operator splitting technique (see e.g. [11]), this in turn allows the analysis and optimization of a complex manufacturing process, like it is described in the next section further below.

Simulated system

Table 1. Properties of the initial particle populations and the particles within the feeds.

Location of particle population with unit number () or Feed symbol	**Initial population in Disperser (1)**	**In Feed 1** $F^{(\mathrm{ext})}_{2,\mathrm{in}}$	**In Feed 2** $F^{(\mathrm{ext})}_{3,\mathrm{in}}$
Particle concentration $n_{\mathrm{F},i}$ / kg (carrier gas)	-	$5\cdot10^{15}$	$2\cdot10^{15}$
Initial number of particles in Disperser / resp. total number of particles inserted by feeds in 28 s	10^{10}	$8\cdot10^{11}$	$3.2\cdot10^{11}$
Initial particle mass in Disperser / resp. particle mass inserted by feeds in 28 s / ng	5.98	6.28	0.07
Mean geometric diameter d_{g} /nm	50	10	3
Geometric standard deviation	1.2	1.46	1.46

We describe a transport algorithm for a particle production process as it is shown in Figure 1. The initial particulate material represented by large particles of the size of ca. 50 nm is dispersed into a carrier gas. These initial particles are mixed in two subsequent units with pulsed aerosol feeds carrying particles of different materials, representing size ranges of around 3nm and 10 nm, the exact properties of the simulated particles can be found in Table 1. In a real application, the particles would grow due to agglomeration and the screen would separate large agglomerated product particles (these are collected by the following filter) from smaller particles (these are directed to a mixer unit for further agglomeration). Hence the agglomeration process is not considered in this work, the screen is operating in a different size range (defined further below) than the one presented in Figure 1.

Transport equations for the carrier gas

The net amount of carrier mass $M_i^{(\mathrm{C})}$ (units kg) changes over time for each unit i and is modeled by the following set of equations:

$$\frac{\mathrm{d}M_i^{\mathrm{C}}}{\mathrm{d}t}=\sum_{j\neq i}\left(f_{i,\mathrm{in}}^{(j)}\cdot M_j-f_{i,\mathrm{out}}^{(j)}\cdot M_i\right)+F_{i,\mathrm{in}}^{(\mathrm{ext})},\qquad \text{with: } f_{i,\mathrm{in}}^{(j)}=f_{j,\mathrm{out}}^{(i)}\ \forall i\neq j, \tag{1}$$

where $f_{i,\text{in}}^{(j)}$ denotes the relative inflow rate (units 1/s) into unit i from unit j and $f_{i,\text{in}}^{(j)}$ the outflow from unit i to unit j. The exact values are listed in Table 2. The pulsed feeds $F_{2,\text{in}}^{(\text{ext})}$ and $F_{3,\text{in}}^{(\text{ext})}$ are described by the following formula:

$$F_{2,\text{in}}^{(\text{ext})}(t) = F_{3,\text{in}}^{(\text{ext})}(t) = \begin{cases} F_{\text{in}}^{(\max)} = 0.01\text{g/sec} = & ,\text{if } \lfloor t/\tau_F \rfloor = 0,2,4,6,\ldots \\ 0 & ,\text{if } \lfloor t/\tau_F \rfloor = 1,3,5,7,\ldots \end{cases}, \quad \text{with: } \tau_F = 4\,\text{sec}. \tag{2}$$

The floor function $\lfloor x \rfloor$ denotes thereby the largest integer number smaller than the rational number x.

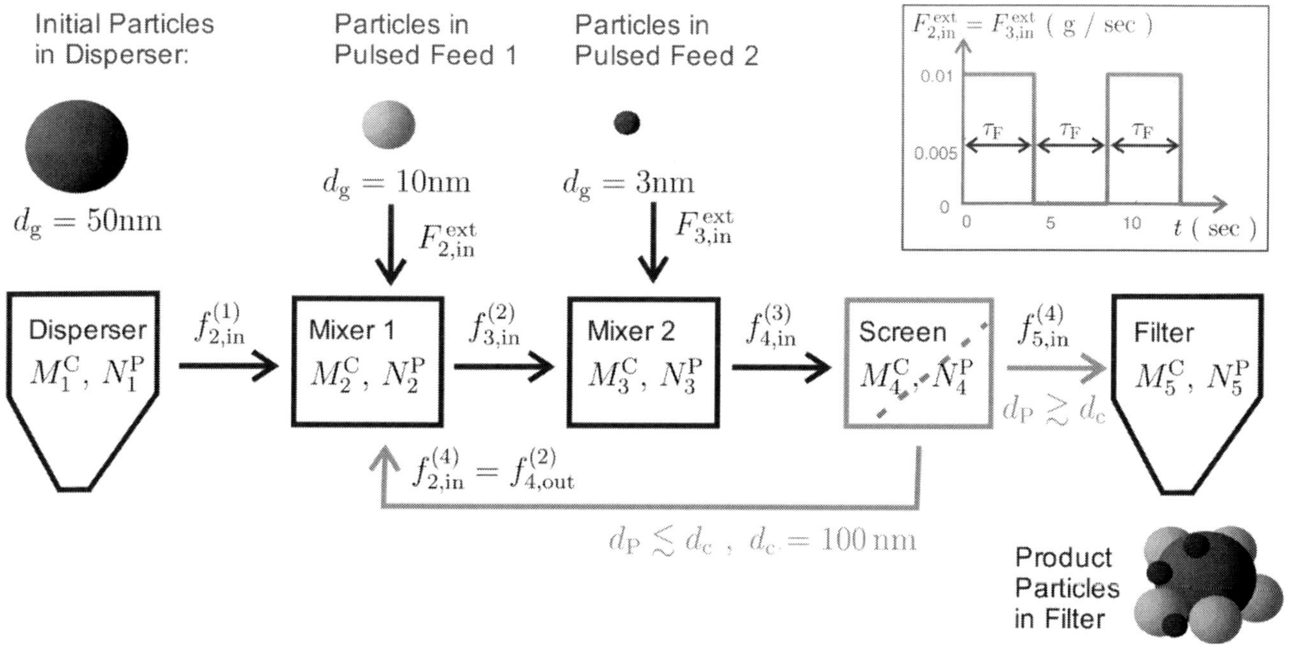

Figure 1. Flow-sheet of the simulated system.

Table 2. Initial Conditions and flow rates for each unit in Figure 1. The properties of the particle populations are listed in Table 1.

Unit **number in equation (1),(3)**	**Disperser** **(1)**	**Mixer 1** **(2)**	**Mixer 2** **(3)**	**Screen** **(4)**	**Filter** **(5)**
Initial particle populations	Yes	No	No	No	No
Initial carrier gas masses $M_i^{C,(0)}$ /mg	100	0.1	0.1	0.1	0
Outflow rates / s^{-1}	$f_{1,\text{out}}^{(2)} = 0.2$	$f_{2,\text{out}}^{(3)} = 1$	$f_{3,\text{out}}^{(4)} = 1$	$f_{4,\text{out}}^{(2)} = 0.7; f_{4,\text{out}}^{(5)} = 2$	No
External pulsed feeds maximal rates / mg /s	-	$F_{2,\text{in}}^{(\text{ext})} = 10$	$F_{3,\text{in}}^{(\text{ext})} = 10$	-	-

Transport equations for particles

We assume that the particles are mixed homogeneously and instantaneously upon arrival in each unit, so that a constant concentration $n_i^{(\mathrm{C})}$ (units / kg) of particles can be found throughout each unit. For the number of particles N_i^{P} (dimensionless number, $N_i^{\mathrm{P}} = n_i^{(\mathrm{C})} \cdot M_i^{\mathrm{C}}$) within each unit, the following equations result for the change of the number of particles due to the transport:

$$\frac{\mathrm{d}N_i}{\mathrm{d}t} = \sum_{j \neq i} \left(f_{i,\mathrm{in}}^{(j)} \cdot N_j - f_{i,\mathrm{out}}^{(j)} \cdot N_i \right) + F_{i,\mathrm{in}}^{(\mathrm{ext})} \cdot n_{\mathrm{F},i} \text{ for } j \neq 4, \tag{3}$$

with the relative flows $f_{i,\mathrm{in}}^{(j)}$, $f_{i,\mathrm{in}}^{(j)}$ and absolute flows $F_{i,\mathrm{in}}^{(\mathrm{ext})}$ used in Equations (1) and (2) and tabulated in Table 1. The values for the particle concentrations $n_{\mathrm{F},i}$ within the feeds are listed in Table 1.

For the description of the screen (unit 4, $j = 4$), it is assumed, that the total outflow of the carrier mass $\left(f_{4,\mathrm{out}}^{(2)} + f_{4,\mathrm{out}}^{(5)} \right)$ corresponds to the correct decrease of the particle number, $N_i \cdot \left(f_{4,\mathrm{out}}^{(2)} + f_{4,\mathrm{out}}^{(5)} \right)$, but that the particles are either found in the stream to the second unit or to the fifth unit, depending on their volume v, resulting in the following modified particle flow rates:

$$f_{4,\mathrm{out}}^{\mathrm{P}(2)}(v) = \left(f_{4,\mathrm{out}}^{(2)} + f_{4,\mathrm{out}}^{(5)} \right) \cdot S(v), \quad \text{and: } f_{4,\mathrm{out}}^{\mathrm{P}(5)}(v) = \left(f_{4,\mathrm{out}}^{(2)} + f_{4,\mathrm{out}}^{(5)} \right) \cdot \left(1 - S(v) \right) \tag{4}$$

, where $S(v)$ is a separation function describing the screening process. The resulting equations (4) have to be used instead of $f_{4,\mathrm{out}}^{(2)}(v)$ and $f_{4,\mathrm{out}}^{(5)}(v)$ in Equation (3) for the screen unit. In this work, the following simplified separation function is applied:

$$S(v) = \begin{cases} 1 & , \quad v < v_{\mathrm{S}} = \pi \cdot d_{\mathrm{S}}^3 / 6, \qquad \text{with: } d_{\mathrm{S}} = 40 \text{ nm} \\ (v_{\mathrm{E}} - v) / (v_{\mathrm{E}} - v_{\mathrm{S}}) & , \quad v_{\mathrm{S}} < v < v_{\mathrm{E}} = v_{\mathrm{S}} + 2 \cdot (\pi \cdot d_{\mathrm{C}}^3 / 6 - v_{\mathrm{S}}), \qquad \text{with: } d_C = 50 \text{ nm}. \\ 0 & , \quad v > v_{\mathrm{E}} = \pi \cdot d_{\mathrm{E}}^3 / 6 \qquad \text{with: } d_{\mathrm{E}} \approx 57.8 \text{ nm} \end{cases} \tag{5}$$

Simulation Methods

Transport of weighted Monte Carlo (MC) particles

The application of weighted MC particles allows to render the PSD in a more precise way [4]. Each MC particle k in the unit i is assigned an additional property, the statistical weight $W_k^{(i)}$. This approach allows alternative formulations to the stochastic determination, which particles are selected to leave a compartment (or unit) and to be inserted into the next one (specific transport events are defined by [12]; [13] allows multiple simulation particles to leave within one time step and formulates binomial probabilities for these events).

If weighted MC particles are simulated, exact calculation of the removed weight becomes possible. The following simple Euler approach summarizes this for the MC particle k in the unit i:

$$W_k^{(i)}(t + \Delta t) = W_k^{(i)}(t) - W_k^{(i)}(t) \cdot \sum_{j \neq i} f_{i,\mathrm{out}}^{(j)} \cdot \Delta t = W_k^{(i)}(t) \cdot F, \qquad \text{with: } F = 1 - \sum_{j \neq i} f_{i,\mathrm{out}}^{(j)} \cdot \Delta t. \tag{6}$$

For each connected unit j which is connected to i, new MC particles are added with the same properties as particle k in the unit i and the statistical weight:

$$W_{\mathrm{new},k}^{(j)}(t + \Delta t) = f_{i,\mathrm{out}}^{(j)} \cdot \Delta t \cdot W_k^{(i)}(t). \tag{7}$$

A scaling factor F can be used to keep track of the depleted statistical weight for all particles, if the flowrates are independent on the particle size, like proposed by [14]. However, in case of a unit operation like the screening process, the flow rates depend on the particle properties (the volume v according to Equation (4)), Equations (6) and (7) have to be evaluated for each individual particle. This might be computationally disadvantageous in the case of a serial implementation, but it is very efficient if a parallel implementation in combination with a graphic processing unit (GPU) is applied. The new particles with the weights $W^{j}_{new,k}$ can be inserted into the simulation by means of parallel merging algorithms or by the random removal of existing particles.

Merging

Following the approach presented in [6], two particles i and j with the statistical weights W_i, W_j and the volumes v_i and v_j are merged with each other to a new particle with W_{new} and v_{new} with the merge error E_m defined via:

$$W_{new} = W_i + W_j, \qquad v_{new} = (W_i \cdot v_i + W_j \cdot v_j)/(W_i + W_j), \qquad E_m = \left(\left(v_i - v_j\right)/\min\left(v_i, v_j\right)\right)^2. \tag{8}$$

While the parallel algorithm in [6] dealt with the problem to include one nucleated particle into an existing particle population by finding merge partners with low statistic weights and the lowest merge errors of a sample, the here presented work merges two entire particle populations with each other. This is done by a large number of parallel comparisons. The algorithm compares for two particles, whether the merging error is smaller than a predefined value E_{min}. The pair is merged if the condition is true. The value E_{min} is increased in the course of the simulation.[1]

Random removal?

The random removal algorithm has been introduced in the framework of a MC simulation of coagulation, in order to ascertain a constant number of simulation particles and avoid the decrease of computational accuracy [15]. We use this algorithm with some modifications for weighted simulation particles (like described in [6]). MC particles of the streams are included into the holdup by removing either the existing particles from the holdup or the particles from the stream and redistributing the removed mass on the rest of the particle population by multiplication with a constant factor. Random numbers are used for the selection of the particles which are to be removed.

Fixed pivot technique

For the validation of the transport code, a fixed pivot technique is used – although defined for the simulation of coagulation [16], it can be easily applied to the given problem. A fine one-dimensional grid of 1000 pivot points is set up, spanning particle sizes from 1 nm (first pivot point) to 10 µm (last pivot point), the sizes of two adjacent pivot points increase by a factor of $10^{(1/12)} \approx 1.028$. The pivot points

[1] An increase of E_{min} by a factor of 4 is done after every 64 parallel comparison steps – if the program is not terminated earlier because of the successful merging of all populations. During each parallel comparison step, 512 MC particles from each stream are compared with 512 MC particles in the hold up. The compared pairs change with each loop iteration, so that after 512 parallel loops (7 increments of E_{min}) each of the 512 MC particles in the holdup unit would have been compared with every particle from every stream, if no particles could have been merged. The data representing the particles is divided in blocks, where each block operates on its 512 particles for each stream and the holdup of the unit (if 10240 MC particles are used for the simulation, then 20 data blocks are used in parallel). We observed, that an initial setting of $E_{min} = 10^{-4}$ leads to the successful merging of all streams long before the end of the theoretically possible 512 loops – the last loop iteration would be done with the largest value for E_{min}, which is: $E_{min} = 1.64$.

are initiated with concentrations corresponding to the particle size distributions defined in Table 1, and for each point the transport equation (3) is solved numerically.

Results

Figure 2 shows exemplary PSDs resulting from the prolonged application of the merging technique as well as the random removal technique. The PSD in the disperser can be approximated by both methods exactly, hence no particles have to be inserted into this unit. It can be seen, that the random removal technique fails to predict the PSDs of the other units correctly. Both simulation techniques preserve the total mass of all particles in all units in the scope of the floating point precision, like it is shown in Figure 3a. The total mass in all units resulting from the pivot method, M_{piv}, is compared in the Figure with the total mass M_{MC} resulting from MC simulations. Similar results can be seen for the number of particles (N_{piv} for the pivot method and N_{MC} for MC particles) in the screen unit, which is shown in Figure 3b. It can be seen, that the accuracy of the simulation can be tuned by the choice of the Euler step size and that this accuracy increases for smaller step sizes. Smaller step sizes lead, however to larger computational times, which are summarized in Table 3.

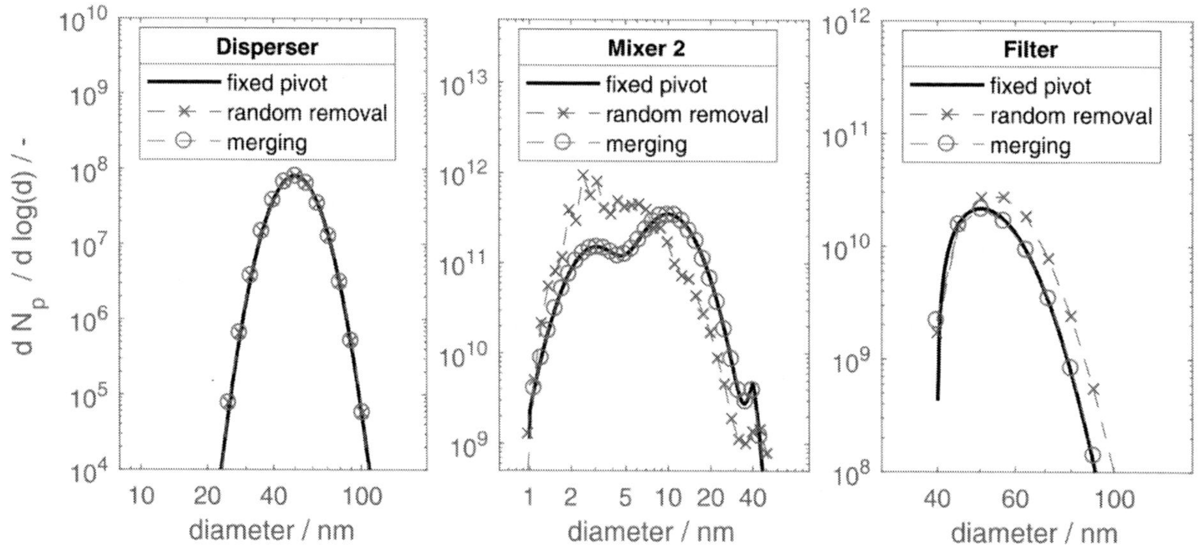

Figure 2. PSDs for 3 units in Figure 1 (the screen operates for particle ranges defined by Equation (5)) for a simulation time of 28 seconds, 10 240 MC particles and an Euler time step of $\Delta t = 0.1$.

Table 3. Computational times for 10 240 MC particles in seconds using a NVIDIA GTX 980 Ti GPU.

Euler step size	$\Delta t = 0.1$	$\Delta t = 0.01$	$\Delta t = 0.001$
Merging (10 parallel simulations)	198.9	2 012.7	19 928.6
Random Removal (100 parallel simulations)	49.5	128.7	1 181.5

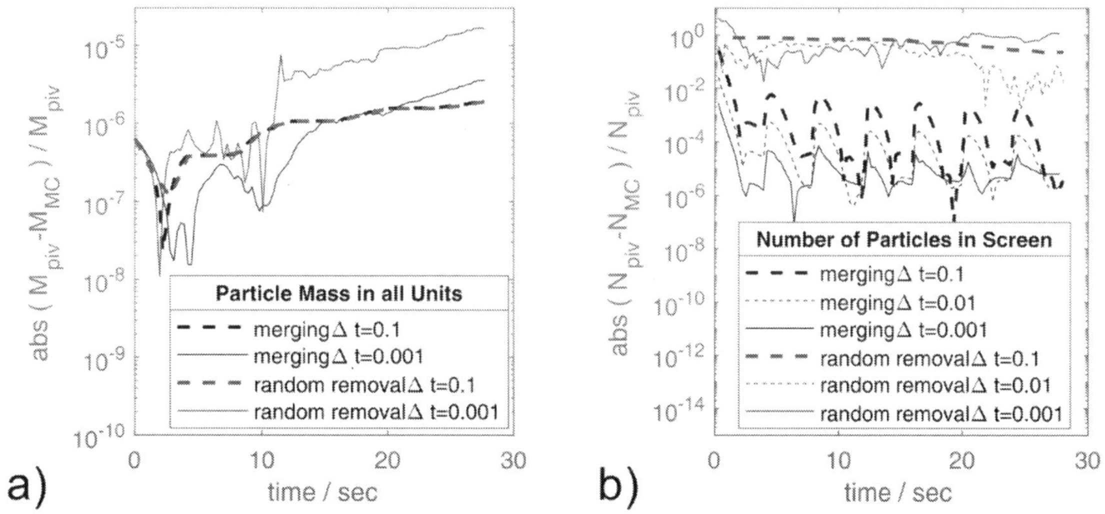

Figure 3. The deviations of the MC simulations from the pivot results for different values of the Euler step Δt. a) the total mass of the system is compared. b) the number concentration in the screen unit is compared. The results are shown for 10240 MC particles.

It should also be mentioned, that the random removal methods exhibit a large amount of statistical noise. The arithmetic standard deviations for the values of the number-concentration of the particles (or the geometric mean diameter or the geometric standard deviations of the PSDs) are higher than 10 % of the arithmetic mean of these values – this very high level of statistical noise disqualifies the here presented random removal technique for many flow-sheet applications.

Conclusions

We introduce and investigate Monte Carlo (MC) simulation techniques for the particulate transport in a network of units (or compartments). The techniques are benchmarked by comparison to simulations based on the fixed pivot method. The application of parallel algorithms, designed to run on graphic processor units (GPUs) has been tested for the simulation of an exemplary production setup with recycle streams, mixers and a screen unit. The simulation of the transport alone poses a challenging task, and it could be shown, that the simulation of a deterministic outflow of particles from single units coupled to a stochastically simulated inflow (simply termed 'random removal' in this work) is not able to reproduce the benchmark results with a high level of accuracy. A deterministic method, a massive parallel merge algorithm has been introduced, and the Euler step size has been identified as a parameter, which allows to tune the accuracy at the expense of the required computational time. This algorithm proves to reproduce the benchmark results with a high accuracy and is a promising candidate for a computationally efficient simulation of a multivariate flow-sheet problem, which also includes other non-linear particle processes, like coagulation, growth, evaporation, breakage, nucleation and others.

Acknowledgement

This work is supported by the DFG (Deutsche Forschungsgemeinschaft) within the priority program SPP 1679 "Dynamic flowsheet simulation of interconnected solids processes".

References

1. Ramkrishna, D. and M. R. Singh (2014), "Population balance modeling: Current status and future prospects", *Annual Review of Chemical and Biomolecular Engineering*, 5, pp. 123–146.
2. Ramkrishna D. (2000) ." Population balances: Theory and applications to particulate systems in engineering", New York,Acad. Press.
3. Xiong, Y. and S. E. Pratsinis (1993), "Formation of agglomerate particles by coagulation and sintering-Part I. A two-dimensional solution of the population balance equation", *Journal of Aerosol Science*, 24(3), pp. 283–300.
4. Zhao, H., Kruis, F. E. and C. Zheng (2009), "Reducing statistical noise and extending the size spectrum by applying weighted simulation particles in Monte Carlo simulation of coagulation", *Aerosol Science and Technology*, 43(8), pp. 781–793.
5. Patterson, R. I., Wagner, W. and M. Kraft (2011), "Stochastic weighted particle methods for population balance equations", *Journal of Computational Physics*, 230(19), pp. 7456–7472.
6. Kotalczyk, G. and F. E. Kruis (2017), "A Monte Carlo method for the simulation of coagulation and nucleation based on weighted particles and the concepts of stochastic resolution and merging", *Journal of Computational Physics*, 340, pp. 276–296.
7. Kotalczyk, G., Skenderovic, I. and Frank Einar Kruis, "A GPU-Based Monte Carlo Technique for the Simulation of Simultaneous Nucleation, Coagulation and Growth Based on Weighted Simulation Particles. In: *Particle Technology Forum 2016 - Core Programming Area at the 2016 AIChE Annual Meeting*. AIChE, 2016, pp. 490–497.
8. Kotalczyk, G., Devi, J. and F. E. Kruis (2017), "A time-driven constant-number Monte Carlo method for the GPU-simulation of particle breakage based on weighted simulation particles", *Powder Technology*, 317, pp. 417–429.
9. Skorych, V., Dosta, M., Hartge, E.-U. and S. Heinrich (2017), "Novel system for dynamic flowsheet simulation of solids processes", *Powder Technology*, 314, pp. 665–679.
10. Hillestad, M. and T. Hertzberg (1986), "Dynamic simulation of chemical engineering systems by the sequential modular approach", *Computers and Chemical Engineering*, 10(4), pp. 377–388.
11. Celnik, M., Patterson, R. I., Kraft, M. and W. Wagner (2007), "Coupling a stochastic soot population balance to gas-phase chemistry using operator splitting", *Combustion and Flame*, 148(3), pp. 158–176.
12. Marshall, C. L., Rajniak, P. and T. Matsoukas (2013), "Multi-component population balance modeling of granulation with continuous addition of binder", *Powder Technology*, 236, pp. 211–220.
13. Irizarry, R. (2012), "Fast compartmental Monte Carlo simulation of population balance models: Application to nanoparticle formation in nonhomogeneous conditions", *Industrial and Engineering Chemistry Research*, 51(47), pp. 15484–15496.
14. Menz, W. J., Akroyd, J. and M. Kraft (2014), "Stochastic solution of population balance equations for reactor networks", *Journal of Computational Physics*, 256, pp. 615–629.
15. Smith, M. and T. Matsoukas (1998), "Constant-number Monte Carlo simulation of population balances", *Chemical Engineering Science*, 53(9), pp. 1777–1786.
16. Kumar, S. and D. Ramkrishna (1996), "On the solution of population balance equations by discretization - I. A fixed pivot technique", *Chemical Engineering Science*, 51(8), pp. 1311–1332.

509616 The Compression and Compaction Behaviour of Pharmaceutical Powders and Their Binary Mixtures

Abstract:

Powder compaction is a very important production process in the pharmaceutical industry due to the prevalence of tablets as solid dosage form. Therefore, various scientific investigations focused on the investigation of the powder compaction process in the past decades. Nevertheless, the physical processes at particle level during powder compression are still not fully understood and this is the reason for the mainly empirical formulation and process development. The missing process understanding can be attributed to the complexity of the powder compaction process due to the high number of influencing parameters on the one hand and different acting micro-processes on the other hand. Exemplary influencing parameters are particle size distribution, particle shape and deformation behaviour of the raw materials as well as compaction stress and velocity [1; 2]. All these parameters affect the contact area and the number of contacts between the particles and hence, the mechanical and structural properties of the tablets.

Furthermore, it is well known that different micro-processes occur during powder compaction [3–6]. At the beginning of the powder compaction and thus, at low compression stresses, particle rearrangement is the main deformation mechanism and causes the reduction of inter-particulate pores due to the filling of large pores. The increase of compaction stress leads to the elastic and plastic deformation as well as fragmentation of single particles, which, in addition influences the contact area and the number of contacts between the particles. Additionally, elastic deformation of the molecule lattice occurs and causes the decrease of solids volume and thus, an increase of solids density with rising compression stress [7–9]. This phenomenon is referred to as solids compressibility. The different deformation mechanisms do not appear successively, but appear in parallel. The proportions of the single mechanisms differ in dependence on the used materials and process parameters and a differentiation between the mechanisms during the process is difficult or rather impossible. At the decompression phase, elastic recovery of the particles takes place and can lead to the destruction of weak inter-particulate bonds and to the occurrence of capping and lamination phenomena [6; 10; 11].

The missing physical process understanding requires the comprehensive characterization of the compressibility and compactibility of raw materials and the derivation of characteristic material parameters enables a more valid prediction of formulation and process development. The compressibility of a powder can be described by the relation between volume or porosity decrease and compaction stress [4]. Instrumented tablet presses enable the determination of this relationship directly during the compaction process (in-die). Various mathematical models for the description of this relation were developed, such as the models of Heckel, of Kawakita and of Cooper and Eaton [12–14]. Nonetheless, these models are mostly empirical and often describe only one part of the compaction process. Additionally, the existing mathematical models do not consider the phenomenon of solids compressibility.

In this work, the compressibility and compactibility of pharmaceutical powders with different deformation behaviour were investigated using in and out-die analysis. The compaction experiments were performed using the compaction simulator Styl'One Evolution (Medel'Pharm, France), which is a single station tablet press instrumented with force and displacement sensors. Furthermore, the compression and compaction behaviour of binary powder mixtures, consisting of an active pharmaceutical ingredient (API) and an excipient, were investigated under systematic variation of the API concentration. Additionally, a new process function for the description of the in-die compression curves was developed based on mechanistic considerations and their applicability for pharmaceutical powders with different deformation behaviour as well as for binary powder mixtures, was examined. It is found that the new process function is applicable for all model materials and well suited for the derivation of characteristic material parameters. The characteristic material parameters can facilitate formulation and process development.

References

[1] F. Podczeck, M. Sharma, The influence of particle size and shape of components of binary powder mixtures on the maximum volume reduction due to packing, Int. J. Pharm. 137, 41–47.

[2] K.C. Tye, C. Sun, G.E. Amidon, Evaluation of the effects of tableting speed on the relationships between compaction pressure, tablet tensile strength, and tablet solid fraction, J. Pharm. Sci. 94, 465–472.

[3] E.N. Hiestand, J.E. Wells, C.B. Peot, J.F. Ochs, Physical Processes of Tableting, J. Pharm. Sci. 66, 1977, 510–519.

[4] H. Leuenberger, B.D. Rohera, Fundamentals of Powder Compression. I. The Compactibility and Compressibility of Pharmaceutical Powders, Pharm. Res. 3, 1986, 12–22.

[5] P. E. Wray, The Physics of Tablet Compaction revisited, Drug Dev. Ind. Pharm. 18, 1992, 627–658.

[6] D. Train, An investigation into the compaction of powders, JPP. 8, 1956, 745–761.

[7] S.N. Vaidya, G.C. Kennedy, Compressibility of 18 Molecular Organic Solids to 45 kbar, J. Chem. Phys. 55, 987–992.

[8] S. Pedersen, H.G. Kristensen, Change in crystal density of acetylsalicylic acid during compaction, S.T.P. Pharm. Sci. 4, 1994, 201–206.

[9] E.V. Boldyreva, High-pressure diffraction studies of molecular organic solids. A personal view, Acta Cryst. A64, 218–231.

[10] S.K. Dwivedi, R.J. Oates, A.G. Mitchell,

[11] J.S.M. Garr, M.H. Rubinstein, An investigation into the capping of paracetamol at increasing speeds of compression, Int. J. Pharm. 72, 117–122.

[12] A.R. Cooper, L.E. Eaton, Compaction behaviour of several ceramic powders, J. Am. Ceram. Soc. 45, 1962, 97–101.

[13] R.W. Heckel, Density-Pressure Relationships in Powder Compaction, Trans. Metall. AIME. 221, 1961, 671–675.

[14] K. Kawakita, K.-H. Lüdde, Some considerations on powder compression equations, Powder Technology. 4, 61–68.

508477 Tablet Disintegration: Advanced Methodology Development

Abstract:

Nowadays, most of the drugs are introduced into the organism in the form of orally administered tablets. It is commonly known that composition, excipients and manufacturing process significantly affect behaviour of the tablets when exposed to aqueous media. significantly affect behaviour of the tablets when exposed to aqueous media. There is a big effort to quantitively and qualitatively describe disintegration, using techniques such as UV-VIS imaging, terahertz time-domain spectroscopy or Raman spectroscopy. In the presented work, there is an overview of used techniques to describe disintegration on binary model tablets made of mixture of ibuprofen and lactose (content of ibuprofen from 0.2 to 0.5 w/w), prepared at different compaction pressures (low, medium and high) are subjected the tests, which should help to identify their dissolution behaviour, mechanical and surface properties, wettability or liquid penetration.

Certified dissolution tests showed that the composition is significant in the most cases (there is a step change in dissolution times between 0.2/0.3 and 0.4/0.5 w/w). Further testing in aqueous environment revealed that there are some properties that are dependent on both composition and compaction pressure.

Measurements by sessile drop method provided information about soaking time and contact angle. Contact angle experiments shows surface tendency to get more hydrophobic with ibuprofen increment. This fact also has minor role in soaking time of liquid to tablet, where major influence is done by manufacturing pressure and thus porosity of tablet.

Experiments based on Texture Analysis exploring the structural properties showed dependence both on compaction pressure and composition as well as on used solution. Acidity plays significant role in tablet strength when there is risk of side reaction with excipient. Overall structure weakening was observed independently on used solution.

Magnetic Resonance Imaging was used for quasi-online monitoring of tablet behaviour in different solution. From experiment can be shown that both higher ibuprofen content and higher compaction pressure slows down disintegration and dissolution processes as well as local saturation of component in solution. There is a surface erosion of tablet observed, when exposed to dissolution medium. Through MRI it can be observed different tablet behaviour such as laminating, disintegration to clusters or dissolution of whole tablet.

The roughness of the tablet surface depends only on the composition and can be evaluated via Scanning Electron Microscopy and subsequent image analysis. Image analysis showed progressive lactose washing out with revealing ibuprofen crystals.

Chemical composition of the tablets can be determined using FTIR microscopy or imaging, where not only the amount but also the morphology of the primary particles on the tablet`s surface can be detected. The size of the primary ibuprofen particles was determined and the differences between two ibuprofen suppliers were clearly detected.

Dissolution and disintegration tests in discriminative conditions showed significant decrease in dissolution rate of ibuprofen, when lactose dissolution was prevented. However, there is an interesting discovery that the tablets with 0.2 w/w ibuprofen are disintegrating also in the case when neither component is dissolving.

These findings lead to the idea of the disintegration can be driven by affinity of the hydrophobic substance to the solvent, which facilitates water penetration and subsequent disintegration and that disintegration occurs due to weakening of inter-particle forces in presence of solution.

509395 Tablet Coating By Injection Molding Technology: Characterization of Coating Materials for Optimal Process and Product Performance

Abstract:

Purpose

ablet coating is one of the most commonly used pharmaceutical unit operations in tablet manufacturing. It renders possible advantages of taste and odor masking, physical and chemical protection, product differentiation, and elegant look to the finished tablet products. Spray coating is the most frequently used film coating process which involves organic and aqueous solvent based polymer systems. The organic solvents can be flammable, costly, and toxic in nature, whereas moisture sensitive drugs cannot be coated with aqueous coating systems. Longer drying time and resulting higher energy costs are other disadvantages of aqueous systems. We have developed a tablet coating process using solvent-free injection molding (IM) technology that can address the limitations experienced by other coating processes. Also, IM technology could be suitable for continuous manufacturing if incorporated with a continuous IM tableting platform. The aims of this research were to 1) optimize coating formulation attributes and process parameters of IM technology, 2) characterize physical properties, process performance and resulting quality attributes of pharmaceutical coating materials, and 3) evaluate the suitability and applicability of IM coating for pharmaceutical tablets.

Methods

Coating formulations (coating polymer and plasticizer in particular ratios) were fed to a vertical, miniature, twin-screw extruder and the generated extrudates were collected through the exit port. Dogbone shaped specimens were created with a microinjector and tested on a universal testing machine. Major mechanical test parameters, like Young's modulus, percentage elongation at the break, toughness, tensile stress at break and tensile strength were determined. A melt flow analysis was conducted using the microinjector with some modifications in the original instrument. Various coating formulations were fed into the heated IM barrel until it was completely filled. A pressure of 600-630 psi was applied by the injection piston onto the coating formulation and the material coming out from heated barrel was then collected in a container to measure the melt flow (g/min). Based on the mechanical properties and melt flow analysis, coating formulations were selected to coat tablets using an in-house built vertical IM machine in a 2-step process. Briefly, the temperature controlled injection barrel was first filled with a coating formulation and the uncoated tablet was placed inside the cavity of the mold inserts. Next, an injection piston applied pressure to the coating formulation and allowed the softened coating material to flow from the barrel to the mold insert. The coating material solidified inside the mold cavities and rendered a smooth coating layer attached to the first half of the tablet surface (step 1). Mold halves were opened, tablet was flipped over and step 2 coating was performed similar as step 1 to obtain fully coated IM tablet. Process parameters (barrel temperature, injection pressure, mold temperature, and cooling time) were optimized for different coating formulations. Coated tablets were stored in various storage conditions and evaluated after 8 weeks to evaluate the coat stability. Drug release from uncoated and coated griseofulvin (GF) tablets was studied, as per GF tablets USP monograph - test 1 dissolution methodology.

Results

Coating formulations suitable for IM process were ranked by evaluating different material properties (Young's modulus, toughness, percentage elongation, and tensile strength/Young's modulus ratio) acquired from the stress-strain analysis. The melt flow property of the coating formulations played a crucial role in IM processing. IM process parameters like injection pressure, barrel temperature, and mold temperature were critical for successful IM coating and were evaluated in depth. Sufficient room humidity (>30% RH) was required while employing polyethylene oxide (PEO) based formulations to avoid immediate cracks, whereas other formulations were insensitive to the room humidity. Tested formulations based on Eudragit E PO and Kollicoat IR did not have the desired tensile properties. Based upon this study, hydroxypropyl peastarch + 30% glycerol, PEO 1,000,000 + 30% PEG and Opadry + 25% glycerol were proven to be viable coating formulations for IM based tablet coating. These formulations had favorable tensile (<700 MPa Young's modulus, >30% elongation, > 95×10^4 J/m^3 toughness) and melt flow (>0.4 g/min) properties, that provided suitable IM coats. These three formulations increased the dissolution time by 10, 15 and 35 minutes respectively (75% drug release) compared to the uncoated tablets (15 minutes). Coated tablets stored in several environmental conditions remained stable to cracking for the studied 8-week time period.

Conclusions

The study confirmed IM as a viable technology for tablet coating. The study serves as a model for product development with specifications of excipients in ranges within the designed acceptance space for optimal product performance.

509487 Quantification of Lubrication and Particle Size Distribution Effects on Tensile Strength and Stiffness of Tablets

Abstract:

We adopt a Quality by Design paradigm to better control the mechanical strength of tablets as a critical quality attribute by understanding the effects of critical process parameters and critical material attributes. To this end, the effect of particle size distribution, lubricant concentration, and mixing time on the tensile strength and stiffness of tablets were studied.

Two grades of lactose, lactose Î±-monohydrate and spray-dried lactose, were selected. Tablets were compressed to different relative densities ranging from 0.8 to 0.94 using an instrumented compactor simulator. We propose a general model, which predicts the elastic modulus and tensile strength envelope that a specific powder can obtain based on its lubrication sensitivity for different particle size distributions. This was possible by introducing a new parameter in the existing tensile strength and elastic modulus models. A wide range of lubrication conditions was explored and the model exhibited a good predictability. The mechanical properties of lactose monohydrate tablets were noticeably dependent on particle size, unlike spray-dried lactose where little to almost no sensitivity to initial particle size was observed. The model is designed in a general fashion that can capture all the possible mechanical integrity behaviors in response to different lubrication conditions and initial particle size. Our model can be extended to all the powders that undergo different deformation mechanisms and is applicable for more complex pharmaceutical formulations.

507254 Evaluating the Influence of Dry Granulation Processing Variables on the Rheological Properties of Granules

Abstract:

Granulation is often employed in the production of pharmaceutical formulations for oral solid dose delivery. It is most often required when constituent materials in a formulation have poor flow properties, enabling these components to be combined to improve process performance and ensure content uniformity.

Dry granulation, based around the use of a roller compactor and integral mill/screen, can have significant advantages over more common wet granulation techniques. Dry granulation can reduce processing costs and can also be used with moisture/solvent/heat sensitive materials. Unlike a wet granulate, the dry granulate does not require additional drying and milling before encapsulation or tabletting.

Dry granulation is not suitable for all types of material, however, there is little research into which properties make a formulation suitable for this process. In addition, there is little to indicate which process parameters produce optimal granulate quality in order to achieve interruption free processing and high quality final products.

In this study an FT4 Powder Rheometer® (Freeman Technology, Gloucestershire, UK) was used to demonstrate how processing parameters influence the properties of a placebo formulation granulate. The granulate was produced using a roller compactor (MiniPactor, Gerteis Maschinen+Processengineering AG, Switzerland) on which the roll gap and compaction force were varied. The dynamic flow, bulk and shear properties of the resulting granulates were then evaluated using the FT4 Powder Rheometer.

The data clearly demonstrates that several rheological properties, known to influence die filling efficiency and Critical Quality Attributes (CQAs) of the resulting tablets, could be controlled by changes in roller compactor settings.

This study emphasises the importance of assessing powder properties using techniques that closely simulate process conditions and how it is possible to identify Critical Process Parameters to optimise a dry granulation process and ensure high final product quality.

507489 Assessment of the Homogeneity of an Alumina Powders Mixture Using Rheological Parameters

Abstract:

This study has been carried out in order to improve the comprehension of powder mixing in the nuclear fuel manufacturing process. Thus, a binary mixture of two Alumina (Al_2O_3) powders, GE15 and CR6, was used to simulate the oxides powders actually involved in the process. Different mixing technics and parameters were tried and compared in terms of homogeneity of the mixtures produced. The concept of intensity of segregation [1] was used for assessing mixture homogeneity through a sampling protocol and classical statistical analysis tools [2]. One of the difficulties of working with these compounds is that, due to the similarities between CR6 and GE15 powders (same chemical composition, same colour, etc.), the direct measure of the composition of a given sample is not possible. Since homogeneity of a mixture is assessed from the variation of the composition between samples, it is important to find an indirect way to measure the composition of an Alumina powders mixture. Therefore, different correlations between measurable parameters (specific surface area, Sauter diameter, cohesion and flow function coefficient) and the composition of the mixtures were found empirically allowing indirect measurement of the composition of a given sample.

The four studied parameters: specific surface area (, measured by nitrogen adsorption using BET method), Sauter diameter (, measured by laser granulometry, dry transportation with air at 0.1 bar), cohesion and flow function coefficient (and , measured by FT4 shear test at 9 kPa on a 10 mL cell), were compared in terms of precision, relevance and feasibility.

It appears that the homogeneity assessment methods involving rheological parameters (cohesion and flow function coefficient) show many advantages compared to the other methods. Not only are they the most accurate correlations found but both parameters are measured simultaneously, meaning that it is possible to apply both methods on the same samples while all the tests (N_2 adsorption, laser granulometry and FT4) are destructive. Those two methods also allow the estimation of the homogeneity at a more interesting scale (around 7 g of powder), indeed, specific surface area and Sauter diameter methods give a homogeneity at a scale of less than one gram which is not relevant for the end-use properties of the final product.

[1] P. Danckwerts, The definition, measurement and some characteristics of mixtures, Applied scientific research. Section A, 3 (1952) 279-296

[2] S. Massol-Chaudeur, H. Berthiaux, S. Muerza, et J. Dodds, A numerical model to identify the structure of a high-dilution powder mixture, *Powder Technol.*, 128 (2002) 131-138

508516 Characterising the Grain Shape: In Search of Size Independent Shape Descriptors

Abstract:

Granular materials contain a large number of grains that differ in size and shape. While the characterisation of the size distribution is typically done using sieve analysis, the characterisation of the shape of the grains remains a difficult task and is one of the major bottlenecks in the characterisation of the granular material. In this study, we attempt to characterise the shape of three different types of grains commonly used in iron and steel making industry, namely pellet, sinter and iron ore lump. While the pellet particles can be considered nearly spherical in shape, imposing the spherical shape on the iron ore lumps and highly irregular sinter particles seems a very poor choice. We consider two different size ranges of pellet particles and four different size ranges for sinter and iron ore lumps. Imaging of the grains is done using an iPhone 7 Plus camera and sufficient number of grains of each type in all the size ranges are considered. We perform the image analysis and obtain various boundary parameters (such as projected area, minor and major axis, perimeter, etc.) of the grains using MATLAB image processing toolbox. In addition, we define two additional boundary parameters to characterise the deviation of the projected area from that of a circle and calculate their values for all the different type of grains considered in this study. Using these length scales, we define a total of nine different shape descriptors and obtain the distribution of these shape descriptors for each type of grain for all the size ranges. Interestingly, for a given type of grains, the frequency distribution curves for these nine shape descriptors turn out to be nearly independent of the size range of the grains. This suggests that the chosen shape descriptors indeed represent properties related to the shape of the grains alone, irrespective of the size of the grains considered.

We investigate the correlation among these shape descriptors by evaluating the values of the correlation coefficients between each pair of the shape descriptors using the Pearson method, Kendall method and Spearman method. Our results indicate that out of the nine shape descriptors considered in this study, five shape descriptors are strongly correlated with each other and only the remaining four shape descriptors need to be considered in characterising the shape of the grains. Out of these four uncorrelated shape descriptors, two shape descriptors, namely the aspect ratio of the grains and the root of form factor, have been considered important by researchers in previous studies. The two other shape descriptors proposed in this study, which are used to characterise the deviation of the projected grain shape from that of a circle, are not measured previously to the best of our knowledge. We compare these four shape descriptors for the pellets, sinter and iron ore lumps having the same sieve size range and show that the shape descriptors for these three different types of grains are different.

We next calculate the values of these four shape descriptors for relatively larger coke particles using 2d image analysis. We also perform 3d image analysis of some of these coke particles and calculate the values of the analogous shape descriptors in 3 dimensions. We investigate if the identified 2d shape descriptors (obtained from 2d imaging) can be correlated to the analogous 3d shape descriptors (obtained from 3d imaging) and explore if it is possible to use the 2d imaging technique to estimate the 3d shape descriptors, thus eliminating the need of 3d imaging.

508543 Investigating Mechanical Properties of Surfactant Films at the Solid-Liquid Interface Using AFM

Abstract:

Corrosion is a well-known problem in the oil and gas sector and also in many other industries. Surfactant based inhibitors have been used for corrosion control in oil pipelines but their inhibition mechanism is still not fully known. Our group is studying surfactant corrosion inhibition mechanism by various characterization techniques (AFM, Electrochemical and Spectroscopic techniques). A better understanding of corrosion inhibition mechanism will lead to the design and development of not only cost-effective but also greener surfactants system for corrosion inhibition purposes.

Atomic force microscopy (AFM) is an effective tool to study the barrier properties of surfactant films at the solid-liquid interface. Determination of mechanical and topographical properties of single and mixed surfactant coatings by atomic force microscopy (AFM) can delineate the role of physical vs. electrochemical barrier by such films in the corrosion inhibition process. . So far, the AFM data, obtained using a closed cell, indicate the effect of surfactant concentration and that of hydrocarbon chains interactions on the mechanical properties of cationic surfactant films. Ultimately, integration of the AFM and electrochemical results will reveal the correlation between the surfactant structures at the solid-liquid interface and their corrosion inhibition efficacy.

508545 Deposition Rate Consequences of the Formation of Multi-Spherule Cluster Aggregates in Gases "Role of Momentum Shielding™"

Abstract:

For engineering flow systems characterized by particulate suspensions in gases, will the formation of *cluster aggregates* (CAs) in the mainstream accelerate deposition rates on immersed targets or confining walls ---or is aggregate formation more likely to suppress capture rates? Recent research on the drag properties of large cluster aggregates (e.g., stimulated in part by the Sorensen (2011) review) has opened the door to answering these important mechanism- (and pressure-) dependent questions. Sub-100 nm "primary" spherules, which may find themselves in even tenuous 'fractal-like' N-spherule clusters, can take advantage of what we call "momentum shielding"---analogous to the drag advantages experienced by birds who elect to fly in formation. We show here that the extent of this "momentum shielding" can be conveniently quantified (in a dimensionless function we call: $S_{mom}(N;Kn_1;\text{aggregate structure}))$, and used to predict the deposition-rate consequences of aggregation in aerosol flow systems when the cluster deposition mechanism is dominated by either: i) isothermal convective diffusion (C-D), ii) thermo-phoresis (TP-) or iii) inertial impaction (II). As expected, we show that for any type of aggregate structure (including 'fractal-like'), momentum shielding increases with cluster size (e.g. N) and gas pressure (via the reciprocal Knudsen number based on spherule radius R_1). We also show how S_{mom} can be calculated for the limiting case of a N>>1 random close-packed-D_f=3 cluster ---which is nearly gas impermeable. For fractal-like aggregates, including DLCAs (D_f=1.8) and RLCAs (D_f=2.1) we have recently shown that the abovementioned function: $S_{mom}(N, Kn_1; D_f)$ is actually sufficient to make useful engineering estimates of CA Brownian diffusion coefficients (Rosner and Tandon (2017b), thermophoretic diffusivities (Rosner and Tandon (2017a) and aggregate "stopping times" (Rosner and Tandon (2017c)----each in terms of their previously known values (D_1, $\alpha_T D_1$, and $t_{p,1}$) for isolated individual spherules in the same environment.

This information now enables instructive engineering calculations of the deposition-rate consequences of spherule aggregation to form large cluster aggregates-----i.e. at the same mainstream spherule volume fraction we can now determine how much aggregation will increase or decrease the total spherule acquisition rate on an immersed target. We present here representative results for each of the abovementioned particle transport mechanisms----i.e.: i) isothermal convective-diffusion (C-D), ii) thermo-phoresis (TP-) or iii) inertial impaction (II)----- including a brief account of the consequences of dealing with Kn1-dependent coagulation-'aged' aggregate size distributions (ASDs) in the mainsteam. We summarize our current re-examination of the canonical problem of inertial impaction on a circular cylinder in Re>>1 crossflow (Rosner and Tandon (2017c), dealing now with CA-"projectiles" but in the absence of spherule loss from the much larger target. We conclude by briefly outlining our planned work on the efficient "inversion" of sampled cluster-ASD-data to reconstruct, using the abovementioned $S_{mom}(N,Kn_1;D_f$ information, what ASD must have existed in the mainstream---at least in the absence of appreciable supercritical aggregate break-up.

References:

1. Rosner, D. E. and P. Tandon (2017a). Knudsen Transition Effects on the Thermophoretic Properties of Fractal-like Aggregates; Implications for Thermophoretic Sampling of High-Pressure Flames, Aerosol Science and Technology (in press).

2. Rosner, D. E. and P. Tandon (2017b). Aggregation- and Knudsen Transition Rarefaction- Effects on Particle Mass Deposition Rates by Convective-Diffusion, Thermophoresis or Inertial Impaction; Consequences of Multi-spherule 'Momentum Shielding', Aerosol Science and Technology (in press).

3. Rosner, D. E. and P. Tandon (2017b). Mainstream Aggregation and Polydispersity Effects for Inertially-Dominated Deposition Rates on Cylinders (or Leading 'Edges') in Crossflow"---- Consequences of 'Momentum Shielding' for the Impaction of Large, "Compact" (2<D_f<3) Multi-spherule Aggregates, Aerosol Science and Technology (prepared for submission, August 2017).

4. Sorensen, C. M. (2011). The Mobility of Fractal Aggregates: A Review, Aerosol Science and Technology. 45: 765-779.

508609 Experimental and Numerical Studies on the Thermal Diffusivity of Packed Powder Beds

Abstract:

Several industrial processes, such as calcination and drying, involve thermal treatment of particles. In order to achieve efficient processes and good quality products, the thermal properties of the particles and that of the bulk powder must be well defined. Previously, some models, such as the Krischer and Maxwell-Eucken models, have been proposed to relate particle thermal properties to that of the bulk powder. Though these models have been shown to work well, it is not very clear how to account for changes in the packing density, moisture content, and arrangement of particles. In addition, the particle thermal properties of most materials are not readily available, which limits the application of these models. To get a better understanding of the relationship between the particle thermal properties and that of the bulk powder and also to quantify the effects of packing density and moisture content on the bulk powder thermal properties, we run several experiments and discrete element method (DEM) based simulations. In both the experiments and simulations, a pipe is filled with particles where the temperature of the pipe is kept constant. We measured the temperature change in the powder bed and compute the actual thermal diffusivity of the powder bed. We tested several types of powders and glass beads in the experiments. The packing fraction, the moisture content, and size of particles were varied in the experiments. In the simulations, in addition to the packing fraction and particle size, the thermal properties of the particles were also varied. In this talk, we will present the details of the experimental and simulation results.

509556 Determination of Size, Size Distribution and Refractive Index of Artificial and Biological Microparticles

Abstract:

For the approval of clinical studies, it is mandatory to validate the production process of artificial oxygen carriers by reliably specifying the properties of the microparticles used like the newly developed blood substitute HbMP-700 (1-3). In particular, besides shape, size and size distribution, the amount of hemoglobin (Hb) and methemoglobin (metHb) content is essential.

We present a method based on the simultaneous measurement of collimated and diffuse transmission spectra in the visible and ultraviolet range and numerical analysis to infer the wavelength-dependent refractive index, size and size distribution of microparticles.

For the validation of the method, we used aqueous suspensions of monodisperse polystyrene microspheres of known concentration, determined by flow cytometric measurements. The related data analysis is carried out for synthetic and experimental data. For homogeneous spheres, Mie theory is applied to compute the solution for the scattering of a plane electromagnetic wave by a particle and to calculate the extinction cross section C_{ext}, i.e. the shadow cast by the particle as a function of the vacuum wavelength λ, the particle size, and the complex refractive index $n(\lambda)$. For dilute suspensions of particles, the measured quantity is an ensemble-averaged extinction cross section C_{ext} . This average is obtained by integrating over the particles' size distribution, which may not be known a priori. From the resulting spectral extinction cross sections, the complex refractive index and the mean and standard deviation of the size distribution can be inferred by means of nonlinear least-squares optimization. In order to reduce the number of free parameters of the problem to a feasible amount, $n(\lambda)$ is represented in a set of harmonic-oscillator-like resonance functions. The refractive index of polystyrene microspheres obtained is in good agreement with literature values for the bulk material, the mean and standard deviation of the distribution of particle *size D* are determined with an uncertainty in the range of a few nanometers. For example, for microspheres of mean diameter mean(D) = 2μm and distribution width std(D) = 9.4nm the uncertainty of the mean is $u\{\text{mean}(D)\}$ = 1.2nm and the uncertainty of the distribution width amounts to $u\{\text{std}(D)\}$ = 0.4nm.

In a second step, we applied the method to characterize HbMP-700 microparticles, approximately featuring a prolate spheroidal shape with 700nm length and 400nm width. Besides shape, size and size distribution, the amount of hemoglobin (Hb) and methemoglobin (metHb) content is essential. The relative concentrations of oxyHb, deoxyHb, metHb are accessible from spectral extinction and diffuse transmission measurements, since the different Hb variants exhibit different wavelength-dependent refractive indices. The method to infer size and refractive index described above can be applied to these non-spherical particles, since the deviation of their light scattering properties from those of spheres turns out to be small. The results based on the optical method will be compared to alternative procedures for HbMP particle characterization with respect to their size and size distribution, i.e. electron microscopy, centrifugation and dynamic light scattering. In addition, x-ray spectrometry will be used to detect the ratio between Fe^{2+} (oxy/dexyHb) and Fe^{3+} (metHb) as well as the iron concentration.

References

1) Novel hemoglobin particles--promising new-generation hemoglobin-based oxygen carriers. Bäumler H, Xiong Y, Liu ZZ, Patzak A, Georgieva R. Artif Organs. 2014 Aug;38(8):708-14. doi: 10.1111/aor.12331.

2) Nonvasoconstrictive hemoglobin particles as oxygen carriers. Xiong Y, Liu ZZ, Georgieva R, Smuda K, Steffen A, Sendeski M, Voigt A, Patzak A, Bäumler H. ACS Nano. 2013 Sep 24;7(9):7454-61. doi: 10.1021/nn402073n.

3) Hemoglobin-based oxygen carrier microparticles: synthesis, properties, and in vitro and in vivo investigations. Xiong Y, Steffen A, Andreas K, Müller S, Sternberg N, Georgieva R, Bäumler H. Biomacromolecules. 2012 Oct 8;13(10):3292-300. doi: 10.1021/bm301085x.

517131 FlowCam Nano ® Provides Counts, Sizes and Images of Nano-and Microparticles: Application to a Therapeutic Protein Pumping Study

Abstract:

Sub-visible particle characterization is becoming a more important method in assessing drug substance and drug product stability. However, the current USP protocol does not require any characterization of particles below 10 Î¼m. This greatly hampers our ability to properly assess protein stability as a function of particulate formation and obscures most sub-visible particles and various morphological properties.

In this study, we assess the capabilities of the new FlowCam Nano from Fluid Imaging Technologies, in conjunction with a pumping study that focuses on particulate formation generated during fill operations. Initial experiments show that the new FlowCam Nano can generate highly resolved images of nanoparticulates, allowing the simultaneous assessment of particulate morphology for nano- and microparticulates. This is extremely useful, as currently unpublished data (for the pumping study) show that as a function of post-pumping agitation, a significant increase in microparticle concentration coincides with a significant depletion of the nanoparticle concentration. This suggests that agglomeration (and probably not nucleation) is the more likely mechanism for larger microparticulate formation.

The ability to characterize the nano- and microparticulate morphology simultaneously, will allow us to better assess the cause of particulate formation (such as observing the transition of nanoparticles to microparticles and seeing if specific morphological properties of larger nanoparticulates are conserved in microparticulates). In addition, the FlowCam Nano is also sensitive enough to assess differences in the colloidal stability of a drug product in different formulations. During filling operations, and accelerated degradation studies, FlowCam Nano showed the differences in size distributions of IVIG in PBS and 0.25 M glycine buffer. Since IVIG is less colloidally stable in PBS, the size distribution varies significantly more per size bin and the overall mean effective spherical diameter is also ~ 600 percent greater after 24 hours of agitation in comparison to the size distribution data of IVIG in glycine. Thus, Flow Cam Nano has capabilities that are very useful in assessing drugs in the substance (e.g. optimizing drug formulations) and product (e.g. assessing particulates post fill & finish operations) phases.

Numerical and experimental estimation on the normal and tangential capillary bridge force adhered to two spheres

Kazuo Murase*, Keisuke Arai, Takato Ootsuka, Daiki Sakamoto and Futa Egawa,
Department of Applied Chemistry, Chuo University
1-13-27, Kasuga, Bunkyo-ku, Tokyo, 1128551, Japan
**hmurase@kc.chuo-u.ac.jp*

Introduction

Liquid bridges extensively occur as a binder liquid adhered to solid surfaces in a variety of engineering applications such as crystal growth, ink-jet printing, material processing and powder technology. Especially the strength of adhesion and cohesion of liquid bridges adhered to particles contribute to the consolidation of an agglomerate in wet granular materials. Estimation of the attractive or repulsive force between particles is important in controlling kinetics of agglomeration and handling of granular materials. Characteristics of a liquid bridge e.g. the equilibrium bridge geometry [1], stability [2], static and dynamic binding force [3-6], rupture energy [7,8], and wetting hysteresis [9] etc., have been experimentally and numerically investigated.

The liquid distributions within a particles bed are classified into four types of categories, i.e. pendular, funicular, capillary and slurry states, according to the liquid contents [10]. The static normal capillary force at pendular liquid state have elaborately been investigated as Fisher [11], toroidal model [8] and the exact solution of Yung-Laplace equation [12]. The dynamic normal capillary forces [13] are experimentally estimated with lubrication theory. The tangential force of a liquid bridge [14] also contributes to the primary factor for the granular consolidation with the shear stress of wet granular materials but have not yet substantially been investigated.

This work aimed to estimate experimentally and numerically the normal and tangential capillary force of a static and dynamic liquid bridge adhered to two spheres. The asymmetry static liquid bridge with the shear stress demand for the exact solution of Young-Laplace equation for the equilibrium liquid drop including the effect of gravity.

1. Theory

1.1 Formulation of a asymmetry static liquid bridge

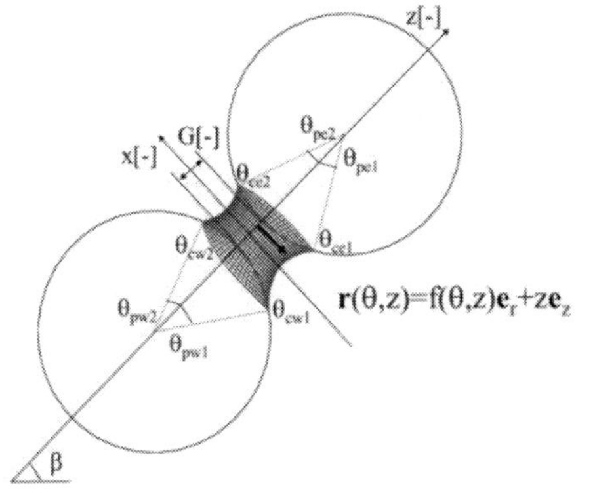

Figure 1 Geometry of a liquid bridge adhered to two spheres with arbitrary inclination in cylindrical coordinate system.

The asymmetry static bridge shape demands for the exact solution of the Young-Laplace equation due to the limit of the toroidal approximation. A prevalent geometry of a liquid bridge adhered to two spheres is drawn in Figure 1. The dimensionless position vector of a liquid surface in cylindrical coordinates is described as $\mathbf{r}(\theta,z) = f(\theta,z)\mathbf{e}_r + z\mathbf{e}_z$ divided by spherical radius, R', which $\mathbf{e}_r$, $\mathbf{e}_z$ are unit basis vectors in r-,z-directions. The symbol (') means the dimensional value. Fig.1 indicates the two dimensional front view with four half-filling (θ_{Pw1}, θ_{Pw2}, θ_{Pe1}, θ_{Pe2}) and contact angles (θ_{Cw1}, θ_{Cw2}, θ_{Ce1}, θ_{Ce2}).

The unit outward normal vector of a bridge surface, $\mathbf{n}_L$, is presented as follows [15],

$$\mathbf{n}_L \equiv \frac{r_\theta \times r_z}{|r_\theta \times r_z|} = \frac{f\mathbf{e}_r - f_\theta \mathbf{e}_\theta - ff_z\mathbf{e}_z}{[f^2(f^2+f_z^2)+f_\theta^2]^{1/2}} \quad ...(1)$$

,where low suffixes (r,θ,z) describe partial derivatives.

The dimensionless effective potential energy, $E=E'/\sigma R'^2$ with Lagrangian multiplier, $K=K'/\sigma R'^2$, on the constrained condition of a constant volume, $V=3V'/4\pi R'^3$, is described as follows [16],

$$E = \int_{-\pi/2}^{\pi/2}\int_{z_{min}}^{z_{max}} 2\left[f^2\left(f^2+f_z^2\right)+f_\theta^2\right]^{1/2} d\theta dz + \frac{Bo}{2}\int_{-\pi/2}^{\pi/2}\int_{z_{min}}^{z_{max}} (z\sin\beta + x\cos\beta) f^2 d\theta dz + K\left[V - \frac{1}{2}\int_{-\pi/2}^{\pi/2}\int_{z_{cp1}}^{z_{cp4}} f^2 d\theta dz\right] \quad ...(2)$$

Bond number (Bo) is a dimensionless number defined by $(\rho_L-\rho_G)gR'^2/\sigma$, where is the ratio of gravitational force to surface tension force equally as Etovoes number (Eoe).

The integral domain of a bridge surface is determined by the trapezoidal section of the upper and lower surface lines, [$z_{w2}<z<z_{e2}$ at $\theta=\pi/2$, $z_{w1}<z<z_{e1}$ at $-\pi/2$] and contact lines, [$z_{w1}<z<z_{w2}$, $z_{e1}<z<z_{e2}$ at $-\pi/2<\theta<\pi/2$] with four contact points settled by four half-filling angles ($\theta_{Pe1},\theta_{Pe2},\theta_{Pw1},\theta_{Pw2}$) in Fig. 1. Two contact lines on the upper and lower spheres were approximated by ellipse lines with parametric aspect ratios, which pass through respectively pairs of two contact points settled with half filling angles, ($\theta_{Pw1},\theta_{Pw2}$) and ($\theta_{Pe1},\theta_{Pe2}$). Filling angles, ($\theta_{PW}(\theta)$, $\theta_{PE}(\theta)$) on the lower and upper spheres have variable properties along the circumferential direction. The formulation of a static bridge shape at equilibrium state results in the extremum problem, which corresponds to the condition that the first order variation equals to zero, $\delta E=0$.

1.2 Estimation of the capillary force of a liquid bridge

The dimensionless mean curvature (H) and contact angles (θ_c) of a bridge surface are respectively calculated by eqs. (3),(4) once the outline of a liquid bridge is numerically evaluated.

$$-2H = \nabla_s \cdot \mathbf{n}_L = \left[(\alpha^2 + f_\theta^2) - f(f^2 + f_\theta^2)f_{zz} - f(1+f_z^2)f_{\theta\theta} + 2ff_z f_\theta f_{\theta z}\right]/\alpha^3 \quad \ldots(3)$$
$$,where\ is\ \alpha^2 = f^2(1+f_z^2) + f_\theta^2$$

$$\theta_c(\theta) = \cos^{-1}(\mathbf{n}_s \cdot \mathbf{n}_L) \quad \ldots(4)$$

,where contact angles also depend on the circumferential angle.

According to the Young-Laplace equation, the dimensionless pressure difference, $\Delta P= \Delta P'R'/\sigma$, across a liquid surface is related with the mean curvature of the surface.

$$\Delta P = (-2H) \quad \ldots(5)$$

The capillary force on a liquid bridge adhered to two spheres must be evaluated only on the contact line as known by the boundary method due to unspecified neck radius of a bridge and asymmetric condition. Dimensionless averaged capillary forces, $F_{Ci} = F'_{Ci}/(2\pi R'\sigma)$ on the upper and lower sphere (i=1,2 respectively correspond to lower, upper sphere) are presented as follows by three terms, which correspond to the surface tension, hydrostatic pressure and buoyancy terms along the longitudinal angle (θ). The sign of the capillary force set to be positive for attractive direction.

$$F_{ci} = \frac{1}{\pi}\int_{-\pi/2}^{\pi/2} \sin(\theta_{Pi})\sin(\theta_{Pi}+\theta_{Ci})d\theta - \frac{1}{2\pi}\int_{-\pi/2}^{\pi/2} \sin^2(\theta_{Pi})\Delta P d\theta \mp \frac{2}{3}\mathrm{Bo}\cdot V_{bi} \quad \ldots(6)$$

, which V_b is sphere volume submerged in a bridge

Finite Element Method is applied for the solution of the $\delta E=0$ equation. Isoparametric elements of a biquadratic polynomial shape function were selected for the integral domain and unknown radius. The set of non-linear algebraic equations for unknown radius at element nodes and Lagrangian multiplier was resolved by the Newton-Raphson method [9].

2. Experiment

Tensile strength of a static and a dynamic pendular liquid bridge

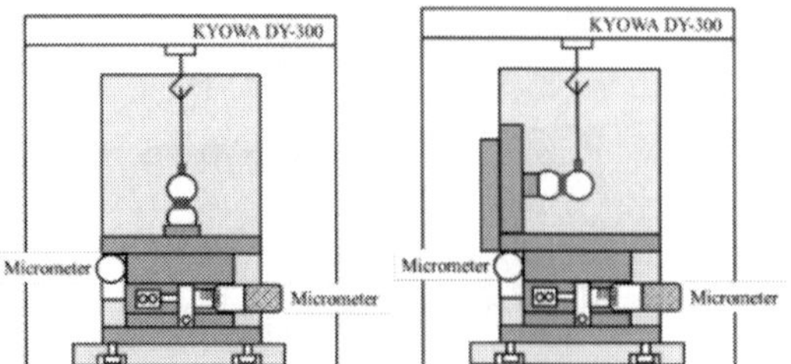

Figure 2 Experimental apparatus for (a) normal and (b) tangential tensile strength of a liquid bridge adhered to two spheres

Figure 2 shows the surface tensiometer (Kyowa interface Science Co., Ltd. DY-300) for measuring (a) the normal and (b) the tangential tensile strength of a water liquid bridge adhered to two stainless steel spheres (radius 3.0mm). The rod attached with one sphere is hanged up to the tensiometer and the other sphere fixed on the displacement stage is moved upward and downward by the tensiometer. The discrete and continuous displacement of the stage are performed for the normal and tangential deformation of a liquid bridge for the static and dynamic states. Two micrometers adjust the appropriate position of the lower spheres for the upper sphere. Water droplets are poured into the spheres aperture and substantially wetted by the displacement of the stage. The tensile strength at static and dynamic state and snapshots of liquid droplets by the digital camera and movie are recorded. The outlines of droplets are evaluated by a digitizer and estimate the half filling angles (θ_{Pe1},θ_{Pe2},θ_{Pw1},θ_{Pw2}) and contact angles (θ_{Cw1},θ_{Cw2},θ_{Ce1},θ_{Ce2}) at four contact points. The sphere configuration (radius, gap), the droplet volume, the half filling angles and the contact lines on spheres initiate parameters of numerical simulation.

3. Results and discussions

3.1 The normal tensile strength of a liquid bridge adhered to two vertical spheres.

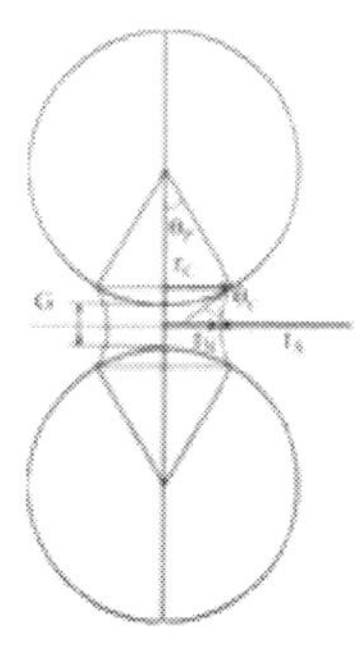

Figure 3 Geometry of toroidal model

The symmetric geometry of a liquid bridge adhered to two vertical spheres without gravity has been extensively investigated by toroidal approximation in Figure 3. The boundary method overestimates the capillary for of a liquid bridge rather than the gorge method but is feasibly and distinctly defined for the asymmetric bridge outline. Symmetric condition simplifies eq.(3) for the mean curvature and eq.(6) for the capillary force into eq.(7),(8) [7,16,17].

$$-2H = \nabla_s \cdot \mathbf{n}_L = \frac{1}{\alpha}\left(\frac{1}{f} - \frac{f_{zz}}{\alpha^2}\right), \textit{where} \quad \alpha^2 = 1 + f_z^2 \qquad \text{...(7)}$$

$$F_{Boundary} = r_C \sin(\theta_{Pi} + \theta_G) - \frac{r_C^{\ 2}}{2}(\Delta P) \mp \frac{2Bo}{3} V_{bi}, \quad r_C = \sin\theta_P \qquad \text{...(8a)}$$

$$F_{Borge} = r_N - \frac{r_N^{\ 2}}{2} \sin\theta_{cpi}(\Delta P) \mp \frac{2Bo}{3} V_{bi} \qquad \text{...(8b)}$$

,where $r_N - r_A = \dfrac{A}{\tan(\pi/2 - \theta_P - \theta_C)} + \sin\theta_P, r_A = \left[A^2 + \{\sin\theta_P - (r_N - r_A)^2\right]^{1/2} A = G/2 + 1 - \cos\theta_P$

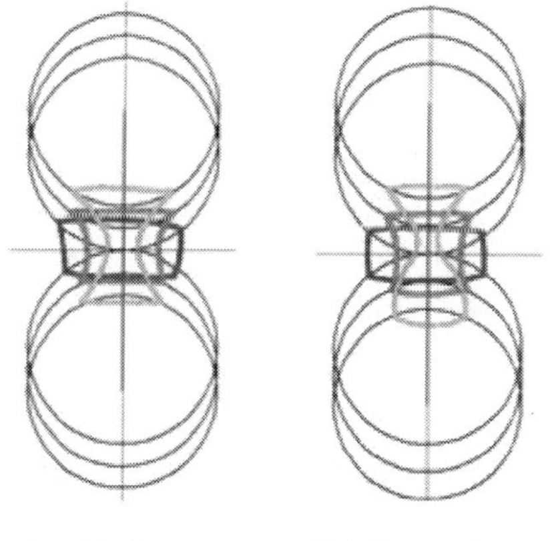

(a) Static state $G=0,\ 0.47,\ 0.91$ (b) Dynamic state $G=0,\ 0.59,\ 1.16$

Figure 4 Experimental droplet outlines at static and dynamic state, Bo=1.21, V=0.088

Figures 4 show symmetric outlines of liquid bridges around the vertical axis respectively at the static and dynamic state. Static state results in the shorter rupture gap and the larger dimensionless normal tensile strength from Figures 5a. Sphere gaps at the peak static capillary force depend on liquid volume. The contact angle on the upper sphere corresponds to the advancing contact angle at the higher Bo number due to gravity effect.

Fig. 4b indicates the downward transition of a liquid immediately before the rupture due to the wettability a liquid. The toroidal approximations over-estimate experimental data at the lower sphere gap but the sphere gaps at the peak dynamic capillary force depend on the velocity of stage displacement. The velocity reduces the peak value of dynamic capillary due to the downward transition of a liquid in Fig. 5b. The stepwise profiles of capillary force are due to the movement sensitivity of tensiometer for the stage displacement.

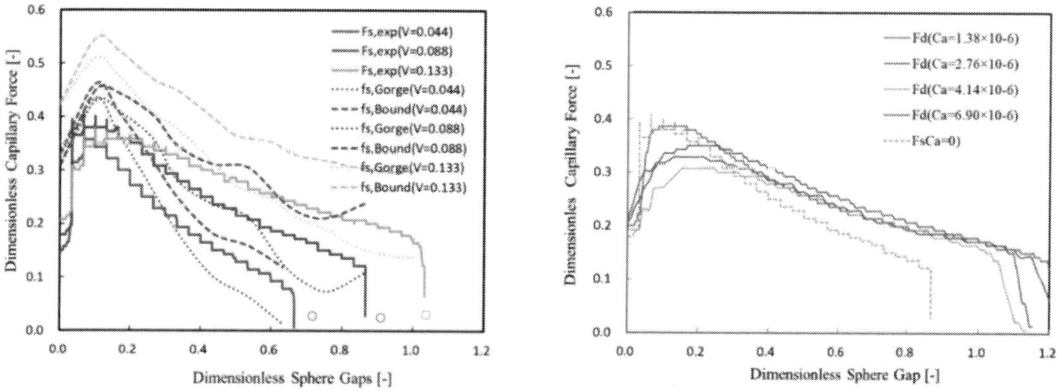

(a) at the static state (b) at the dynamic state

Figure 5 Effect of sphere gap on dimensionless capillary force

3.2 Estimation of boundary condition at contact line by LBM

The Young's equation describes the force balance among three interface tensions (solid-liquid, solid-gas and liquid gas) at three phase contact point and also defines the relationship between interface tension and contact angle. The solution of the Young-Laplace equation at the static state demands for one unknown boundary condition of either a contact angle or a contact line. The simulation for the wettability on three dimensional contact line demands for the elaborate process. Our experiment determines the contact lines on spheres under the constant liquid volume and is utilized for the boundary condition. Besides the perfect wettability the smooth transition of three phase contact lines is unexpected due to the advancing and receding contact angle. Lattice Boltzmann Method (LBM) is expected to estimate the boundary condition at static and dynamic state. The geometry of a liquid bridge is numerically estimated by FEM with half-filling angles predicted by 2 dimensional Shan & Chen model of LBM, which is used by D2Q9 model for transition of distribution function. The estimation for the liquid geometry by FEM has been verified. Figures 6 show the predicted liquid outlines by LBM and FEM. at a given liquid volume.

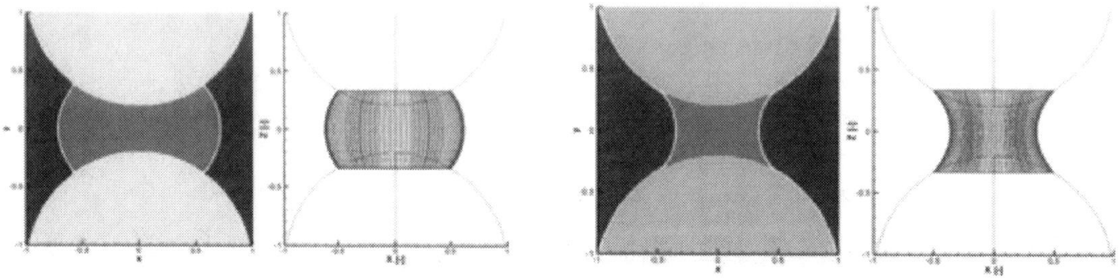

(a) Repulsive interparticle force | (b) Attractive interparticle force

Figure 6 Snapshots of liquid geometries at static state by LBM and FEM

Figures 7 show snapshots of dynamic liquid outlines at a given liquid volume up to the rapture state by LBM. The transition of boundary conditions, i.e. filling and contact angles, is smoothly described. Gravity and imperfect wettability, i.e. advancing and receding contact angles, are excluded in Figs. 7 but the effect of gravity on the liquid outline is avairable.

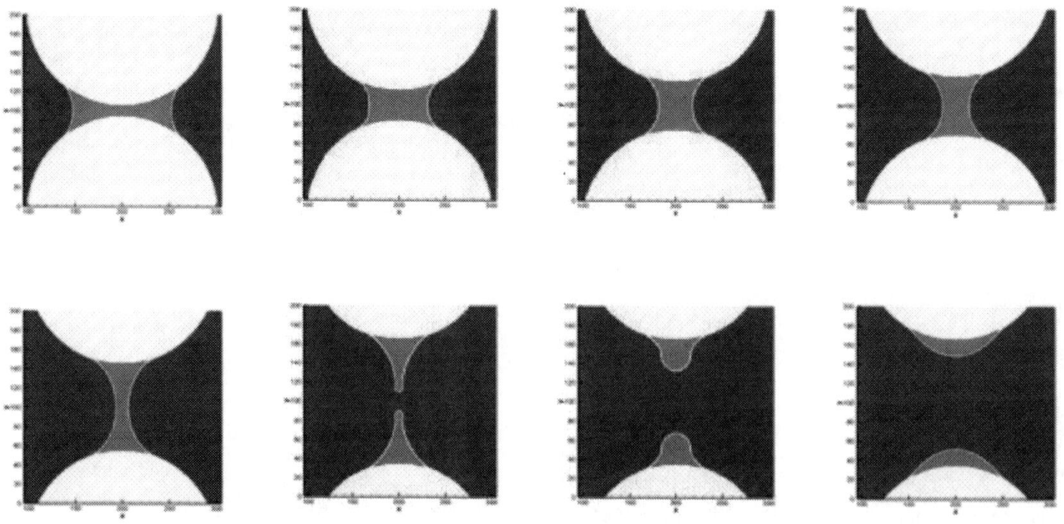

Figure 7 Snapshots of snapshots of dynamic liquid outlines at a given liquid volume

3.3 The tensile strength of a liquid bridge with shear deformation.

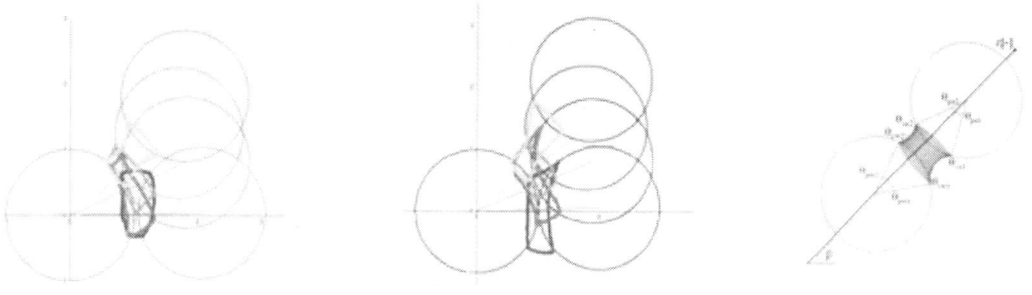

(a) Static state (G=0, 0.59, 0.98) (b) Dynamic state (G=0, 0.59, 0.98)

Figure 8 Liquid outlines at static and dynamic state (Bo=1.21, V=0.088)

Definitions of half-filling and contact angles.

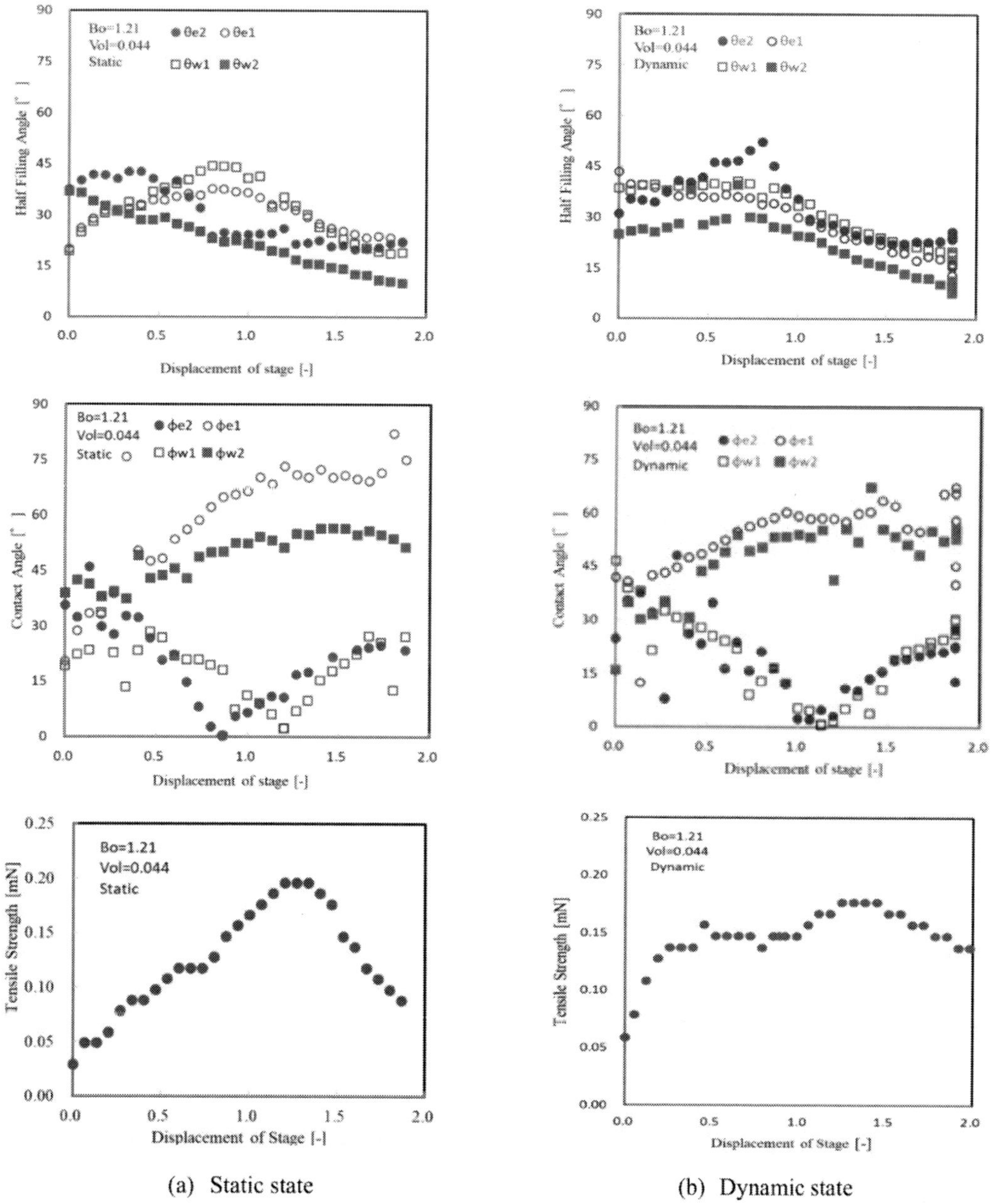

(a) Static state (b) Dynamic state

Figure 9 Profiles of half-filling angles, contact angles and tensile strength , Bo=1.21, V=0.088

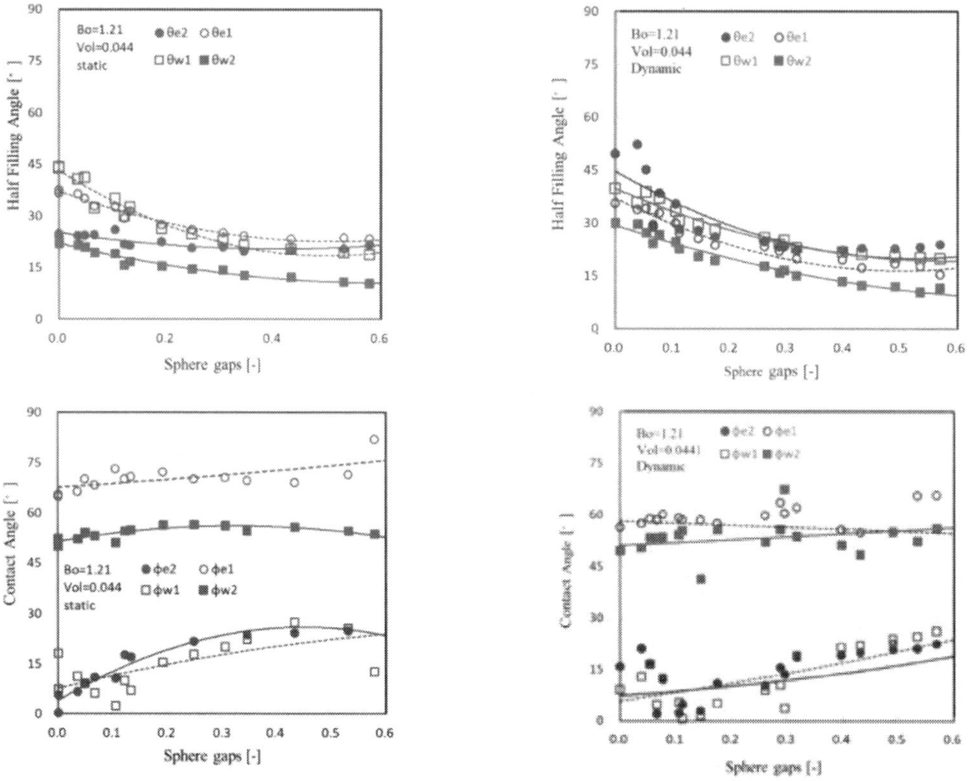

(a) Static state (b) Dynamic state

Figure 10 Profiles of half-filling angles, contact angles with sphere gaps , Bo=1.21,V=0.088

Figures 8 show four types of liquid outlines at static and dynamic state. Magenta and green plots are drawn respectively at the extreme values of half filling angles, contact angles and tensile strengths in Figures 9. During the lower stage displacement two spheres keep at the contact state and the half-filling angles reached to the peak values. Half-filling angles sifted with displacements of a stage have the peak values except the static half-filling angle at the top point of the lower sphere (θ_{Pw2}), which monotonously decrease with stage displacement. The increase and decrease of contact angle respectively corresponds to the advancing and receding contact angle. The contact angles at the bottom point of the lower sphere (θ_{Cw1}) and at the top point of the upper sphere (θ_{Ce2}) transit from the receding to the advancing contact angle due to the infinitesimal values. Figures 10 redraw the half-filling and contact angles with respect to sphere gap. Contact angles contribute to the peak tensile strength, which is measured as resultant force (F_T) of normal (F_n) and tangential (F_t) tensile strength in Figure 11. The normal tensile strength conventionally corresponds to the capillary force. The stage displacement at the peak tensile strength minimize the sum of half-filling and contact angle due to the positive hydrostatic pressure across liquid surface. The larger dynamic tangential strength occurs during the sphere contacting due to the asymmetric deformation. The adhesion force of liquid to sphere at the static state contribute to the larger peak tensile strength.

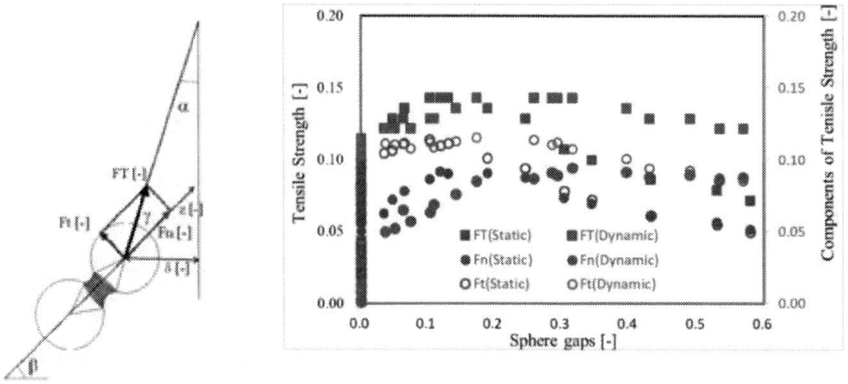

Fig.11 Normal and tangential components of tensile strength at static state

3.4 Numerical simulation at the static state

The solution of Young-Laplace equation demands for one unknown necessary condition on the contact point, i.e. half-filling or contact angle. Liquid surfaces simulated by FEM are drawn with solid lines in Figures 12. The simulation reveals the asymmetric liquid geometry with half-filling and contact angles along longitudinal angle. Disagreements of liquid interfaces results from the approximation of contact lines on spheres. The tangential strength influences depend on the mean curvature. Especially the second figure at the peak half-filling angle has the negative hydrostaic pressure in the direction of relative sphere displacement and positive pressure in the oppsite direction.

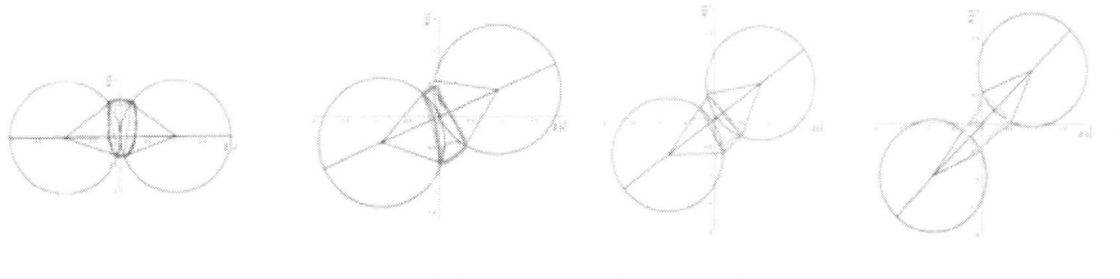

At the initial geometry At the peak halffilling angle At the peak tensile strength At the maximum shpere gap

Figure 12 Numerical estimation at the static state

Figure 13 displays transitions of measured normal tensile strengths ($F_{exp,n}$) and simulated capillary force (F_{cal}) to sphere gap. The initial droplet is attached to the above aperture in the direction of sphere movement due to gravity. The stage displacement drifts parts of liquid volume into the lower aperture of spheres. The sphere on displacement stage keep contacting with the other sphere hang up to the tensiometer up to the peak tensile strength. Fcal and FcalB are numerically estimated by outlines of liquid surfaces respectively with Young-Laplace equation and boundary toroidal model. The sign of hydrostatic pressure with deformation of liquid droplets results in the repulsive capillary force. The normal tensile strength corresponds to the capillary force during the range of symmetric liquid bridge. The angle of beta is defined by the inclination of center distance of a sphere pair. The lower normal tensile strength than the capillary force results from the orthogonal component of capillary force at the low range of beta.

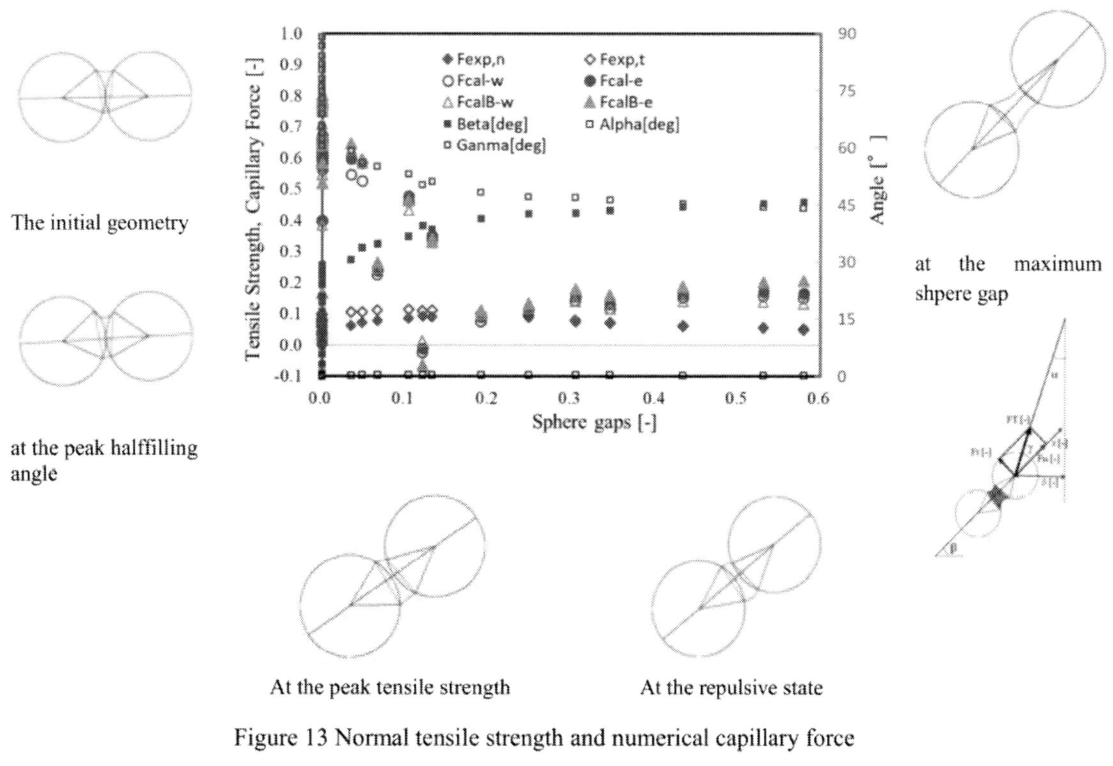

Figure 13 Normal tensile strength and numerical capillary force

Conclusions

Liquid bridges adhered to particles are deformed with the normal and tangential stress in wet granulator. The interparticle force with asymmetric droplet by shear stress must be estimated by the three dimensional solution of Young-Laplace equation. A surface tensiometer and numerical simulation contribute to the normal and tangential tensile strength of liquid bridge adhered to two spheres.

1. The normal strength of symmetric liquid bridge with a surface tensiometer is verified by the conventional toroidal model and the solution of Young-Laplace equation.
2. Sphere gaps at the peak capillary force in the vertical sphere configuration depend on liquid volume and displacement velocity of sphere.
3. The lattice Boltzmann Method is expected to numerically predict the boundary conditions on the contact lines, which are half-filling angle and contact angle at the static and dynamic state.
4. The tensile strength depends on transition of contact angles, which correspond to advancing or receding angles. The larger sum of contact and half-filling angles than the orthogonal angle corresponds to the positive hydrostatic pressure across liquid surface.
5. The normal tensile strength is related with the solution of Young-Laplace equation by FEM. The tangential tensile strength depends on the liquid transition within droplets and the variable mean curvature.

References

[1] K.Murase, H.Shibata, Y.Yasuda and S.Toyama, *Jpn J. Multiphase Flow,* **5**, 322(1991)
[2] A.E.Saez and R.G.Carbonell, *J. Colloid Interface Sci.,* **140**, 408(1990)
[3] D.N.Mazzone, G.I.Tardos and R.Peffer, *Powder Tech.*, **51**, 71(1987)
[4] C.D.Willet, M.J.Adama, S.A.Johnson and J.P.K.Seville, *Langmuir*, **16**, 9405(2000)
[5] M.J.Adama, S.A.Johnson, J.P.K.Seville and C.D.Willet, *Langmuir*, **18**, 6180(2002)
[6] R.K.Niven, *J Petroleum Sci. Eng.*, **52**, 1(2006)
[7] O.Pitois, P.Moucheront and X.Chateau, *Eur. Phys. J.B*, **23**, 79(2001)
[8] S.J.R.Simons, J.P.K. Seville and M.J. Adams, *Chem. Eng. Sci.*, **49**, 2331(1994)
[9] C.D.Willet, M.J.Adama,S.A.Johnson and J.P.K.Seville, , *Powder Tech.*, **130**, 63(2003)
[10] S.M.Iveson, J.A.Beathe and N.W.Page, *Powder Tech.*, **127,** 149(2002)
[11] R.A.Fisher, *J. Agri. Sci.*, 16,492(1926)
[12] F.M.Orr, L.E.Scriven and A.P.Rivas, *J. Fluid Mech.*, **67**, 723(1975)
[13] O.Pitois, P.Moucheront and X.Chateau, *J. Colloid Interface Sci.*, **231**, 26(2000)
[14] U.El Shamy and T.Groeger, *Int. J. Numer. Anal. Meth. Geomeh.*, **32**, 1763(2008)
[15] F.M.Orr, R.A.Brown and L.E.Scriven, *J. Colloid Interface Sci.,* **60**, 137(1977)
[16] K.Murase, T.Mochida, Y.Sagawa and H.Sugama, *Adv. PowderTech.*, **19**, 349(2008)
[17] C.D.Willet, M.J.Adama,S.A.Johnson and J.P.K.Seville, *Handbook of powder technology*, **11**, chap.28, ed. A.D.Salman, M.J.Hounslow and J.P.K.Seville,Elsevier (2007)

NUMERICAL SIMULATION OF WIRE-PLATE ELECTROSTATIC PRECIPITATOR - EFFECT OF PARTICLE CONCENTRATION

Ding Yang[1], Baoyu Guo[2], Jun Guo[1], Aibing Yu[2]*

[1] *Fujian Longking Co. Ltd, Longyan, Fujian, 364000, China*

[2] *Department of Chemical Engineering, Monash University, Clayton VIC 3800, Australia*

Introduction

Electrostatic Precipitators (ESP) have been widely used to separate fine particles from large scale industrial exhaust gases, such as coal fired power plants. In-depth knowledge and careful design optimization are critical to meeting the increasingly strict emission standard in terms of $PM_{2.5}$ emission limit, e.g., 5-10 mg/Nm^3. Numerical modelling has been employed as a powerful tool in this area. Complicated interactions among electric field, gas flow and particle transport are summarized by Adamiak [1]. To model such a system, various levels of simplifications have been made in published papers.

Among them, Lagrangian particle tracking technique is a widely accepted approach. The mass concentration of ash particles in coal-fire flue gas is typically 10-50 g/Nm^3, corresponding to volume fraction in the order of 10^{-5}. Such a small value is often regarded as sufficiently dilute particle concentration in numerical modeling of multiphase flows. For the case of ESP, complicated interactions among the e-field, gas flow and charged particles are involved, even if the deposited dust layer is ignored. Full coupling between the continuous gas phase and discrete particle phase is computationally expensive to achieve a converged steady solution. To avoid such a high cost, the model for ESP was simplified as so called one-way coupling in many previous works, so that the electric field is solved without the particle phase (Guo et al. [2,3]; Farnoosh et al. [4]). This simplification may be acceptable in the end stage of ESP, but will introduce significant error in the early collection stage. Limited studies reported in the literature have considered the particle charges as a source term in the Poisson's equation of electrostatics. For example, Choi and Fletcher [5] predicted a reduced current for the particle-loaded case (d_p = 1-5μm) compared with the particle-free case, but the corona ion charge density somehow shows an increasing trend along the stream. When fine dust particles are treated, the large specific surface area of the particle phase, combined with reduced collection efficiency, implies high space charge density in the collection region. Its effect was understood as a partial suppression of the corona current [6]. As was experimentally proved by Podlinski et al. [7], space charge of charged particles can contribute to the EHD flow pattern and particle transport. In present work, particle space charge is considered through two-way coupling and interesting results are generated in terms of ESP performance as particle loading changes.

Integrated ESP Modelling

Mathematical Equation

Coupled electric field, gas-particle flow and particle charging processes are simulated using

Eulerian-Lagrangian method in Computational Fluid Dynamics (CFD). Major governing equations of vector format are listed below for completeness. For electric field, Poisson equation of potential

$$\nabla^2 V = -(\rho_{ion} + \rho_c)/\varepsilon_0 \quad \mathbf{E} = -\nabla V \tag{1}$$

Current continuity equation

$$\nabla \bullet ((b\mathbf{E} + \mathbf{U}_f)\, \rho_{ion} - D\nabla \rho_{ion}) = S_c \tag{2}$$

where ρ_{ion} is corona charge density, b the ion mobility, D is ion diffusivity, ρ_c is particle space charge density, evaluated by averaging the charges of particles that pass through a control volume (cell),

$$\rho_c = \frac{1}{V_{cell}} \sum_{n=1}^{N} \dot{n}_p q_p \Delta t \tag{3}$$

Where $\dot{n}_p$ is the number flow rate of a parcel of particles tracked, Δt is the residence time in the cell. The source term, S_c, in Equation (2) is transfer rate of ion charges to particles, which is not considered at the moment, i.e., $S_c = 0$. For the gas flow, the momentum equation is,

$$\begin{aligned} \nabla \bullet (\rho_g \mathbf{U}_g \otimes \mathbf{U}_g) = \nabla(p + \tfrac{2}{3}\rho_g k) + \nabla \bullet [(\mu + \mu_t)(\nabla \mathbf{U}_g + (\nabla \mathbf{U}_g)^{\mathrm{T}})] + \rho_{ion}\mathbf{E} \\ -\frac{3\pi n_p d_p \mu}{\mathrm{C_C}}(1 + \mathrm{Re}_p^{0.687})(\mathbf{U}_g - \mathbf{U}_p) \end{aligned} \tag{4}$$

For the particle flow, the force balance of a single particle is

$$m_p \frac{d\mathbf{U}_p}{dt} = \frac{3\pi n_p d_p \mu}{\mathrm{C_C}}(1 + \mathrm{Re}_p^{0.687})(\mathbf{U}_g - \mathbf{U}_p) + \tfrac{1}{6}\pi d_p^3 \mathbf{g}(\rho_p - \rho_g) + n_e q_e \mathbf{E} + \mathbf{F}_{dis} \tag{5}$$

k-ω SST turbulence model is chosen for gas flow, and the dimensionless Lawless charging rate equation is used for particle charging [8].

Model Conditions

An integrated ESP model developed comprises two parallel modules or submodels, namely, electric field model and coupled gas-particle flow model. One-way coupling between the two modules is used earlier is upgraded to two-way in the present work. The electrostatic field caused by charged particles in suspension is considered in the solution of electric field. The two submodels are coupled through a small under-relaxation factor, in views of strong interaction between the electric field and particles flow, and the stochastic nature of the discrete particle phase. In this way all the volume averaged quantities related to the particle phase (such as the particle space charge density) are smoothed out over time in the simulation.

A single channel, 4-wire and flat plate configuration is considered. The dimensions are 3.5mm in wire diameter, 0.24m for wire-wire spacing, 0.2m for wire-plate spacing. The plate length is 1.2m. The simulation domain is defined as 1.74m long by 0.4m wide box with the 4 holes inside. The grid is smoothly refined at places where higher gradient in electric quantities is expected, such as the proximity of the wires and middle of the adjacent wires. Other base case conditions are as follows.

- Applied voltage on the wire electrode: Va = -60kV. The charge density on the wire surface is determined using an injection law combined with Peek equation (Guo et al.[9]).
- Gas temperature: 120°C. As the inlet boundary condition for the gas phase, the velocity is set as 1 Nm/s; relative turbulence intensity 3.7%, turbulent length scale $0.3D_h = 0.24$m.
- Dust particle size is mono-dispersed at d_p=2.5 μm. Particle density 2000 kg/m^3; ash dielectric constant of κ_p=6.6, Particle loading at the inlet, $(\rho_p)_{in}$, is varied with default value of 10 g/Nm^3.

Typical Results

Electric Field without Particles

When the inlet/outlet boundaries are moved sufficient long distance away from the electrodes in the simulation domain (referred to as "natural boundary condition"), the electric field around the boundary wires show an asymmetric distribution pattern that differ from that around the inner wires. Electric field-lines/current can go out of the limit of the collection region (Figure 1). Electric field intensity on the collection plate is of interest as it is considered a factor to determine the upper limit of operation before spark-over [10]. Normally the field intensity on the plate is highest at the point that faces exactly the wire center with the least distance. This maximum field intensity on the plate is 3.80×10^5 V/m. However, when the natural boundary condition is used, the highest value (1.06×10^6 V/m) predicted in field intensity appears at the endpoint of the plate due to its sharpness. This is a likely point that would trigger electric spark-over prematurely. Practically such a sharp edge should be avoided. Generally, the field strength at the middle wires is smaller than that at a boundary wire, due to the counteraction of adjacent wires. Then the corona charge density at the boundary wires should be greater than that at the inner wires (Figure 2), which is similar to the simulation result of Kasdi [11].

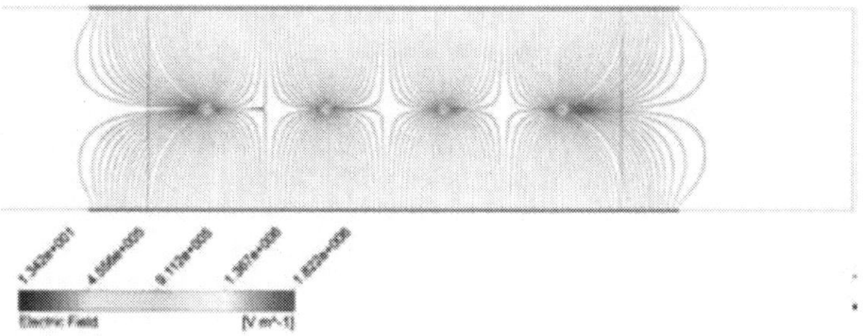

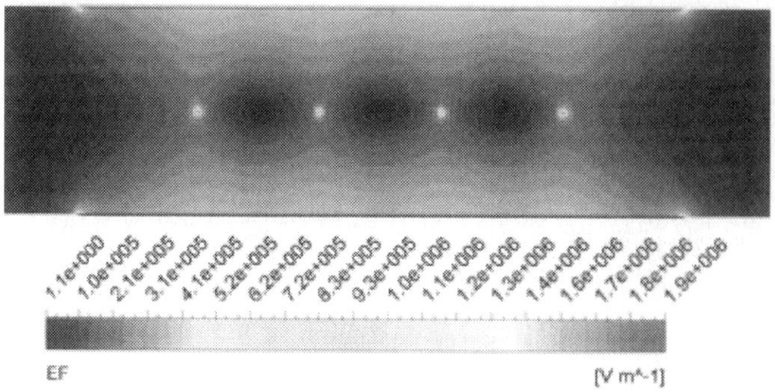

Figure 1. Field streamlines and intensity distribution (no particles).

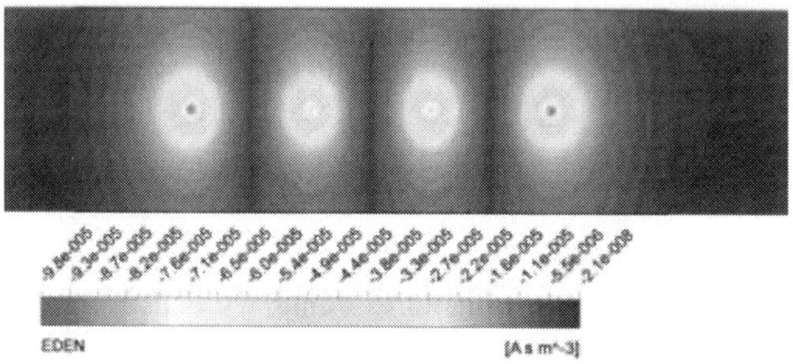

Figure 2. Corona charge density (no particles).

Effect of Particle Space Charge

Particles are randomly injected from the inlet plane with uniform probability in locations and no-slip for velocity (same as the gas velocity). In using random number generator, it is ensured that the injection points are not repeated for every coupling cycle. Figure 3 shows sampled particle trajectories with color scaled to particle charge. Figure 4 shows a pattern of particle deposition rate on the plate. It is seen that the end points of the plate, particularly the downstream one, are special spots to accumulate more dust.

Figure 3. Particle trajectories with colour scaled to particle charge.

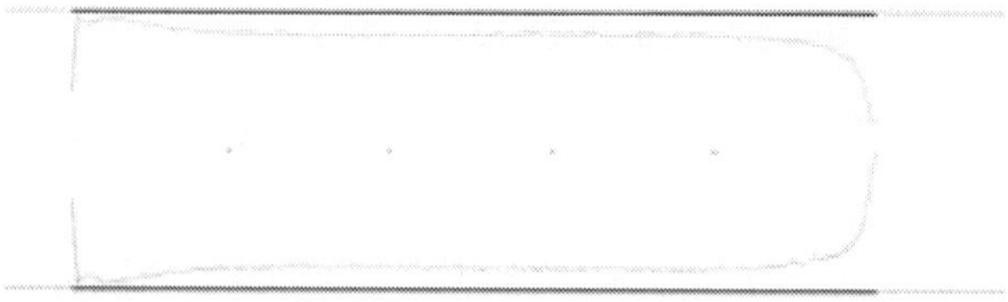

Figure 4. Indicative pattern of particle deposition rate on the plate.

Figure 5 shows the averaged space density of charges carried by particles. Highest space charge density appears near the first wire electrode, where electric current density is highest. It diffuses gradually from the centre line with particles drifting toward the plate while moving forward. Due to the radiating nature of the field-lines around each wire, the lateral component of Coulomb's force in the region close to the centre line may not be as large as that off the centre line. Not collected charged particles of negative polarity are conveyed towards the outlet, lowering the potential downstream outside the collection region (Figure 6), causing distortion of the overall electric field distribution.

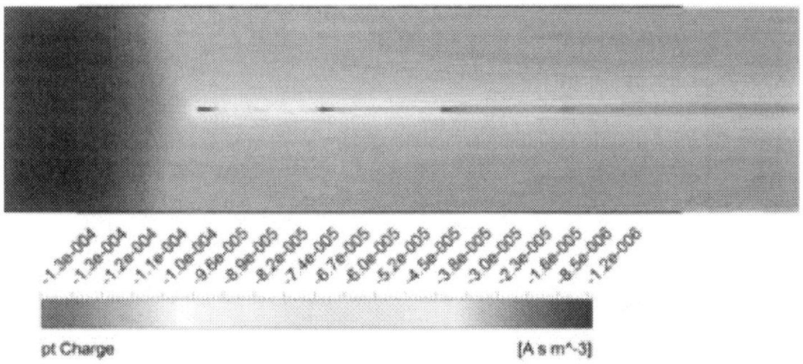

Figure 5. Distribution of space charge density of particles.

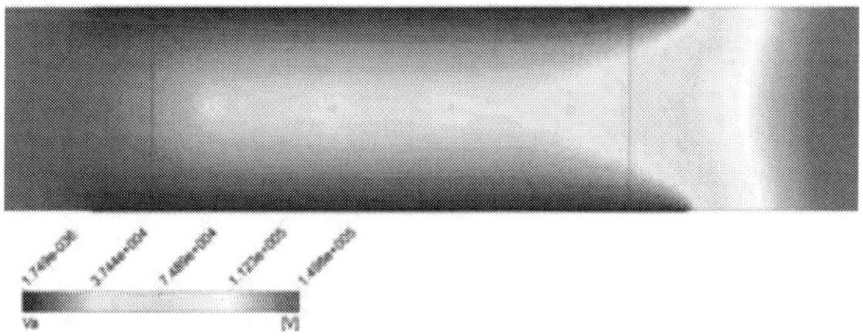

Figure 6. Distribution of electric potential (in absolute value).

Effect of Particle Loading

On Electric Current

Figure 7(a) shows the mean corona current, based on unit length of wire. As can be seen, the corona current drops quickly as particle loading increases, although a fixed voltage is applied. This is caused by the secondary electrostatic field created by the space charges of particles that cancels out the primary electric field at the wire surface and subsequently suppressed the corona generation.

On Particle Collection Efficiency

Predicted particle collection efficiency, as particle loading changes, is given in Figure 7(b). The collection efficiency, increases initially with particle loading (<3 g/m^3), peaks at particle loading of 4-5 g/m^3 and then decreases as particle loading increases further. When ignoring particle space charge, the electric field would be independent of particle loading. This is the case for very low particle loading, where the fractional collection efficiency is constant, or in other words, the quantity of dust collected as well as that penetrated increases proportionally with particle loading.

The observed enhancement of collection by a slight amount of particles is due to the redistribution of electric field so that the near-plate and volume-averaged field intensity happens to increase, which favor the charging of most particles. The asymptotic lower limit of the curve should correspond to the situation when the corona current is so small that turbulent dispersion dominates over the drift force.

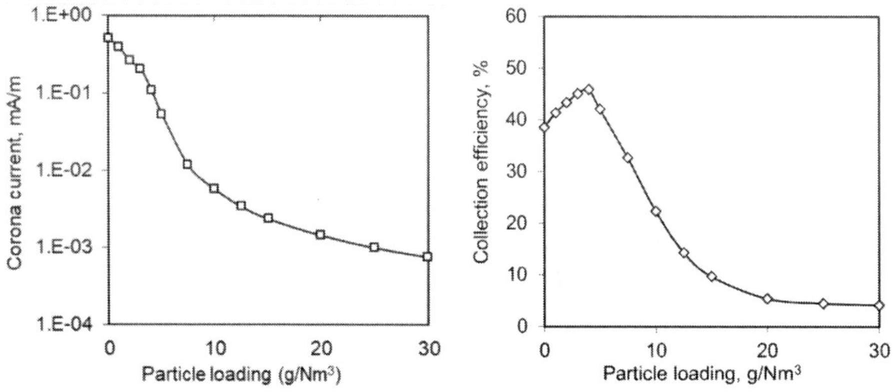

(a) (b)

Figure 7. Mean corona current (a) and collection efficiency (b) as function of particle loading.

On Electric Field Distribution

Figure 8 visualizes the predicted corona charge density and field-lines, which shows how the electric field distribution evolves with particle loading. As the electric field is distorted starting first from the end side, corona density at 3 of the 4 downstream wires is suppressed faster than the 1st wire. When the particle loading is above 7.5 g/m^3, all wires except the first one stop emitting. Meanwhile, reversed field-lines appear around the last wire, and the affected region expands as the loading increases further. Otherwise the field-lines become smoother with increasing loading. Since the field-lines are extended to the outlet, some escaped particles can be pulled back into the collection region, a factor that contributes to significantly greater deposition rate around the end-point of the plate.

Overall, the particle charges affect the particle collection adversely, as they tend to suppress the corona emission or reduce the corona current. Particularly the penetrated particles, which carried significant charges, have prolonged effect as they float along. The corona at the downstream electrodes even extinguish unless the applied voltage is raised individually. When highly charged, more particles precipitate to the plate, reducing the number of particles floating ahead (or less space charge density). The charges carried by these remaining particles exert a negative feedback to the electric field and particle charging, which compromises the collection efficiency.

Electric Quenching

Corona quenching is a common phenomenon in ESP when fine particles are treated. Riebel et al. [12] experimentally studied the corona discharge characteristics of ESP with glycerol aerosol, which found corona onset voltage increases and the corona current significantly decreases after the glycerol steam was injected into the system. Under fixed voltage in the present simulation, corona currents from all the wire electrodes drop with increasing solid loading. The downstream coronas (from the second one) will eventually extinguish at some point. An interesting phenomenon arises when the particle loading increases above 10 g/Nm^3, i.e., the electric streamlines in a small zone around the last wire reverse direction locally. This implies that a charged particle, once falls accidently within the zone, could be trapped and attracted to the same wire, as seen from the particle trajectories in Figure 9. Thus this wire is subject to heavier dust deposition.

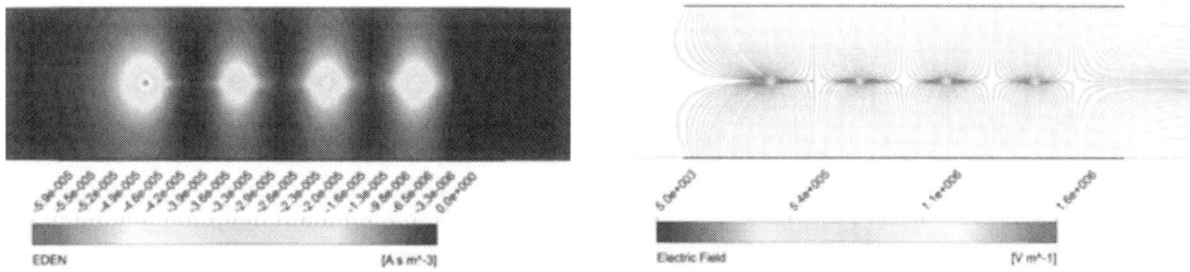

$(\rho_p)_{in}$ = 3 g/Nm^3

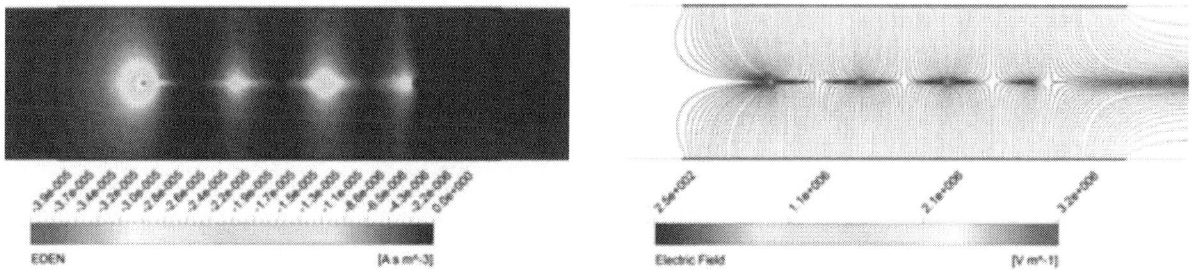

$(\rho_p)_{in}$ = 5 g/Nm^3

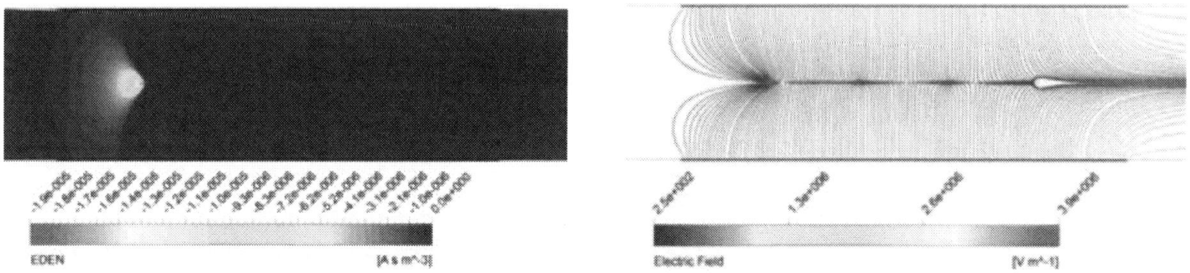

$(\rho_p)_{in}$ = 10 g/Nm^3

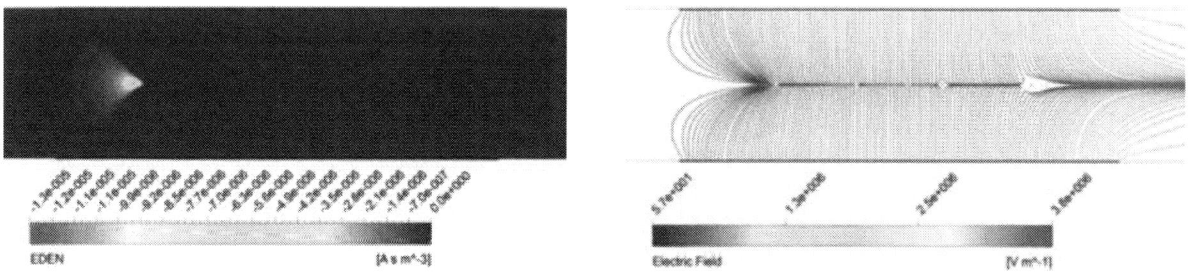

$(\rho_p)_{in}$ = 20 g/Nm^3

Figure 8. Visualizations of corona charge density and field streamlines for different particle loadings.

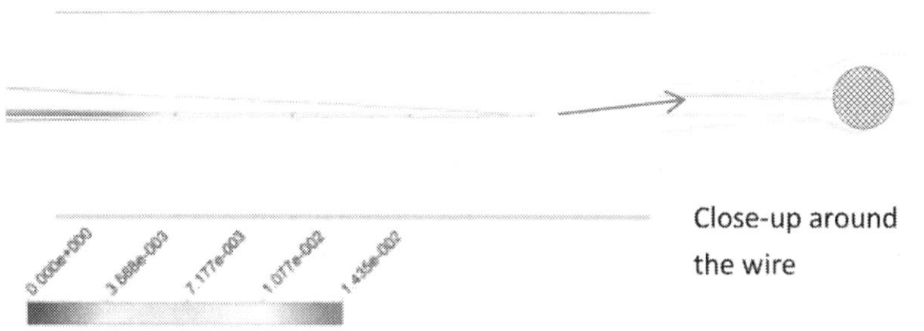

Figure 9. Trajectories of particles falling on the corona wire ($(\rho_p)_{in} = 20$ g/Nm^3).

Concluding Remarks

CFD based ESP model has been advanced, with an enhanced coupling between electric field and particle flow. So the model can be readily applied to the early collection stage of the ESP, where dust particle loading is too high to neglect. Space charge of particles is considered, and its significance is clarified in a broad range of particle mass concentration in the gas. Dust loading significantly affects the dust collection efficiency for fine particles considered. Maximum collection efficiency exists at a critical particle loading that depends on particle size. However, as particle loading increases, collection efficiency decreases and eventually approaches asymptotically to a lower limit. The phenomenon of electrical quenching is observed, where downstream corona is extinguished or one of the emitters is polluted when the particle loading is high.

References

1. Adamiak K. (2013), "Numerical models in simulating wire-plate electrostatic precipitators: A review", *J. Electrostatics*, 71, pp.673-680.
2. Guo B.Y., Yu A.B. and Guo J. (2014), "Numerical modelling of electrostatic precipitation: effect of gas temperature", *J. Aerosol Science*, 77, pp.102-115.
3. Guo B.Y., Yu A.B. and Guo J. (2015), "Numerical modelling of ESP for design optimization", *Procedia Engineering*, 102, pp.1366-72.
4. N. Farnoosh, K. Adamiak, G.S.P. Castle (2010), "3-D numerical analysis of EHD turbulent flow and mono-disperse charged particle transport and collection in a wire-plate ESP", *J. Electrostatics*, 68, pp.513-522.
5. Choi B.S. and Fletcher C.A.J. (1998), "Turbulent particle dispersion in an electrostatic precipitator," *Applied Mathematical Modelling*, 22, pp.1009-1021.
6. Parker K.R., Applied Electrostatic Precipitation, Blackie Academic & Professional, 1997.
7. Podlinski J., Niewulis A., Mizeraczyk J. and Atten P. (2008), "ESP performance for various dust densities", *J. Electrostatics*, 66, pp.246-253.
8. Lawless P.A. (1996), "Particle charging bounds, symmetry relations, and an analytic charging rate model for the continuum regime", *J. Aerosol Sci.*, 27, pp.191-215.
9. Guo B.Y., Guo J. and Yu A.B. (2014), "Simulation of the electric field in wire-plate type electrostatic precipitators", J. Electrostatics, 72, pp.301-310
10. White H.J. (1974), "Resistivity problems in electrostatic precipitation", *J. Air Pollut. Control Assoc.* 24, pp.313–338.
11. Kasdi A. (2016), "Computation and measurement of corona current density and V–I characteristics in wires-to-plates electrostatic precipitator", *J. Electrostatics*, 81, 1-8.
12. Riebel U., Radtke R. and Loos R. (2002), "An experimental investigation on corona quenching", *J. Electrostatics*, 54, pp.159-165.

Triboelectric Charge of Spherical Glass Particles against Metal Pipeline

HoSu CHOI [1], Teruo SUZUKI [2] and Kwangseok CHOI [3,]*
[1] Christian Academy in Japan, [2] Kasuga Denki Inc.,
[3] National Institute of Occupational Safety and Health, Japan
**choiks@jniosh.go.jp*

Introduction

Insulating powders are used in many ways such as in the food, pharmacy, and chemical industries. Usually, these powders are highly charged when transported, loaded, dried, and mixed, due to repeated contact and separation between powder particles and the equipment. This results in the powders acquiring electrostatic charges. If not dissipated quickly, electrostatic charge buildup can cause inter-particle cohesion and adhesion to operating equipment, causing a production efficiency decrease. Incendiary electrostatic discharges such as spark discharges, cone discharges, and propagation brush discharges can also take place during industrial operations and may lead to fires and even explosions. Riedewald reported on an industrial fire in a pharmaceutical powder transfer operation [1]. A large number of studies involving electrostatic charge generation on powders, as well as electrostatic discharges and hazards during operation, have been reported. For example, Kaialy reviewed the main factors affecting the electrostatic charging of pharmaceuticals such as particle resistivity, size, shape, and surface roughness [2]. Gibson also reviewed industrial hazards related to electrostatic charges and discharges [3]. Our previous studies [4–6] investigated the electrostatic charging of polymer powder in mixing processes and the ignitability of aluminous, magnesium and polymer powders due to electrostatic discharges involving the effect of the ratio of N2 in the air on the ignitability.

In order to easily measure and confirm the electrostatic charge level of powders, a novel tribocharging system that utilizes the Faraday cup is presented, and the charging behavior of the sample powders is investigated in this paper.

Experimental Apparatus and method

Figure 1 shows the spiral air tribocharge apparatus presented in this study. The apparatus was divided into four main parts: the air supply part, the powder supply part, the tribocharge part, and powder collection part. The air supply part consisted of a compressor (Hitachi, PO3.7MA) that included an auto dryer (CKD, F3-217199), air controller, electronic valve, and switch. In this study, dried air was used because moisture in the air may have an effect on the electrostatic characteristics of the powder sample. The powder supply and tribocharge parts consisted of a powder port, a spiral air flow device, and a pipeline. The spiral air flow device (photo 1) was attached in front of the sample powder port in order to create spiral air inside the pipeline. The spiral air flow device had 8 holes that were 6 mm in diameter and that opened at a slant (60 degrees), which is very different from conventional models [7, 8, 9]. The straight pipeline was made up of stainless steel metal and was 1300 mm in length and 35 mm in diameter. In order to prevent external noise, the overall powder supply and tribocharge parts were shielded. The powder collection part was made up of a cyclone-type Faraday cup, unlike conventional models [7, 8, 9]. The cyclone-type Faraday cup had a separate metal container (30 mm in diameter) for easy measurement of the amount of collected powders. The other devices included two electrometers (Kethley,

6517A), a data logger (Graphtec, GL900), an electrostatic ionizer (Kasuga Denki Inc., KD-510W), an electrical balance (Shimazu, BL220H), and a timer.

The typical operating procedure for the tribocharge test is summarized as follows:
(1) Adjust the air pressure using the air pressure valve.
(2) Load 1 g of the sample powder into the powder port.
(3) Eliminate the initial charge of the sample powders using an electrostatic ionizer.
(4) Reset the electrometers (Q mode) connected to the pipeline and cyclone.
(5) Switch the air flow on, and run the sample powder through the pipeline where friction occurs.
(6) Measure the indication value when it stabilizes.
(7) Measure the amount of powder collected in the Faraday cup.
(8) Clean out the pipeline and cyclone part using an air spray in preparation for the next test.

The charge to mass ratio, q (nC/g), of the sample powder was obtained by dividing the charge, Q (nC), by mass, m (g), of the sample powder loaded into the Faraday cup:

$$q = \frac{Q}{m} \quad (1)$$

This experiment is mainly to determine how much the charge to mass ratio, q (nC/g), of the sample powders is affected by the size of the powder and the amount of air pressure. The cyclone-type Faraday cup and the pipeline were suspended electrically with insulators. All experiments were conducted in ambient conditions (air temperature: 22 ℃; relative humidity: 55 %).

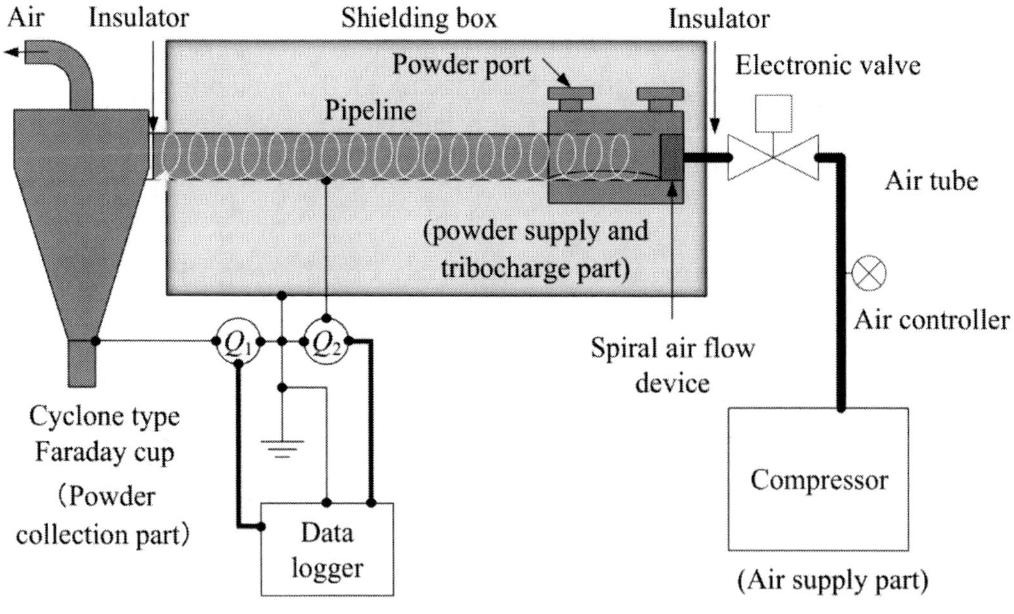

Fig. 1 Schematic of the spiral air tribocharge apparatus used in this study.

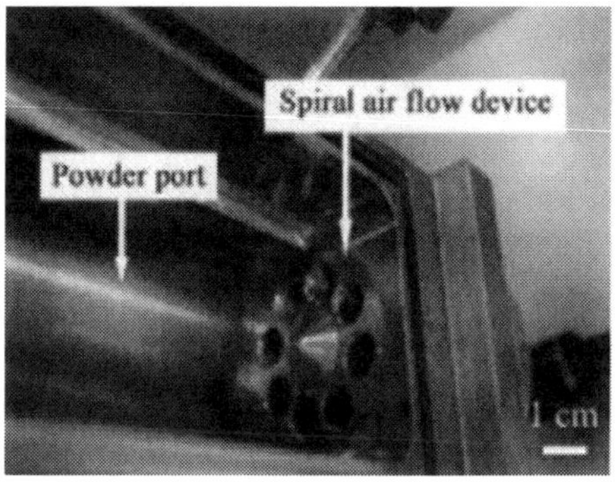

Photo 1 Structure of the powder port and spiral air flow device.

Sample powders

Glass beads are extremely good light reflectors and thus are widely used as the main component of the coating material for light-reflecting adhesive tape, background paint for traffic control sign boards, and paint for road signs. Four different grades of commercial glass beads were used in this experiment, as shown in Table 1. The particle size was determined using a laser diffraction system (LDA Win 1.21, dry type). Hereafter, they will be called by their abbreviation shown in Table 1. The analysis of the images from microscopy (Neoscope, JCM-500) revealed that the G-4 powder particles were spherical, as shown in photo 2. The other glass bead particles were the same shape as G-4.

Table 1. List of powder specimens used in experiments.

Specimen	Abbreviation	Diameter [μm]		
		D_{10}	D_{50}	D_{90}
Glass beads 1	G-1	22	62	85
Glass beads 2	G-2	93	131	175
Glass beads 3	G-3	130	180	241
Glass beads 4	G-4	162	225	294

D_X: X% volume average.

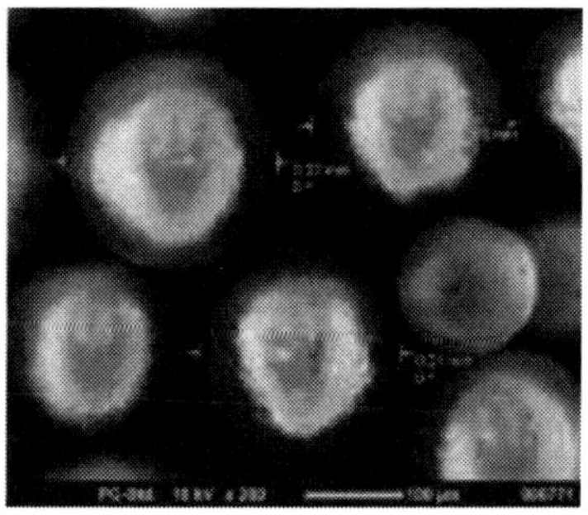

Photo 2 Shape of G-4 powder used in this experiment.

Experimental results and discussion

Figure 2 shows the typical change of charge detected over time when the G-3 particle was used and the air pressure was 0.15 MPa. It should be noted here that when no particle was used and only air was passed through the pipeline, the charge was detected as almost zero. The charge Q1 of the powder was found through the cyclone-type Faraday cup and then was confirmed through the charge Q2 obtained from the pipeline. As the powder particles passed through the pipeline for about 0.82 seconds, as seen in the Fig. 2, the charge clearly increased. Q1 and Q2 had almost the same trend and absolute values. This result confirmed the reliability of the test and showed that there was no disturbance by external noise.

In this study, all of the glass beads (G-1 to G-4) were negatively charged when stainless steel metal was used as the contact surface. The charge polarity of the materials relates to the work function (or ionization potential) of each material. The contact potential difference between two substances is defined by the difference between their fermi levels, which correlate with the work function. In general, materials with higher work functions tend to appropriate electrons from materials with lower work functions. This point will be carried out in our further study.

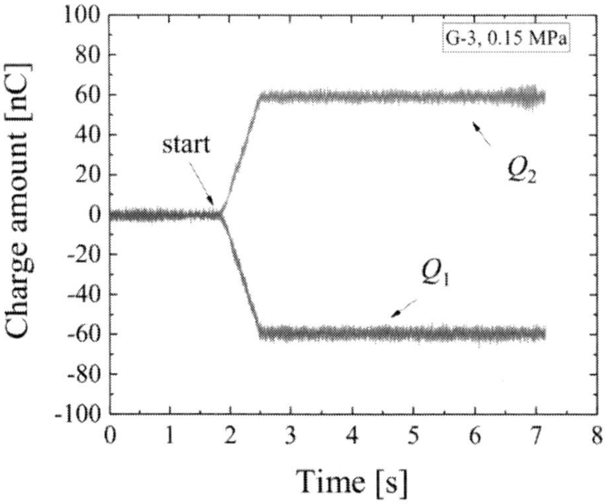

Fig. 2 Change of charge detected over time.

Figure 3 shows the charge to mass ratio of glass beads as a function of applied air. The range of applied air was between 0.1 MPa and 0.3 MPa in this study. When the applied air pressure was below 0.1 MPa, it was most likely not enough to send all of the powder through the pipeline. Each experiment was repeated five times, and the average values were reported. This graph does not contain an error bar, but these were about 10%. The results showed that the charge to mass ratio of both powders increased as the applied air increased. Although not shown in Figs. 2 and 3, the increase in air pressure also causes the slope of the graph to increase significantly. This can be explained by the fact that with increased pressure of applied air, the particles collide against the surface of the pipeline with more strength [9]. In order to investigate in more detail, the velocity of air at approximately 10 mm from the inside of the pipeline wall was measured using an anemometer (Kanomax, Anemomaster-6115). The approximate velocity of air flow was 9.5 m/s at 0.1 MPa, 12 m/s at 0.15 MPa, 15 m/s at 0.2 MPa, 20 m/s at 0.25 MPa, and 22 m/s at 0.3 MPa. It could be clearly seen that as the air pressure increased, the air velocity also increased. Another reason for this may be that because the air was applied in a spiral form, the particles may have collided more frequently against the pipeline when the air flow was increased. This exception will be investigated in detail in the future.

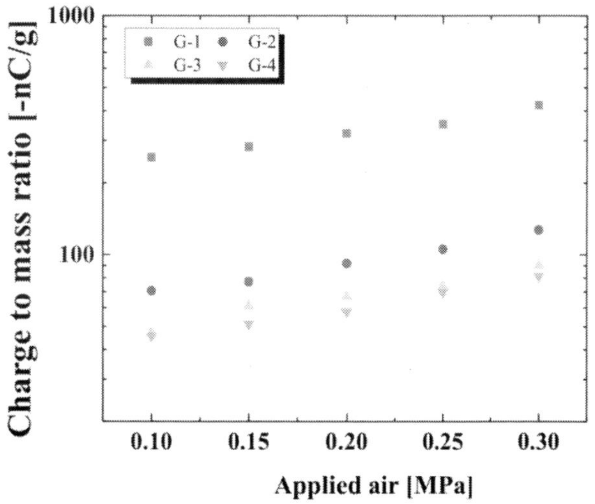

Fig. 3 Charge to mass ratio of glass beads as a function of applied air.

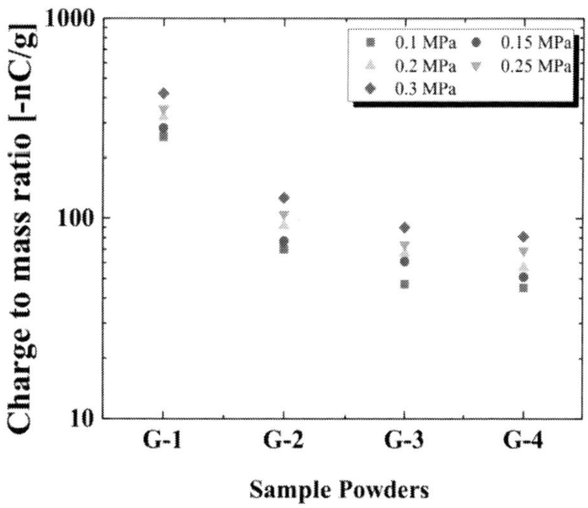

Fig. 4 Charge to mass ratio as a function of different glass bead sizes.

Figure 4 shows the charge to mass ratio as a function of different sizes of glass beads. The size distribution of each glass bead in this study was described in Table 1. The results showed that the charge to mass ratio of the glass beads increased as the powder size decreased. That is to say, smaller particles had a larger q than the larger ones. The total surface area of all of the smaller beads combined is greater than that of the larger ones, which results in a high charge to mass ratio development in the tribocharge. The value range of the q of the glass beads obtained in this study was about 40 nC/g to 100 nC/g. These values agreed with those obtained from the micronizing and pneumatic transfer process [10]. These values were high enough to generate incendiary electrostatic discharges during handling in industrial processes. This suggests that if glass beads are used with solvents in the cleaning process, they must be handled carefully, as incendiary discharges can ignite the solvents even though the glass beads themselves are nonflammable material.

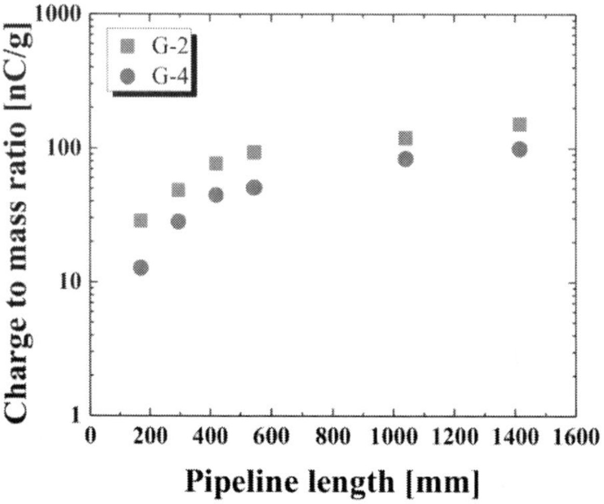

Fig. 5 Charge to mass ratio as a function of pipeline length.

Figure 5 shows the charge to mass ratio as a function of pipeline length. Six different pipeline lengths, *d*, were used: 170, 295, 420, 545, 1040, and 1415 mm. The glass beads used in this experiment were G-2 and G-4, and the air pressure was 0.35 MPa. Since the pipeline length changed with each test, it was difficult to attach the Faraday cup to the end of the pipeline. Therefore, in this study, only the value of q_2 was measured. As seen in Fig. 2, q_2 had the almost same value as q_1 but had an opposite sign. The length of the pipeline was measured starting from beginning of the powder port because the inside structure of the powder port was the same as that of the pipeline. Each experiment was repeated five times, and the average values were reported. The results showed that q_2 increased as the length of the pipeline increased for both sizes of glass beads. This can be explained by the fact that as the length increases, the time the glass beads come in contact with the pipeline also increases. The increase rate at the beginning section of the pipeline was especially large. It can also be seen through Fig. 5 that overall, G-2 had a larger q_2 than G-4 because of the total surface area. This confirms the reliability of the data shown in Fig. 4.

In addition, both beads seemed to reach saturation when the pipeline length was 1415 mm. It was compared to the maximum charge to mass ratio, q_c [nC/g] calculated from the well-known equation shown below [11]:

$$q_c = |\sigma_{max}|/(D \cdot \rho), \quad \text{-------------------------(2)}$$

where σ_{max} [kg/m^3] is the maximum surface charge density (27 μC/m^2); D [μm] is the particle diameter; and ρ is the particle density (1.5×10^3 kg/m^3).

In this study, the particle sizes of G-2 and G-4 were 131 μm and 225 μm, respectively, as shown in Table 1. The calculated maximum values q_c and experimental maximum values q_t are compared as shown in Table 2. The calculated and experimental values for both G-2 and G-4 agreed well with each other. Therefore, it can be seen that both G-2 and G-4 reached saturation.

Table 2. Calculated maximum values and experimental maximum values of the charge to mass ratio.

	q_c [nC/g]	q_t [nC/g]
G-2	137	151
G-4	80	99

Conclusion

In this paper, a novel tribocharging system that utilizes the Faraday cup was presented and the electrostatic charge behavior of glass beads was experimentally investigated. The results were as follows:
1) For all of the glass beads, the charge clearly increased as the glass bead particles passed through the pipeline. The charge amount of the glass beads and the charge amount obtained from the pipeline had almost the same trend and absolute values. This result confirmed the reliability of the test and showed that there was no disturbance by external noise.
2) All of the glass beads were negatively charged when stainless steel metal was used as the contact surface. The value range of the charge to mass ratio of the glass beads obtained in this study was about 40 nC/g to 100 nC/g.
3) The charge amount of the glass beads increased as the applied air increased. It could also be seen that as the air pressure increased, the air velocity also increased.
4) The charge amount of the glass beads increased as the powder size decreased due to the total surface area of the beads.
5) The charge amount of the glass beads (G-2, G-4) increased as the length of the pipeline. The increase rate at the beginning section of the pipeline was especially large.

Acknowledgments
The authors gratefully acknowledge Mr. Naoto Nogera for his assistance during the test.

References

1. F. Riedewald (2012): "A fire in a secondary pharmaceutical powder transfer operation," *Process safety progress*, 31, 4, pp. 390-392.
2. W. Kaialy W, "A review of factors affecting electrostatic charging of pharmaceuticals and adhesive mixtures for inhalation," *Int.J.Pharm*. Vol.503, pp.262-276.
3. N. Gibson (1997), "Static electricity — an industrial hazard under control," *J. Electrostatics,* Vol. 40-41, pp. 21-30
4. K. Choi, K. Choi, K. Nishimura (2015), "Experimental study on the influence of the nitrogen concentration in the air on the minimum ignition energies of combustible powders due to electrostatic discharges," *J. of Loss Prevention in the Process Industries*. Vol.34.pp.163-166.
5. K.S. Choi, H.Sakasai, and K. Nishimura (2015), "Experimental study on ignitability of pure aluminum powders due to electrostatic discharges and Nitrogen's effect," *J. of Loss Prevention in the Process Industries*. Vol.35.pp.232-235.

6. K.S. Choi, H. Sakasai, and K. Nishimura (2015), "Minimum ignition energies of pure magnesium powders due to electrostatic discharges and Nitrogen's effect," *J. of Loss Prevention in the Process Industries.* Vol.41.pp.144-146.
7. N. Masui, and Y. Murata (1981), "charging tendency of the flowing powder through a metal tube (II), " *The five Annual meeting of Institute of Electrostatic Japan,* pp. 244-247.
8. S. Watano, S. Saito, and T. Suzuki (2002), "Feedback control of powder electrostatic charge in a pneumatic conveying process," *J. Soc. Powder,Japan*, Vol. 39, pp. 496-502.
9. P. Boschung, M. Glor (1980), "Methods for investigating the electrostatic behavior of powders, J. of elelctrostatic," Vol. 8, pp. 205-219.
10. IEC50404(2004), "Electrostatics- Code of practice for the avoidance of hazards due to static electricity," p. 66.
11. M.Glor (1988), "Electrostatic hazards in powder handling," *Research studies press LTD.*, p.26.

509266 Spray-Drying of a Layered Double Hydroxide Nanosuspension

Abstract:

Over the past few years, layered double hydroxide (LDH) has attracted particular interests in both academia and industry. LDH is 2D anionic clay of brucite-like layered structure with various organic or inorganic anions located between layers. The diverse composition of LDH, and its associated physical and chemical properties open up new possibilities in various applications. LDHs are successfully applied as precursors in synthesis of supported catalysts, sorbents, carriers delivering bioactive molecules, matrices containing active pharmaceutical ingredients for control release, electrochemical biosensors, etc. LDHs are mainly synthesized in a form of nanosuspensions. The application of nano-sized sorbents or catalysts in the liquid-phase processes requires a special technique to separate them from the final product. An uncontrollable agglomeration of solid nanoparticles makes their industrial application impossible in a dry phase. Moreover, the toxic effects of nanoparticles on human health and ecosystem have been discussed in the literature.

Utilization of larger micron-size agglomerates of nanoparticles is one of the alternative ways to safely handle nanoparticles in the dry phase on the industrial scale. Spray drying is one of the well-known techniques that have been widely applied for production of microspheres from nano-sized particles. Using this technique, it is possible to control the size distributions as well as the morphology of spray-dried agglomerates. However, there are only a limited number of publications related to spray drying of LDH nanosuspensions. Moreover, spray-dried LDH agglomerates of rather wide size distribution, low surface area and irregular shape were reported in the literature.

In this study, we explore the effect of process parameters, such as spraying air flow rate, drying air temperature, slurry concentration and slurry feed rate on the size distribution, morphology, specific surface area and pore volume of LDH agglomerates fabricated by spray drying of nanosuspensions.

The round-shaped agglomerates with a diameter of more than 2 and less than 15 micrometers were obtained by spray drying of suspensions of nanoparticles with median diameter of 171 nm without addition of dispersant and binder. Our results confirmed that the spraying air flow rate has the dominant effect on the agglomerate median size and the width of agglomerate size distribution as well as on the agglomerate specific surface area. The small agglomerates with narrow size distribution were collected during drying at the high flow rate of spraying air. Our results also indicate that lowering the temperature of drying air and using the dilute nanosuspensions promote generation of small agglomerates.

The specific surface area of spray-dried LDH agglomerates is in the range of 49 to 101 m^2/g. The agglomerates of high specific surface area were obtained at low flow rate of spraying air, high slurry feed rate of concentrated slurry and using low temperature drying air. These results also confirm that drying at low drying rate yields to formation of agglomerates of high surface area. Similar trends were observed in terms of the variation of total pore volume with drying process parameters.

Thus, by adjusting the spray drying conditions, it is possible to produce agglomerates of LDH nanoparticles satisfying the product requirements.

513829 Impact of Spheroidization of UO_2 Powders on the Filling of Press Molds

Abstract:

In industry, the use of powders is very common. This underlines the interest in understanding the flow of powders in process lines (hopper, silos, transportâ€¦) and finding the link between powder physical characteristics and its rheological properties. This study deals with the flowing properties of a UO2 powder in a powder metallurgy process for nuclear fuel fabrication. A reproducible filling of the press molds is required to avoid mechanical grinding of the sintered pellets.

A powder is characterized by its particle size distribution but also by the morphology of particles. Morphology has a major impact on the cohesion and then on the flowability of powders [1]. UO2 powders are generally made up agglomerates which are composed of aggregates, themselves formed by crystallites (Figure 1). Agglomerates are primarily governing the powder flowing behavior. Their characteristics sometimes do not allow them to meet the expectations of the process, especially for the filling of the press molds.

For this reason, in a powder manufacturing process, a spheroidization operation (Figure 2) could precede the shaping of pellets to improve the filling properties.

In this study we observe the influence of particles morphology on the rheological properties of UO2 powders. Whereas there is no standard and universal method to characterize the flow of powders [2], there is direct (shear cells) and indirect methods (dynamic flowability measurements like permeability). We use both methods: the indirect methods based on qualitative measures of powder cohesion, the direct methods to determine the powder cohesion for a specific consolidation state through the Freeman shear cell [3]. While a shear cell measurement permits to determine the powder flow in quasi-static system, a permeability measurement shows the flow index in dynamic system. We identify the link between theses measurements and the quality of the filling.

References:
[1] Xiaowei Fu and others, «Effect of Particle Shape and Size on Flow Properties of Lactose Powders», Particuology, Advances in Characterization and Modeling of Particulate Processes, Apr. 2012.
[2] P. Juliano, B. Muhunthan, and G. V. Barbosa-Cánovas, « Flow and shear descriptors of preconsolidated food powders », J. Food Eng., vol. 72, no 2, p. 157-166, Jan. 2006.
[3] J. Schwedes and D. Schulze, « Measurement of flow properties of bulk solids », Powder Technol., vol. 61, no 1, p. 59-68, Apr. 1990.

508203 Modelling a Twin Screw Granulator Using the Discrete Element Method

Abstract:

Wet granulation is a process used to create larger stable agglomerates (granules) from fine powders. This has many desirable outcomes such as improving flowability, compactibility, and homogeneity. Granulation is commonly employed in the food, pharmaceutical, detergent, and fertilizer industries, but despite its wide adoption it is often inefficiently operated, with high recycle ratios in continuous processes and high rejection rates in batch processes. Wet granulation is the most common type of granulation and, in pharmaceutical applications, it is a critical step in tablet manufacturing that affects the uniformity and compactibility of the final dosage form.

Traditionally batch granulation was the favoured granulation process method in the pharmaceutical industry due to the challenges faced from adopting a continuous approach, such as cost or inability to monitor product quality reliably. Since the introduction of Quality by Design (QbD) and Process Analytical Technology (PAT) by the FDA in 2003 there has been a move towards new manufacturing techniques such as continuous processing.

Although tremendous efforts have been made to gain scientific insight into the granulating process, a fundamental understanding of wet granulation is still lacking due to the complexity of the mechanisms involved. With twin screw granulation becoming a popularly employed method of wet granulation, an in-depth understanding of particle enlargement in the granulating process is necessary in order to improve the quality of the final product without the need for large-scale Design-of-Experiment studies.

Despite the extensive experimental research carried out for twin screw granulators in recent years, there has been little computational work carried out. This paper employs the Discrete Element Method (DEM) to study a 25 mm diameter, GEA ConsiGmaâ„¢ 1 twin screw granulator with a typical 60Â° forward configuration for kneading elements. The DEM simulations were conducted using the commercial code EDEM with a DEM contact model developed for cohesive solids implemented through an API. The contact model is based on an elasto-plastic contact with adhesion and uses hysteretic loading and unloading paths to model the elastic-plastic contact deformation. In these simulations, the adhesion is used to capture the effect of the binder liquid without the complication of modelling the liquid directly. The adhesion parameter is a function of the plastic contact overlap. The model has previously been shown to be able to predict the stress-history-dependent behaviour depicted by a flow function of the material. In this study the effect of particle size, gap tolerance and cohesion on the residence time distribution and local solid fraction at different elements in the granulator is assessed. The level of cohesion used is related back to uniaxial test results which can be used to measure the flowability of a cohesive granular solid.

508216 Hybrid Multiscale Modelling of a Twin Screw Granulator

Abstract:

Wet granulation is a process used to create larger stable agglomerates (granules) from fine powders. This has many desirable outcomes such as improving flowability, compactibility, and homogeneity. Granulation is commonly employed in the food, pharmaceutical, detergent, and fertilizer industries, but despite its wide adoption it is often inefficiently operated, with high recycle ratios in continuous processes and high rejection rates in batch processes. Wet granulation is the most common type of granulation and, in pharmaceutical applications, it is a critical step in tablet manufacturing that affects the uniformity and compactibility of the final dosage form.

Modelling processes such as wet granulation is not straightforward, but may be achieved using techniques such as Population Balance Modelling (PBM) and Discrete Element Method (DEM). PBM groups particles into different classes based on their properties and uses empirically derived differential equations to describe key processes such as aggregation or breakage. However, although PBM tracks the size and number of particles over time, it cannot account for the distribution of the particles and the effect of this local variability.

By contrast, DEM simulations track individual particles, and as such, phenomena such as segregation or liquid migration can be studied. DEM simulations also consider the effect that the geometry and its dynamics have on the particles' behaviour, which is something that PBM cannot account for directly. However, DEM simulations of fine materials can be computationally expensive and this often results in reduced numbers of particles being used to make the simulation computationally tractable. This is usually achieved through increased, unrealistic particle sizes or the simulation of isolated local zones without considering adjacent zones which may affect the behaviour.

Traditionally, large-scale Design-of-Experiment (DoE) studies would be needed to estimate key rate parameters for use in PBM modelling of granulation and this can be both costly and time-consuming. DEM simulations provide an alternative method to estimate these parameters. By coupling the two simulation methods, one can utilise the key benefits of each method to better simulate complex processes.

This paper presents a hybrid approach to modelling twin screw granulation via DEM–PBM coupled simulations to study a 25 mm diameter, GEA ConsiGma™ 1 twin screw granulator with a typical 60° forward configuration for kneading elements. The DEM simulation was conducted using the commercial code EDEM with a DEM contact model developed for cohesive solids implemented through an API. The PBM simulation is carried out with the commercial code gFORMULATE which is built on top on the advanced modelling platform gPROMS. A bi-directional coupling is implemented with information shared between the two simulation tools at specified intervals. Preliminary results for this coupled approach are presented and discussed.

509577 Blocking Rules for Discharging Granular Materials from a Flat Bottom Hopper

Abstract:

Hoppers and silos are widely used in storing granular materials in various industries, such as agricultural, chemical, food and pharmaceutical industries. Three key aspects need to be considered when hoppers are used to store powders and grains: 1) stress analysis; 2) hopper blockage ; 3) mass flow rate. A hopper must be designed strong enough to sustain the stress generated from the bulk materials during storage and discharge. On the other hand, a hopper must be designed properly so that the stored bulk materials can be discharged, i.e. without blockage and formation of ratholes and arches. Furthermore, the mass flow rate needs to be predicted and controlled. In this study, a combined experimental and numerical study was performed to determine the blocking rule for discharging granular materials from a flat bottom hopper. Granular materials of various sizes were considered and the minimal orifice size through which the bulk materials can be discharged was determined experimentally using a Flodex tester and numerically using DEM. The dependency of the minimal

orifice size on particle size was explored. An semi-empirical model was also developed. Using this model, the critical orifice size at which the flow will stop, i.e. blockage takes place, can be determined from the mass flow rate data. It was found that excellent agreement between the numerical results and experimental measurements was obtained. Both experimental and numerical results showed that, even for spherical particles, layered dead zones could be formed during the discharge from the flat bottom hopper. In addition, mono-layered dead zones were developed when large particles were used. Furthermore, the developed blocking rules were also verified experimentally and numerically.

508310 Characterization of Mesoscopic Structure in Cohesive Powder, Neat or Blended, By X-Ray Computed Tomography and Prediction By the Discrete Element Method

Abstract:

Understanding and predicting structural characteristics of cohesive powder systems is a technical challenge in many powder technology processes and applications. The discrete element method (DEM) [1] is a promising approach to model the structure and flow behavior of cohesive granular systems. Advances in experimental techniques such as X-ray computed tomography (XRCT) have now enabled characterization of fine cohesive powder systems with sub-micron resolution. Combined with standard bulk powder characterization methods (e.g. density), this technique allows for mesoscopic structural characteristics of cohesive powder systems to be compared with DEM simulation. In this study, DEM and XRCT are used to characterize the mesoscale structure, under controlled consolidation conditions, of two moderately cohesive 'model' (roughly spherical) powders with distinct material properties. The DEM simulations are based on a constitutive elasto-plastic model [2] that makes use of the JKR theory [3] to account for cohesive interparticle forces. Density and void structure obtained from DEM simulation are compared to characterization results obtained from XRCT image analysis. XRCT and DEM simulation approach is applied to examine structures of binary mixtures of the two model powders. The ability of DEM simulation and the contact model to accurately predict structural characteristics of single component and binary granular systems is discussed.

References

[1] P. A. Cundall and Otto D.L. Strack. A discrete numerical model for granular assemblies. Geotechnique, 29(1):47–65, 1979.

[2] C. Thornton and Z. Ning. A theoretical model for the stick/bounce behavior of adhesive, elastic-plastic spheres. Powder technology, 99(2):154– 162, 1998.

[3] K.L. Johnson, K. Kendall, and A.D. Roberts. Surface energy and the contact of elastic solids. In Proceedings of the Royal Society of London A: Mathematical, Physical and Engineering Sciences, volume 324, pages 301–313. The Royal Society, 1971.

511638 Numerical Research of Hydrodynamics in Gas-Solid Micro Fluidized Beds

Abstract:

Due to the advantages of less energy consumption, more safety and good operability, the miniaturization of fluidized bed, or named micro fluidized bed, has been extensively studied by a lot of researchers. The Micro Fluidized Bed Reaction Analyzer (MFBRA) was adopted by the Institute of Process Engineering (IPE), Chinese Academy of Sciences (CAS), to study the gas-solid reaction kinetics under isothermal differential condition[1, 2]. In the MFBRA, the hydrodynamics of micro fluidized beds play a vital role on the heat and mass transfer and which is a key to the accurate measurement of gas-solid reaction kinetics. In order to optimize and design the MFBRA, numerous experiments have been conducted to study the hydrodynamics of micro fluidized beds[3, 4]. Unfortunately, those works mostly focused on the bed pressure drop and minimum fluidization velocity, nevertheless, the detail information, such as local solid volume fraction and particle velocity profile in micro fluidized beds are reported rarely. It is because the existing experimental techniques, such as, optical fiber and capacitance probes are hard to measure those local parameters without disturbing the flow field.

The objective of this work is to study the hydrodynamics of micro fluidized beds and to investigate the isothermal characteristics during kinetics testing of graphite powder combustion in MFBRA. In this study, Discrete Particle Model (DPM) and Two-Fluid Model (TFM) are selected to investigate micro fluidized beds, respectively. The DPM is carried out by the commercial software EDEM-Fluent Coupling simulations. And for TFM, we used the Eulerian multiphase model in the ANSYS Fluent. From our numerical simulations, the detailed hydrodynamic characteristics of micro fluidized beds are explored systematically by comparing different column sizes and superficial gas velocity. Meanwhile, it is found that both DPM and TFM have the ability to model the MFBRA under practical operation condition with high superficial gas velocity. In the term of isothermal characteristics in micro fluidized beds, the temperature field distribution as well as the heat transfer between heat-carrier and graphite powder are explored by the combination of simulations and experiments.

Reference:

1. Xu, G., *Micro fluidized bed reaction kinetics analyzer. Supported by instru-ment development project of Chinese Academy of Sciences, China, Project numberY2005014. (in Chinese).* 2005.
2. Yu, J., et al., *Isothermal differential characteristics of gasâ€"solid reaction in micro-fluidized bed reactor.* Fuel, 2013. **103**: p. 29-36.
3. Liu, X., G. Xu, and S. Gao, *Micro fluidized beds: Wall effect and operability.* Chemical Engineering Journal, 2008. **137**(2): p. 302-307.
4. Guo, Q.j., Y. Xu, and X. Yue, *Fluidization Characteristics in Micro-Fluidized Beds of Various Inner Diameters.* Chemical Engineering & Technology, 2009. **32**(12): p. 1992-1999.

507312 An Improved Bubble-Based Drag ModelÂ for Accurate Coarse-Grid Two-Fluid Modeling of Geldart a Powder Bubbling Fluidization

Abstract:

An improved bubble-based energy minimization multi-scale (EMMS) drag model was formulated in which the original bubble-based EMMS drag model at macroscale bed level was extended to microscale sub-grid level for accurate coarse-grid two-fluid modeling (TFM) of Geldart A powder bubbling fluidization. At both the macroscale and microscale levels, the physically heterogeneous structures were modeled by three pseudo phases, i.e., bubble phase, emulsion phase, and inter-phase, where conservation and stability condition equations were applied. Through such scale and phase decompositions, the improved bubble-based EMMS drag model was formulated through a two-step macro-to-micro scheme, where the obtained drag coefficient is related to phase volume fractions, velocities, and accelerations within each coarse grid. Modeling results show that TFM combined with the improved bubble-based EMMS drag model can accurately reproduce the heterogeneous structures for a laboratory-scale Geldart A powder bubbling fluidized bed. At relatively coarse grid, the predicted axial and radial distributions of solid concentration agree very well with experimental measurements. This indicates that our proposed improved bubble-based EMMS drag model incorporated TFM is promising for accurate and economical simulation of engineering-scale Geldart A powder bubbling fluidization.

507860 Numerical Simulation on Fine Particle Transport Behaviour in Electrostatic Precipitators

Abstract:

Electrostatic precipitators (ESPs) are a major air pollution control device to remove particles from treated gas. Extensive research has been done on several aspects such as electrostatics, fluid dynamics, charging mechanism, and particle dynamics. The overall particle-collection efficiency of a modern industrial wire-plate ESP exceeds 99%. However, the removal efficiency of submicron particles by conventional ESPs is as low as 70â€"80% owing to the low charge carried by fine particles and the increase in mobility with decreasing size. Submicron particles possess a high superficial area, thus favouring toxic heavy metals and other pollutants. Therefore, fine particles are collected with much lower efficiency and further research is required to improve this aspect.

In this work, a numerical study on the charging and transport behaviour of fine particles has been carried out based on wire-plate electrostatic precipitators (ESPs) with multiple wire electrodes. The effect of the applied wire voltage, inlet height, and wire distance on particle charging and transport behaviour, and the influence of the precipitator structure on particle trapping are analysed in detail. The 2-D numerical model involves electric field, space-charge density, gas flow, and particle trajectories. The fluid-flow velocity is higher than 1 m/s in this study and the EHD flow has a negligible. A Lagrangian approach for particle transport is utilized with the trajectories of individual particles tracked by solving their equation of motion. The electric field and gas flow are calculated using the finite-volume method. The flow simulation is done using the commercial software FLUENT with other parts coded as user-defined functions. The following main results can be drawn:

(1) It was observed that the Y-displacement of the particles exhibited a universal increase for all particle sizes, especially for larger particles, at low injection positions (Y < 25 mm). The larger the particle size, the higher the Y-displacement of particles because larger particle sizes improve both the particle charging level and electric field strength. No significant improvements were found for submicron particles.

(2) Increasing the spacing of the discharge electrode within a certain range and therefore increasing the electric field, it was observed that the particle-trapping effect improved and the outlet level exhibited a universal increase for all particle sizes, especially for large particles.

(3) The ESP models M1 and M2 were used to investigate their influence on the outlet height. Except for particles with a diameter of 2 μm, all other particle sizes showed an increased outlet height in M2 compared to that in M0. In addition, the different discharge-electrode arrangements, especially in M2, exhibited significant effects on the charging and transport behaviour of particles with a diameter of 1 μm.

(4) Particles with an inlet height of 5 and 10 mm were studied in M1 and M2 precipitators. The changed discharge-electrode arrangement improves particle trapping, especially in M2. This is because of the electron trajectories being closer to the discharge electrodes (M2), giving the particles more charge and leading to a greater electric field.

507879 Lattice Boltzmann Simulations of Porous Particulate Flows

Abstract:

The comprehensive understanding of the particle-fluid interaction is very important for fluid-particles system, and thus many studies have been conducted for the particle-fluid system[1-3] of solid impermeable particles. In the real industrial fields such as the catalyst process in chemical engineering[4], the particles usually possess the porous structure which will have effects on the fluid-particles interaction[5-6]. However, there has been little open literature discussing the particle-fluid system of porous permeable particles. Therefore, the objective of the present study is to investigate the effect of the porosity and permeability on the porous particulate flows which are formulated by the volume-averaged governing equations[7] in terms of intrinsic phase average velocity. The lattice Boltzmann equation model (LBE)[8] is used to solve the governing equations. Firstly, we validate the LBE model by the numerical results existing in the literature for the flow around and through one porous particle. Then, the porous particles settling against gravity in a quiescent fluid, i.e., the DKT process[9] for porous particles, is investigated. The effects of the porosity and Darcy number on the porous particulate flows are studied numerically in details. Also, we present the simulation results of the fluidization of 512 porous particles. Our results demonstrate the significance of the porous structure of particles for the particles dynamics behaviors.

Reference

[1] Ladd, A.J.C., 1994. Numerical simulations of particulate suspensions via a discretized Boltzmann equation Part I. Theoretical foundation. J. Fluid Mech., 271, 285-310.

[2] Hill, R.J., Koch, D.L., Ladd, A.J.C., 2001. The first effects of fluid inertia on flows in ordered and random arrays of spheres. J. Fluid Mech. 448, 213-241.

[3] Van der Hoef, M.A., Beetstra, R., Kuipers, J.A.M., 2005. Lattice Boltzmann simulations of low-Reynolds-number flow past mono- and bidisperse arrays of spheres: results for the permeability and drag force. J. Fluid Mech. 528, 233-254.

[4] Zhao, Y.F., Li, H., Ye, M., Liu, Z.M., 2013. 3D numerical simulation of a large scale MTO fluidized bed reactor. Ind. Eng. Chem. Res. 52, 11354-11364.

[5] Noymer, P.D., Glicksman, L.R., Devendran, A., 1998. Drag on a permeable cylinder in steady flow at moderate Reynolds numbers. Chem. Eng. Sci. 53, 2859-2869.

[6] Zhu, Q.Y., Chen, Y.Q., Yu, H.Z., 2014. Numerical simulation of the flow around and through a hygroscopic porous circular cylinder. Comput. Fluids 92, 188-198.

[7] Wang, L., Wang, L.-P., Guo, Z.L., Mi, J.C., 2015. Volume-averaged macroscopic equation for fluid flow in moving porous media. Int. J. Heat Mass Trans. 82, 357-368.

[8] Chen, S.Y., Doolen, G.D., 1998. Lattice Boltzmann method for fluid flows. Annu. Rev. Fluid Mech. 30, 329-364.

[9] Fortes, A.F., Joseph, D.D., Lundgren, T.S., 1987. Nonlinear mechanics of fluidization of beds of spherical particles. J. Fluid Mech. 177, 467-483.

508370 Modeling of a Novel Multi-Particle Collision Model for Gas-Solid Flows

Abstract:

Fluidized beds are widely applied in the chemical, petrochemical, metallurgical, environmental and energy industries due to their favourable mass and heat transfer characteristics. In past few decades computational fluid dynamics has proven to predict hydrodynamics of fluidized beds. DPM (Euler-Lagrange) [1] and TFM (Euler-Euler) [2] are two most common modelling approaches for gas-solid flow in fluidized beds. In TFM the solid phase itself is treated as a continuum phase interpenetrating with the gas phase. Since the concept of individual particles has disappeared, the effects of particle-particle interactions in the TFM has to be incorporated in a statistically averaged way, via the solid phase pressure and viscosity. TFM is preferred model to study large scale fluidized beds but model suffer from high numerical diffusion. And the KTGF theory in TFM assume elastic collisions and is limited only to spherical particles [2]. In DPM, the gas phase is considered as the continuous phase whereas the particle trajectories are obtained by integrating Newtonâ€™s equations of motion. Particle-particle and particle-wall interactions are explicitly taken into account using various physical models such as the soft sphere model. DPM proves to be accurate modelling of gas-particle flow. However, lagrangian models quickly become computationally too expensive because of the large number of particle interactions. Therefore coarse graining of discrete particles is an emerging model that can handle large number of particles effectively. The new coarse-grained model developments are of vital importance for the advancement of large-scale modelling of gas-particle flows that are flexible to extended for polydispersity, and non-spherical particles. In this study a novel lagrangian multi-particle collision (MPC) method is developed for predicting dense particulate-fluid flow.

The MPC solver is developed in OpenFoam [5], where the fluid phase is treated as a continuum and the solids are modelled by lagrangian particles. The MPC model grew out of the multiphase particle-in-cell (MP-PIC) model [3], that employs a fixed Eulerian grid, and particles are traced in a lagrangian fashion, including a parcel approach where each simulated particle represents a large collection of real particles. Each parcel follows Newtonâ€™s equations of motion, where inter-phase momentum transfer is accounted through mapping and interpolating quantities back and forth between parcel locations and the eulerian grid. Particle collisions in MP-PIC are not fully resolved but are derived from particles stresses that are theoretically calculated from Harris and Crighton [4] stress closure. In the MPC model, we implemented a functional form for the particle stress tensor where the coefficients are obtained computationally by means of DPM simulations. The idea is to extend the model from spherical particles to non-spherical particles, for which theoretical particles closures are currently not available in the literature. The viscous stress and random forces are typically ignored in MP-PIC model due to their insignificant effect [3]. Importantly, we incorporated viscous stress and random forces representing scattering of particles by collisions, which are obtained from DPM simulations by measuring the actual variance. Details of model development and validation is presented in this study.

Results from MPC model are compared with MPPICFoam and, with DPM simulations from DEM-LIGGGHT [6], and MFiX [7] code respectively. Pseuduo-2D fluidized beds of dimension 0.01x0.15x0.45 m is considered for comparison with experiments [8]. Comparison is made on the basic of pressure drop, void fraction, bubble dynamics and solids motion. We observed that MPC is in good agreement with DPM, and experiments, when compared to MPPICFoam. Unrealistic over-packing of solids is predicted in MPPIC due to limitation of Harris and Crighton [4] model and the velocity field updates, whereas in MPC these limitations overcomes and no over packing of particles is predicted. Solid circulations are in good agreement with DPM results, where solid moving above in the center and downward near to the wall. Time-average porosity plots revels the path of bubble motion and is in close agreement with results of DPM. In conclusion a novel multi-particle collision model is developed and validated to study gas-solid fluidized beds. The model is efficient to handle large number of particles. In future study, we plan to extend the model for dense flow of non-spherical particles.

Acknowledgments

The authors thank the European Research Council for its financial support under its consolidator grant scheme, contract no. 615906 (NonSphereFlow).

References

1. N.A. Patankar, D.D. Joseph, 2001. Modeling and numerical simulation of particulate flows by the Eulerian Lagrangian approach, International Journal of Multiphase Flow, Volume 27, Issue 10, October 2001, Pages 1659-1684.
2. Gidaspow, D., 1994. Multiphase flow and fluidization continuum and kinetic theory description, Academic Press, Boston.
3. Andrews, M.J., O'Rourke, P.J., 1996. The multiphase particle-in-cell (MP-PIC) method for dense particulate ows, Int. J. Multiphase Flow 22, 379-402.
4. Harris, S.E., Crighton, D.G., 1988. Solutions, solitary waves and voidage disturbances in gas-fluidized beds, J. Fluid Mech. 195, 437.
5. OpenFOAM, 2015. The Open Source CFD Toolbox, Ver. 3.0.0. ESI Group, Paris, France.
6. https://www.cfdem.com/liggghtsr-open-source-discrete-element-method-particle-simulation-code
7. https://mfix.netl.doe.gov/mfix/
8. M. Goldschmidt, 2001. PhD thesis, Hydrodynamic Modelling of Fluidised Bed Spray Granulation.

508490 Dpm Analysis of Large Fluidized Catalytic Cracking (FCC) Reactors

Abstract:

DPM Analysis of Large Fluidized Catalytic Cracking (FCC) Reactors

Abstract:

Computational Fluid Dynamic (CFD) modeling is a powerful tool for equipment design and development and is widely used at UOP, particularly for the reactor and regenerator design of FCC process unit. Through CFD modeling, a technical solution was developed for a large capacity FCC process unit purposing an elevated amount of propylene production.

A key challenge in FCC process design is how to optimize the size of the reactor vessel, which is a function of the number and size of reactor cyclones (catalyst separationdisengagers). UOP developed a new cyclone cluster configuration via CFD study of the reactor internals, including UOP's VSS™ riser termination device and reactor cyclones, which reduce the reactor size significantly compare to the conventional design, while maintaining effective catalyst / vapor separation and mechanical reliability. The objective of the current CFD study is to evaluate the catalyst separation efficiency and performance of the new riser termination device technology, which accommodate more cyclones in a smaller reactor vessel.

To better understand the flow pattern, flow distribution and the gas residence time inside the UOP VSS™ , the analysis was broken down into two sections. Part 1 focused on single gas phase only and in part 2 the particles behavior is studied using the Discrete Phase Model (DPM) in Fluent.

The single gas flow analysis results showed that the flow in the VSS chamber section has the same rotational pattern as in conventional FCC/VSS units, where the catalyst will separate from the gas products. Also the gas flow distribution at each chamber arm is uniform, which is critical to catalyst separation efficiency.

The DPM analysis confirmed an equal catalyst particle distribution through all the cyclones and a high efficiency separation in the riser termination device. Having this information from CFD allows for the new design to move forward.

509623 Catalytic Reactor Design Using Multiphase CFD

Abstract:

Catalytic Reactor Design Usng Multiphase Cfddimitri Gidaspow the 2011 Book , "Computational Gas-Solids Flows and Reacting Systems ", Edited By S. Pannala, M. Syamlal and Tom O'Brien ,Engineering Science Reference, Shows the Progress We Have Made in Using Multiphase CFD in Modeling Fluidized Bed Chemical Reactors and Other Multiphase Systems, Such As the Pyroplastic Flow Emitted from Volcanoes. in This Presentation the Design of Two Types of Catalytic Reactors Will be Described: A Fischer_ Tropsch Slurry Bubble Column Reactora Polymerization Reactor the CFD Design of a New Slurry Bubble Column Reactor Was Already Described By Gidaspow ,Et Al in Chemical Engineering Science 134,784-799 (2015).in This Presentation the Design Will be Extended to Obtain 100, 000 Barrels of Diesel per Day, Twice That of the Largest Commercial Reactors. the Production of Synthesis Gas Using Oxygen or Carbon Dioxide in Catalytic Fluidized Beds Needed for the Fischer-Tropsch Reactor Will Also be Discussed Briefly. for the Polymerization Reactor Our CFD Code Described in " Computational Techniques " By D. Gidaspow and V. Jiradilok , Nova Science, 2009 Was Modified for Polymerization Reactors. Such Reactors and Their CFD Modeling Are Described in a Chapter in Pannala's Et Al Book from Professor Rodney Fox's Group. They Point out That the Fine Particle (80 microns) Catalyst Radial Distribution, Is a Remaining Issue the Computations to be Presented Will Show That the Catalyst Will Have a High Concentration at the Reactor Walls for Various Operating Conditions. This Leads to a High Rate of Reaction at the Reactor Walls and May Result in Melting of the Polymers and Sheet Formation , Forcing Reactor Shut-Down. in Some of Our Simulations There Is Large Bubble and Slug Formation Similar to Those Mentioned in Fox's Chapter. Acknowledgement : I Thank Dr. Mayank Kashyap of SABIC for a Valuable Discussion and His Papers " Application of Multiphase Flow CFD in the Gas Phase Polymerization Process " Fluidization XV. May 24,2016 and a Related Paper Presented at Doe NETL, July 2016

509627 Simulation of Large Particle Turbulent Fluidization in Riser Reactors By Coarse Grain DEM-CFD

Abstract:

The complex two-phase flow in the riser is an important element of many industrial circulating fluidized bed reactors. In some cases, e.g. gas-solid polymerization, relatively large particles are transported upward in a turbulent regime, in which the high gas velocity results mostly from heat removal requirements. Strong solids acceleration, radial and axial velocity profiles for both the gas and solid, intense clustering and inhomogeneities make it impossible to reliably predict the riser head or local bed density profile. Detailed numerical simulations proved capable of showing some of the peculiarities of the flow, but limitations in the scale, lack of grid independence, need for special drag corrections and sub-grid models still hinder confident use of such tools in design and scale-up, particularly in industrial applications characterized by particle size and Reynolds numbers that are not well covered by the available scientific literature.

In the present contribution, coupled Discrete Element Method-Computational Fluid Dynamics (DEM-CFD) simulations are carried out using a coarse graining technique for the solid particles. According to this method, small number of particles are grouped into a computational particle (*parcel*) and represented by it as a spherical "lump," to reduce the computational cost of the simulations and allow lab-scale systems to be represented in full detail. The investigation focuses on the influence of the coars-graining degree, i.e. the number of real particles represented by a *parcel*, on the simulated flow hydrodynamics both in terms of its advantages (a quantification of the computational savings) and disadvantage, primarily the real capability of the method to resolve the behaviour observed and produce results in line with experimental findings. The gas and solid motion are studied, with the aim to characterize the clustering phenomena observed during the up-flow and its effect on the pressure drop. Assuming prescribed gas and solid inlet velocity fields, the potential tendency to show core-annulus flow, the intensity of solids clustering and the local bed density within the riser are examined at different number of particles per *parcel*.

509640 Numerical Study of Particles Shape Effects on Solid-Liquid Fluidizations

Abstract:

Numerical study of particles shape effects on solid-liquid fluidizations

Esmaeil Abbaszadeh Molaei, Zongyan Zhou*

Laboratory for Simulation and Modelling of Particulate Systems, Department of Chemical Engineering, Monash University, Clayton, VIC 3800, Australia

Solid-liquid fluidizations have been observed in many industries due to their high heat and mass transfer rate, high operation flexibilities, and reduced back mixing of fluid and solid phases. In most of these multiphase operations, particle properties such as size, density, and shape may experience significant changes during the process because of attrition, coalescence, comminution or chemical reactions. The difference in particle properties can result in significant segregation problem in liquid fluidized beds. To gain an understanding of solid-liquid fluidisation behaviour, most existing studies, either experimentally or numerically, have focused on solid flow behaviour with mono-sized particles or binary and ternary mixtures with different particle sizes or densities. In spite of substantial studies on the hydrodynamics of liquid fluidized beds, the effects of particles shape on the solid-liquid flow behaviour are still less reported. Except for two physical experiments, the previous studies considered particles as spheres, and few studies have been remarked on liquid fluidization of non-spherical particles. However, particles are generally irregular in most of the processes such as ore beneficiations using liquid-solid fluidized beds separator. Particle shape can cause different flow behaviour in liquid fluidizations rather than that observed for ideal spherical particles. Nonetheless, the answers to some fundamental questions such as how and why particle shape affects flow phenomena are not clear, and hence still poorly understood.

Therefore, in this work, a combination of computational fluid dynamics (CFD) and discrete element model (DEM) [CFD-DEM] is used to examine the effects of particle shape on the flow characteristics in liquid fluidizations. In the simulations, ellipsoidal particles are used as they can represent a wide range of particle shapes from oblate to prolate particles. Different aspect ratios are employed, and results are analysed mainly in terms of particle flow patterns, pressure drop, and bed expansion. In addition, force and particle orientation analysis is also conducted to examine the effect of particle shape and explain the occurrence of entrainment phenomenon.

The results show that ellipsoidal particles intend to enter to freeboard region and entrainment phenomenon occurs, and consequently, the interface between the bed and freeboard cannot be observed clearly. In addition, the pressure gradient variation with bed height is uniform for spheres; however, this uniformity is replaced with â€œdistorted parentheses-shapeâ€ profile for disc-shape and elongated particles. Finally, bed expansion analyses show that spherical particles have the lowest bed expansion. On the other hand, disc-shape particles intend to expand more significantly than spherical and elongated particles. This causes the entrainment phenomenon which has been found to be due to the effects of particles shape on particles terminal velocity.

* Corresponding author. Tel: 61 3 9905 0846; email: zongyan.zhou@monash.edu.

514318 CFD Investigation on Gas-Solids Flow and Heat Transfer in Two Fluidized Catalyst Cooler

Abstract:

Our previously experimental work showed that the internal circulation of solids in an annular catalyst cooler (ACC) had an ability to enhance the heat transfer between bed and wall, comparing with the base catalyst cooler (BCC). The current work simulated in detailed hydrodynamics of gas-solids flow and heat transfer to advance our understanding of the ACC by using the two-fluid model. The effective particle diameter d_pâŽ was determined to 170 Î¼m and the specularity coefficient Ï† was 0.3 in the 3D simulations. The catalyst cooler can be divided to four distinct regions including draft tube region, annulus region, bottom region and particle diffluence region according to the distributions of solids volume fraction and solids velocity. The effect of gas-solids flow on bed-to-wall heat transfer was compared in the different regions of the BCC and ACC. And the effects of operational gas velocity and bed structure were discussed. The results showed that the proper selection of the ratio of gas velocities and the position of the heat transfer tube were needed to form a stable internal solids circulation in the ACC. The ACC had a combined hydrodynamic feature of up-flow and down-flow catalyst coolers with bigger solids volume fraction and smaller particle resident time, which are beneficial for improving the heat transfer between solids and wall.

Hydrodynamics and mixing characteristics of a new-type particle mixer

Mengxi Liu, Chunxi Lu, Zhenliang Meng

State Key Laboratory of Heavy Oil Processing, China University of Petroleum, Beijing 102249, PR China

E-mail: mengxiliu@sina.com (M. Liu)

Abstract:

In order to further enhance the particle mixing in the internally circulating fluidized bed, especially along the radial direction, a new-type particle mixer with slots on the draft tube was proposed in the present work. The hydrodynamics of the mixer was investigated by using a multi-scale computational fluid dynamics with the EMMS drag. The prediction in terms of cross-sectionally averaged solid holdup and solid velocity were compared to experimental data for validation. The results show that in draft tube a low pressure region can be observed near the slots and in annulus the pressure in the vicinity of the slots seems larger than that in the region between the adjacent slots, which induces horizontal particle immigration and mixing. Nearly 25.4% to 29.7% of gas in the draft tube bypasses and flows through the slots into the annulus. The ratio of the mass flux of the tracer gas bypassing into the annulus to that of total tracer gas is approximately equal to 0. With increase of the superficial gas velocity in the draft tube from 0.3 m/s to 0.45 m/s, the total solid mass flux increases from 4.87 kg/s to 10.86 kg/s, while the mass flux through slots decreases from 2.69 kg/s to 0.52 kg/s.

1. Introduction

The characteristics of particle flow and mixing have a considerable influence on the mass and heat transfer, as well as the product quality. According to the previous study, the mixing degree of particles along the axial and radial direction is quite different [1-5]. Mostoufi and Chaouki [4] discovered that the diffusivities varied from 3.3×10^{-3} m^2/s to 5.6×10^{-2} m^2/s along the axial direction and 2.6×10^{-4} m^2/s to 1.5×10^{-3} m^2/s along the radial direction. Berruti et al. [6] also revealed that the particle mixing rate for radial direction was an order of magnitude smaller than for the axial direction. Therefore, the weak particle flow along the radial direction greatly limits the mixing ability of a fluidized bed. Internally circulating fluidized bed (ICFB) has been widely recognized as an efficient fluidized bed reactor, in which orderly and macroscopic particle circulation occurs. Compared with the conventional fluidized bed, ICFB offers multiple particle circulation between the draft tube and the annulus, which leads to relatively strong particle convection, diffusion and shear, and thereby improves the particle mixing performance along the radial direction [7]. In order to further enhance the particle mixing for some processes in which a uniform particle mixing is required, a new-type particle mixer with slots on the draft tube was proposed in the present work.

With increase of the computational power, the computational fluid dynamics (CFD) provides an additional tool to predict particle flow and mixing in the local region, which is difficult with the experimental methods. In particular, the gas-solid flow in a gas-solid fluidized bed can be accurately described by the energy minimization multi-scale (EMMS) model [8]. In this paper, the hydrodynamics of the mixer was investigated by using a multi-scale CFD based on the EMMS drag model [9, 10].

2. Experimental apparatus, Materials and Operating conditions

The structure of the particle mixer is shown in Fig.1. The mixer has a height of 5 m and an inner diameter of 286 mm. A draft tube with a height of 1 m and an inner diameter of 215 mm is coaxially mounted into the column. The four rectangular slots of 80×40 mm are symmetrically opened on the draft tube at a height of h=860 mm, in order to strengthen the particle radial flow. A gas distributor plate and a ring distributor are installed below the draft tube and the annulus respectively. The height of the gap between the gas distributor plate and the draft tube is 60 mm. For the sake of the convenience, the mixer is divided into six zones: the pre-mixing zone, feed influence zone, draft tube zone, gas-solid separation zone, annulus zone and the bottom zone. A perforated plate is mounted at the bottom of the pre-mixing zone, which is connected with a central downcomer, extending into the draft tube. Particles are first mixed in the pre-mixing zone and then the pre-mixed particles flow downward through the central downcomer and into the feed influence zone. In the feed influence zone, the pre-mixed particles are mixed with the particles moving upward in the draft tube, flow upward and circulate between the draft tube and the annulus. In the annulus, particles move downward to the bottom. Then a part of particles flow out of the mixer and the other part flow horizontally into the draft tube and recirculate.

The solid particles used in the experiment were FCC catalyst (ρ_p=1498 kg/m^3), with an average diameter of 76 μm. Air (ρ_g=1.225 kg/m^3) was introduced into the mixer to fluidize the particles. The superficial gas velocity ranged from 0.3 to 0.45 m/s in the draft tube (based on the cross-sectional area of the draft tube) and remained at 0.05m/s in the annulus (based on the cross-sectional area of the annulus). The mass flux of cold particles remained at 13.8 kg/m^{2}s. As we only investigated the influence of slots on the flow and mixing of particles in this paper, no hot particles were induced into the mixer.

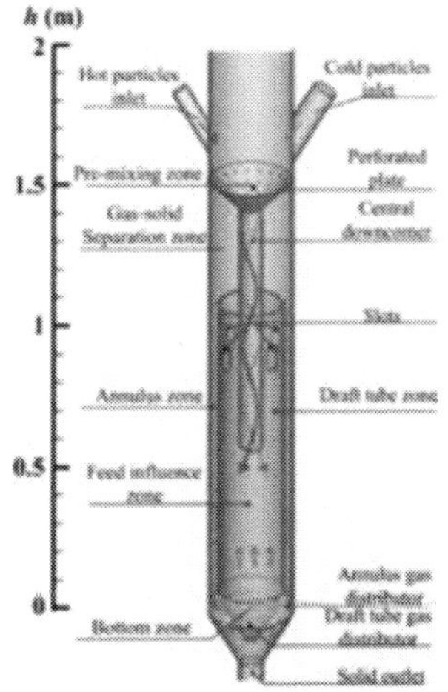

Fig. 1 Schematic diagram of the new-type particle mixer

3. CFD model and simulation method

The hydrodynamics of the mixer was performed using the Eulerian- Eulerian of Ansys Fluent

®13.0. The governing equations [11] include a set of mass and momentum equations for each phase, as shown in Table 1.

A 3-D mesh containing 580,000 cells was generated by using Gambit®2.4. The no-slip boundary was assigned for the gas phase and the partial-slip boundary condition developed by Johnson and Jackson [12] was used for the solid phase. In order to ensure a steady-state operation of the mixer, simulations were first performed for 30 s, and then the time-averaged data was analyzed for another period of 10 s. Other parameters settings in the model are shown in Table 2.

Table 1 Governing equations

Gas-phase Continuity equation	$\frac{\partial}{\partial t}(\varepsilon_g \rho_g) + \nabla \cdot (\varepsilon_g \rho_g \boldsymbol{u}_g) = 0$
Solid-phase Continuity equation	$\frac{\partial}{\partial t}(\varepsilon_s \rho_s) + \nabla \cdot (\varepsilon_s \rho_s \boldsymbol{u}_s) = 0$
Gas-phase momentum equation	$\frac{\partial}{\partial t}(\varepsilon_g \rho_g \boldsymbol{u}_g) + \nabla \cdot (\varepsilon_g \rho_g \boldsymbol{u}_g \boldsymbol{u}_g) = -\varepsilon_g \nabla p + \nabla \cdot \boldsymbol{\tau}_g + \varepsilon_g \rho_g \boldsymbol{g} - \beta(\boldsymbol{u}_g - \boldsymbol{u}_s)$
Solid-phase momentum equation	$\frac{\partial}{\partial t}(\varepsilon_s \rho_s \boldsymbol{u}_s) + \nabla \cdot (\varepsilon_s \rho_s \boldsymbol{u}_s \boldsymbol{u}_s) = -\varepsilon_s \nabla p - \nabla p_s + \nabla \cdot \boldsymbol{\tau}_s + \varepsilon_s \rho_s \boldsymbol{g} + \beta(\boldsymbol{u}_g - \boldsymbol{u}_s)$
Granular temperature equation	$\frac{3}{2}[\frac{\partial}{\partial t}(\varepsilon_s \rho_s \Theta_s) + \nabla \cdot (\varepsilon_s \rho_s \Theta_s \boldsymbol{u}_s)] = (-p_s \boldsymbol{I} + \boldsymbol{\tau}_s) : \nabla \boldsymbol{u}_s + \nabla \cdot (\kappa_s \nabla \Theta_s) - \gamma_s - 3\beta\Theta_s$
Gas-phase stress	$\boldsymbol{\tau}_g = \varepsilon_g \mu_g [\nabla \boldsymbol{u}_g + (\nabla \boldsymbol{u}_g)^T] - \frac{2}{3}\varepsilon_g \mu_g (\nabla \cdot \boldsymbol{u}_g)\boldsymbol{I}$
Solid-phase stress	$\boldsymbol{\tau}_s = \varepsilon_s \mu_s [\nabla \boldsymbol{u}_s + (\nabla \boldsymbol{u}_s)^T] + \varepsilon_s (\lambda_s - \frac{2}{3}\mu_s)(\nabla \cdot \boldsymbol{u}_k)\boldsymbol{I}$
Solid-phase pressure	$p_s = \varepsilon_s \rho_s \Theta_s [1 + 2(1+e_s)\varepsilon_s g_0]$
Solid-phase shear viscosity	$\mu_s = \mu_{s,col} + \mu_{s,kin} + \mu_{s,fr}$
Where	$\mu_{s,col} = \frac{4}{5}\varepsilon_s \rho_s d_p g_0 (1+e_s)\sqrt{\Theta_s / \pi}$, $\mu_{s,kin} = \frac{10\rho_s d_p \sqrt{\Theta_s / \pi}}{96\varepsilon_s (1+e_s) g_0}[1 + \frac{4}{5} g_0 \varepsilon_s (1+e_s)]^2$, $\mu_{s,fr} = \frac{p_s \sin\phi}{2\sqrt{I_{2D}}}$
Radial distribution function	$g_0 = [1 - (\varepsilon_s / \varepsilon_{s,max})^{1/3}]^{-1}$
Diffusion coefficient	$\kappa_s = \frac{150\rho_s d_p \sqrt{\Theta_s / \pi}}{384(1+e_s) g_0}[1 + \frac{6}{5} g_0 \varepsilon_s (1+e_s)]^2 + 2\rho_s \varepsilon_s^2 d_p (1+e_s) g_0 \sqrt{\Theta_s / \pi}$
Collisional energy dissipation	$\gamma_s = \frac{12(1-e_s^2)\varepsilon_s^2 \rho_s g_0 \Theta_s^{3/2}}{d_p \sqrt{\pi}}$
Heterogeneity index	$H_d = \frac{\beta_e}{\beta}$, $\beta_e = \beta H_d$
Drag based homogeneous structure	$\beta = \begin{cases} \frac{3}{4} C_d \frac{\varepsilon_s \varepsilon_g \rho_g \lvert \boldsymbol{u}_g - \boldsymbol{u}_s \rvert}{d_p} \varepsilon_g^{-2.65}, \varepsilon_g > 0.8 \\ 150 \frac{\varepsilon_s^2}{\varepsilon_g} \frac{\mu_g}{d_p^2} + 1.75\varepsilon_s \frac{\rho_p}{d_p} \lvert \boldsymbol{u}_g - \boldsymbol{u}_s \rvert, \varepsilon_g \le 0.8 \end{cases}$
Where	$C_d = \begin{cases} 0.44 & \mathrm{Re}_s \ge 1000 \\ \frac{24}{\mathrm{Re}_s}(1 + 0.15\mathrm{Re}_s^{0.687}), & \mathrm{Re}_s < 1000 \end{cases}$, $\mathrm{Re}_s = \frac{\varepsilon_g \rho_g d_p \lvert \boldsymbol{u}_g - \boldsymbol{u}_s \rvert}{\mu_g}$

Table 2 Simulation Parameters

Gas-solid model	Eulerian-Eulerian
Solids Pressure	Lun et al.
Particle-wall specularity coefficient	0.5
Gas-wall friction coefficient	No-slip
Pressure-Velocity Coupling	Simple
Gradient discretization	Least squares cell based
Momentum discretization	Second-order upwind
Volume fraction discretization	Quick
Max. number of iterations per time step	40
Pressure outlet (atm)	1
Inventory (kg)	56
Incipient solid holdup	0.6

4. Results and discussion

4.1 Model validation

Fig. 2 gives the distribution of cross-sectionally averaged solid holdup and particle velocity in the mixer. As shown in Fig. 2(a), the solid holdup is larger in the bottom zone (h=0.04 m) and due to the accelerated upward motion of particles, the solid holdup decreases with increase of the axial height (when $h \leq 0.16$ m). The cross-sectionally averaged solid holdup is nearly constant in the draft tube and gas-solid separation zone of $0.4 \leq h \leq 1.25$ m. The cross-sectionally averaged particle velocity is shown in Fig. 2(b). The particle velocity is smaller in the bottom zone and keeps almost invariable in the draft tube and the gas-solid separation zone. Compared with the experimental data, the particle velocity seems slightly higher, while the solid holdup in the simulation agrees reasonably with the experimental results.

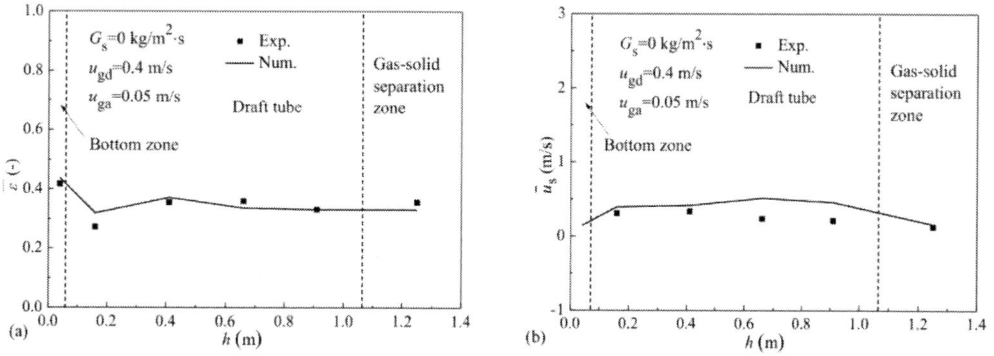

Fig. 2 Distribution of cross-sectionally averaged solid holdup and particle velocity as a function of axial heights

4.2 Pressure distribution near the slots

Fig. 3 shows the time-averaged contours of static pressure on the cross-section of h=0.9 m, the cross-section on the middle of the slots. It is seen that in the draft tube a low pressure region can be observed near the slots, smaller than the pressure in the region between the adjacent slots. This pressure drop forced particles inside the draft tube to move towards the slots and pass through them, which significantly facilitates the mixing of the particles along the radial direction. In the annulus the pressure in the vicinity of the slots seems larger than that in the region between the adjacent slots, inducing horizontal particle immigration and mixing.

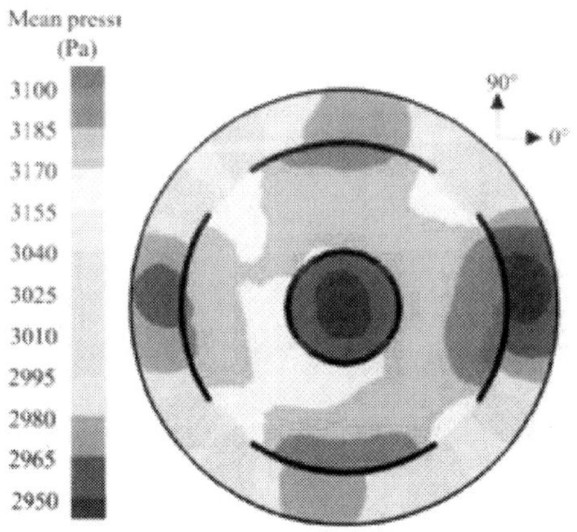

Fig. 3 Time-averaged contours of static pressure at the cross-section of the slots
(u_{gd} =0.4 m/s, G_s=13.8 kg/m^{2}s, h=0.9 m)

4.3 Gas bypassing

In the present work, the gas bypassing behavior was investigated by CFD simulation, and a tracer gas having identical physical properties with air was defined. After the mixer running stably, the tracer gas was introduced into the mixer from the gas distributor plate to replace the original gas. A cross section (h=0.06 m) on the exit of the annulus was selected and the mass flux of the tracer gas through the cross section was registered. The results are shown in Fig. 4. f_B refers to the ratio of the mass flux of the tracer gas through the cross section (h=0.06 m) to that of total tracer gas. It is seen that f_B varies between -0.05 and 0.05 with a frequency of 10 Hz, indicating an unstable gas bypassing from the draft tube to annulus. This suggests that the gas bypassing into the annulus was finally brought out of the annulus by the circulating particles. The average f_B is approximately equal to 0, much smaller than that in a conventional ICFB (usually within 15%) [13], because massive particle circulation in the mixer makes it difficult to bypass from the draft tube to annulus.

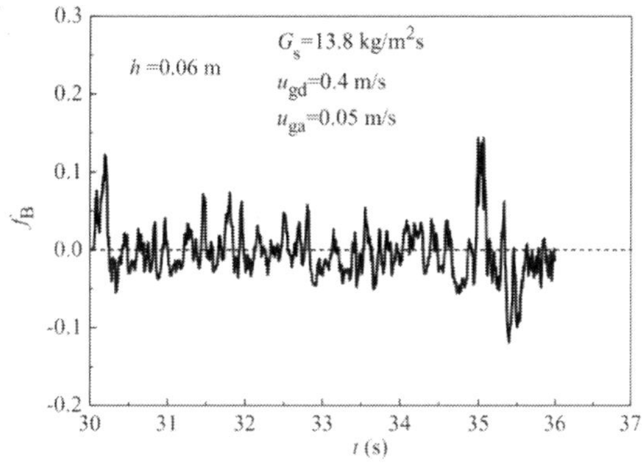

Fig. 4 Gas bypassing from draft tube to annulus in the bottom zone

As discussed previously, in the mixer there exists a pressure gradient from the draft tube to the vicinity of the slots, and then the annulus. Such a pressure difference leads to horizontal flows of particles and gas from the draft tube to the slots and then to the annulus. In the mixer the upside and downside of the slots is located at h=0.94 m and 0.86 m, respectively. Mass flux of the gas bypassing through the slots was obtained by monitoring the gas flux through the lower cross section (h=0.7 m) and the upper cross section (h=1 m) in the draft tube. Fig. 5 shows the gas mass flux under different operating conditions. The total gas mass flux in the draft tube increases with increasing gas velocity in the draft tube. With increase of the gas velocity from 0.3 to 0.45 m/s, the mass flux of the gas bypassing through the slots increases from 0.157 to 0.221 kg/m^{2}s. But the mass flux of bypassing gas does not increase linearly with the gas velocity. It is also seen that the mass ratio of the gas through the slots to that in the draft tube drops from 29.7% to 25.4%.

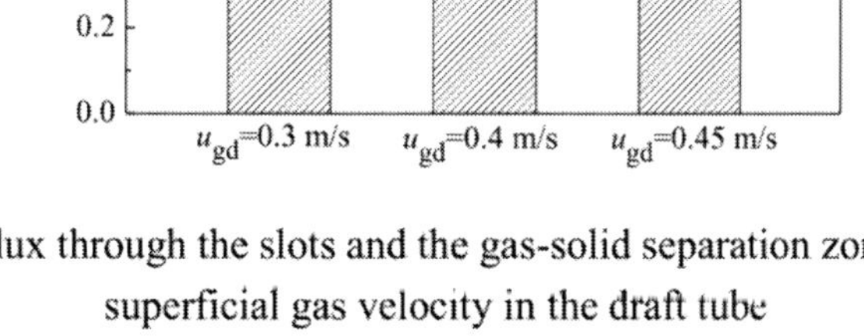

Fig. 5 Gas mass flux through the slots and the gas-solid separation zone as a function of superficial gas velocity in the draft tube

4.4 Particle mass flux

Fig. 6 shows the particle mass flux in the mixer. It is seen that the mass flux increases with increase of the superficial gas velocity in the draft tube. The mass flux through the slots accounts for about 50% of the total mass flux when u_{gd}=0.3 m/s. Because the axial particle velocity rose with increase of the superficial gas velocity from 0.3 m/s to 0.45 m/s in the draft tube, making it difficult to change the direction of the particles. Meanwhile, the radial particle velocity above the draft tube increases. Consequently, with the increase of the superficial gas velocity in the draft tube , the mass flux through the slots considerably decreased, from about 55.3% of the total mass flux (about 2.69 kg/s) at u_{gd}=0.3 m/s to only 4.8% (about 0.52 kg/s) at u_{gd}=0.45 m/s. Meanwhile, the total particle mass flux in the mixer increased from 4.87 to 10.86 kg/s as u_{gd} increased from 0.3 to 0.45 m/s.

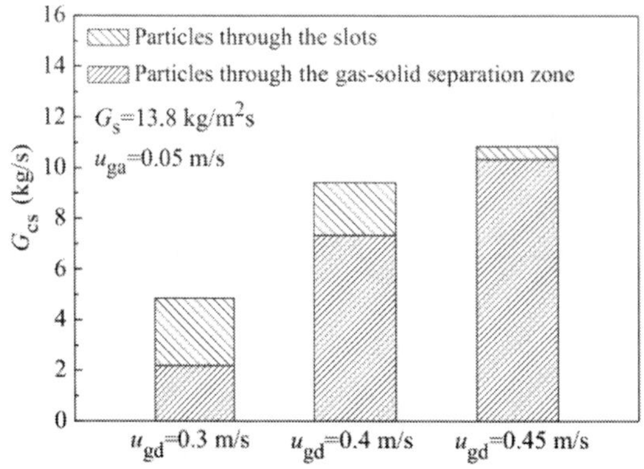

Fig. 6 Particle mass flux through the slots and the gas-solid separation zone as a function of superficial gas velocity in the draft tube

5. Conclusions

The characteristics of particle flow and mixing have a considerable influence on the mass and heat transfer, as well as the product quality. In order to further enhance the particle mixing in the ICFB, especially along the radial direction, a new-type particle mixer with slots on the draft tube was proposed in the present work. The hydrodynamics of the mixer was investigated by using a multi-scale computational fluid dynamics with the EMMS drag. The prediction in terms of cross-sectionally averaged solid holdup and solid velocity were compared to experimental data for validation. The results show that in draft tube a low pressure region can be observed near the slots and in annulus the pressure in the vicinity of the slots seems larger than that in the region between the adjacent slots, which induces horizontal particle immigration and mixing. Nearly 25.4% to 29.7% of gas in the draft tube bypasses and flows through the slots into the annulus. The ratio of the mass flux of the tracer gas bypassing into the annulus to that of total tracer gas is approximately equal to 0. With increasing the superficial gas velocity in the draft tube from 0.3 m/s to 0.45 m/s, the total solid mass flux increases from 4.87 kg/s to 10.86 kg/s, while the mass flux through slots decreases from 2.69 kg/s to 0.52 kg/s.

Notation

G_s	solid mass flux (kg/m^2 s)	G_{cg}	gas mass flux in the mixer (kg/m^2 s)
G_{cs}	solid mass flux in the mixer (kg/s)	h	height (m)

p	pressure (Pa)	R	radius of the reactor (mm)
r	radial position of measurement points (mm)	t	time (s)
u_{ga}	superficial gas velocity in the annulus (m/s)	u_{gd}	superficial gas velocity in the draft tube (m/s)
$\bar{u}$	cross-sectionally averaged axial particle velocity (m/s)		

Greek

ε_g	voidage	ε_s	solid holdup
$\bar{\varepsilon}_s$	cross-sectionally averaged solid holdup	ρ	density (kg/m^3)

References

[1] F. Niklasson, H. Thunman, F. Johnsson, B. Leckner, Estimation of solids mixing in a fluidized-bed combustor, Ind Eng Chem Res, 41 (2002) 4663-4673.

[2] L.H. Shen, M.Y. Zhang, Effect of particle size on solids mixing in bubbling fluidized beds, Powder Technol, 97 (1998) 170-177.

[3] L.H. Shen, M.Y. Zhang, Y.Q. Xu, Solids Mixing in Fluidized-Beds, Powder Technol, 84 (1995) 207-212.

[4] N. Mostoufi, J. Chaouki, Local solid mixing in gas-solid fluidized beds, Powder Technol, 114 (2001) 23-31.

[5] D.Y. Liu, X.P. Chen, Lateral solids dispersion coefficient in large-scale fluidized beds, Combust Flame, 157 (2010) 2116-2124.

[6] F. Berruti, D.S. Scott, E. Rhodes, Measuring and Modeling Lateral Solid Mixing in a 3-Dimensional Batch Gas Solid Fluidized-Bed Reactor, Can J Chem Eng, 64 (1986) 48-56.

[7] M.M. Fang, K. Luo, S.L. Yang, K. Zhang, J.R. Fan, Computational Fluid Dynamics-Discrete Element Method Investigation of Solid Mixing Characteristics in an Internally Circulating Fluidized Bed, Ind Eng Chem Res, 52 (2013) 7556-7568.

[8] J. Li, M. Kwauk, Particle-fluid Two-phase Flow: the Energy-minimization Multi-scale Method, Metallurgical Industry Press, Beijing, 1994.

[9] Z.S. Shi, W. Wang, J.H. Li, A bubble-based EMMS model for gas-solid bubbling fluidization, Chem Eng Sci, 66 (2011) 5541-5555.

[10] K. Hong, Z.S. Shi, W. Wang, J.H. Li, A structure-dependent multi-fluid model (SFM) for heterogeneous gas-solid flow, Chem Eng Sci, 99 (2013) 191-202.

[11] D. Gidaspow, Multiphase flow and fluidization: Continuum and Kinetic Theory Descriptions, Academic Press, Boston, 1994.

[12] P.C. Johnson, R. Jackson, Frictional–collisional constitutive relations for granular materials, with application to plane shearing, J. Fluid Mech. 176 (1987) 67–93.

[13] B.H. Song, Y.T. Kim, S.D. Kim, Circulation of solids and gas bypassing in an internally circulating fluidized bed with a draft tube, Chem Eng J, 68 (1997) 115-122.

514498 Power Spectral Density Analysis of Pressure Signal in 18 m Circulation Fluidized Bed Riser

Abstract:

Studies on the gas-solids flow characteristics using Power Spectral Density based on the pressure data from different axial locations in a riser of diameter 80 mm and height 18 m. Results show that the pressure signal fluctuation is mainly affected by gas-solids interaction, which included gas-solids friction, particle-particle collision and particle-particle aggregation. With the increasing solids circulating rate, frequency of particle interactions including particle-particle collision and particle-particle arrogation become strong and the pressure fluctuation also increased. It can be seen that the pressure signal fluctuation intensity decreased gradually with the increasing axial position of the riser. The low frequency but high power phenomenon nearly disappears in the fully developed zone near the top of the riser. Based on the results, it can be concluded that the gas-solids interaction is stronger at the bottom of riser. On the other hand, the gas-solid interaction is more homogeneous, and the pressure fluctuation is relatively weak in the fully developed zone.

514514 Characteristics of Pressure Fluctuations in Particle-Transport Inclined Pipe of a Circulating Fluidized Bed

Abstract:

In a large circulating fluidized bed device, the characteristics of pressure fluctuation in a particle-transport inclined pipe under different solid circulating flux were investigated based on measuring the dynamic pressure of gas-solid two-phase flow by using FCC equilibrium catalyst as the particles. The experimental results showed that with increase of particle mass flux, the flow patterns of FCC particles above and below the butterfly valve in the inclined pipe both successively changed, and the fluctuation amplitude of dynamic pressure increased. The standard deviation of the dynamic pressure below the butterfly valve possessed a linear relation with the particle mass flux, which could be used to characterize the real-time dynamic pressure in the inclined pipe. By wavelet analysis, it is indicated that there exists a main frequency caused by inclined pipe flow structures in the pressure fluctuations above and below the butterfly valve, besides, there is a secondary frequency in the pressure fluctuation below the butterfly valve caused by gas turbulence and gas-solid interactions. With the increase of particle mass flux, the wavelet energy ratio of the secondary frequency gradually decreased. The fluctuation pressure caused by discharge in the inclined pipe is a low frequency pressure and the excitation source of inclined pipe vibration.

508323 Dynamic Modeling of Attrition and Reactions in Circulating Fluidized Bed Reactors

Abstract:

Fluidized bed reactors are widely applied in chemical processes due to the excellent gas-solid contact. A problem that occurs in fluidized beds is the mechanical stress that particles are exposed to during the fluidization, which causes particle attrition. Additionally, fluidized beds are often operated in batch mode, which will cause ever-changing operating conditions with time. This makes fluidized bed modeling especially challenging, because the distributed particle properties, like size, wetness or reaction state, have to be tracked over the whole course of the simulation.

At the Institute of Solids Process Engineering at Hamburg University of Technology, novel process simulation software named DYSSOL is developed, particularly tailored for solids processes. The software is based on the flowsheeting concept. This means that chemical unit operations, like for example chemical reactors or separators, are put into process modules, which are interconnected into a process flowsheet and then mathematically solved. To make dynamic simulations available, new dynamic models had to be proposed for the involved equipment. The models themselves are based on empirical and semi-empirical correlations for hydrodynamics and kinetic particle reaction models are integrated.

The dynamic flowsheeting software was then used to simulate a system of interconnected fluidized bed reactors, which are used for Chemical Looping Combustion of solid fuels. This process is especially interesting, because heterogeneous reactions take place. Further, the process is a reactor-regenerator system causing the reactive solid to be oxidized and reduced and, hence changing its conversion state. Attrition plays a major role for the cost effective operation of the process.

In a small lab scale bubbling bed reactor, the jet and bubbling attrition rate constants for the solid oxygen carrier were experimentally determined under ambient conditions. In a small stand-alone cyclone the cyclone attrition was investigated. Furthermore, the reaction kinetics of the solid material with fuel gases was experimentally determined. All this information was included into the flowsheet simulation of a pilot scale circulating fluidized bed reactor, which is actually operated in our institute to deliver validation data.

The dynamic flowsheet simulation of the pilot plant, which was enhanced with kinetic and attrition information of smaller scale facilities, was conducted over 1000 seconds. A fuel load change was simulated and the chemical conversion was tracked. Over the whole simulation, the attrition and small particles leaving the reactor system was also covered. When comparing the simulation with experiments it could be shown that the gas conversion was sufficiently calculated and the conversion state of the oxygen carrier tracked correctly. The attrition modeling, however, with its parameters measured had to be adapted, so that it fitted the outlet particle flows measured under reaction conditions.

The financial support of DFG (Deutsche Forschungsgemeinschaft) within the priority program SPP 1679: â€œDynamic simulation of interconnected solids processesâ€ is gratefully acknowledged; grant number HA 6935/2-1.

514599 Analysis of FCC Cyclone Fault Diagnosis Technology Based on Particles Information

Abstract:

Cyclones are key equipment in fluid catalytic cracking unit (FCCU), which are used for separating the catalyst particles from oil gas or flue gas. Hence, it is very necessary to carry out condition monitoring and fault diagnosis for the running cyclone. The properties of particles vary at different positions in the particle circulation loop, such as size distribution, morphology and element composition, etc. In consequence, the particles contain much information about operation. Once the cyclone operates abnormally, the information of run-off catalyst particles appear abnormal too, and there exists a corresponding relationship between them. So, taking the catalyst particles as the carrier of fault information, this paper put forward cyclone fault diagnosis technology by extracting catalyst particle information, such as particle packing density, particle size distribution, particle micrographs etc, to build the relationship model between the cyclone fault and catalyst particle information.

514600 The Multi-Hole Throttling Effect and the Erosion Characteristics of the High Pressure Natural Gas

Abstract:

Based on Euler-Lagrange multi-phase flow numerical simulation method, the single hole and multi-hole throttling process of the high pressure natural gas was studied by using the RNG k-$\hat{I}\mu$ turbulent model, the Sutherland Viscosity Law compressible fluid viscosity model, and the advanced erosion model numerical simulation. The calculated mass flow variation law in the throttle choke is comparable with the theoretical calculations, which verified the adopted models and calculating methods. And then the throttling effect and the erosion characteristic of the high pressure natural gas was studied to check its specific law with different throttle holes by using the advanced erosion model. The results showed that, high pressure natural gas can reduce pressure and expansion after flowing through the single hole and multi-hole throttle choke, in which lead a high-speed jet region came up near its outlet. But the maximum jet velocity of multi-hole throttle choke was significantly lower than that of the single hole. At the same time the multi-hole uniform jet may reduce the vortex entrainment and the wall erosion damage, in which can be easier to mix the cold core and achieve stable pressure recovery at the role of the multi-jet mixing effect. Therefore, the multi-hole throttling can effectively restrain the erosion damage and the gas hydrate formation probability.

514601 Numerical Simulation of Flow Field In a Gas Pipe Distributor of the FCCU Regenerator

Abstract:

The flow field in the pipe distributor used in regenerator of FCCU (Fluid Catalytic Cracking Unit), was analyzed based on the numerical simulation. Firstly, the distributing gas performance of the pipe distributor was investigated. The result shows that the pressure increases along the branch pipe of the pipe distributor and the axial flow rate decreases in the branch pipe, which leads to the nozzle flow rate increasing gradually. Furthermore, an obvious bias flow was found in the inlet of branch pipe, which brings some nozzle flow rate less than the designed flow rate and causes an uneven gas distribution in the fluidized bed as well. Meanwhile, the momentum equation with variable mass flow was applied to analyze the flow in the pipe distributor. The flow type in the branch pipe belongs to the "momentum exchange controlling model". The pressure in the branch pipe always increases from the inlet to the end, and the pressure distribution is not consequentially uniform. The uneven gas distribution of the pipe distributor results in fluidized gas non-uniformity in the bed and erosion in the branch pipe and nozzle. Finally, according to the pressure characteristics in fluidized bed system, relationship between the pressure drop of branch pipe and pressure drop of nozzles, as well as of the critical nozzle pressure drop and the designed nozzle pressure drop, were discussed. The gas distributor performance could be optimized by means of the structural improvement of branch pipe.

514602 Full-Loop Simulation of Gas-solids Flow in Circulating Fluidized Bed of FCC System

Abstract:

Circulating fluidized beds are widely used in petrochemical, coal, metallurgy, and other industrial fields. A typical circulating fluidized bed (CFB) includes the riser, gas-solid separator, downer and solids flow control device. To obtain detailed information and get better understand of gas-solids flow in CFB, computational fluid dynamics (CFD) had been applied extensively in many systems such as circulating fluidized bed combustion (CFBC) systems and fluid catalytic cracking (FCC) systems. In most studies, the riser, the core part in CFB system, was isolated as a single simulation object by applying proper boundary conditions. However, there are some limitations of riser-only simulation, without consideration of dynamics in solids circulation rate and effects of other part of CFB system. Therefore, the full-loop simulation is a more promising way to describe complex gas-solids interaction in CFB by treating the whole CFB system as a simulation object. Nowadays, most of current full-loop simulations were focused on CFBC systems, there was little work about FCC system, with different type of particles and recycle mechanism from CFBC system.

In this work, both full-loop and riser-only simulations were carried out by TFM approach to investigate a pilot-scale gas-solid circulating fluidized bed of FCC system which has been set up in our group. The simulation condition was chosen according to experiments with different superficial gas velocity, valve opening and fluidizing air flow rate. Detailed comparison between full-loop and riser-only numerical simulations has been conducted with respect to the hydrodynamics inside the riser, which showed full-loop simulation had advantages on predicting dynamics of solids circulation rate and the gas-solids flow behavior. Distribution of pressure, solids holdup and velocity in system was investigated which was helpful to understand the hydrodynamics in CFB. The effects of operating condition were then studied, which revealed the system stability mechanism.

506758 Immobilization of Sulfur-Oxidizing Bacterium, *Thioalkalivibrio Sp.* D301 on Magnetic Nanaoparticles and Biodesulfurization

Abstract:

There is an urgent need to develop desulfurization methods for halo-alkaline wastewater. *Thialkalivibrio sp.* D301, an obligatory chemolithoautotrophic, haloalkaliphilic, sulfur-oxidizing bacterium, was used in the removal of sulfide in wastewater. We developed a process to immobilize D301 cells by using superparamagnetic Fe_3O_4 nanoparticles (NPs) under halo-alkaline conditions. Fe_3O_4 NPs were synthesized by coprecipitation of ferric and ferrous salts in the presence of N_2 gas. TEM image shows the spherical shape of Fe_3O_4 NPs with an average diameter of 10 to 15 nm. The particles had superparamagnetic properties. The saturation magnetization (Î´$_s$) of the Fe_3O_4 NPs was 73.985 emu g$^{\text{âˆ'1}}$ with zero remanence and coercivity. Analysis of zeta potentials and hydraulic diameters showed that the Fe_3O_4 NPs and bacterial cells interact mainly through the nano-size effect. The NPs-coated cells have the same sulfur oxidization activity as that of free cells. Experiment results showed that the immobilized cells still retained >90% of their sulfur oxidization activity after six times of reuse. The cells coated with Fe_3O_4 NPs did not interact through electrostatic attraction. The mechanism of NPs adsorption mainly involved the nano-size effect. Magnetic immobilization is a convenient technique for HA-SOB immobilization under halo-alkaline conditions and is a promising technology for large-scale industrial applications.

507944 Simulation of Bulk Solids and Granular Systems By Using Combined Discrete Element Models

Abstract:

Granular materials can be found in many industrial processes from chemical to energy, agriculture, mining, pharmacy and many other fields. They are almost always opaque, making the acquisition of experimental data below the surface difficult. DEM (Discrete Element Method) simulations are an effective way to study the physics of granular materials because they can simulate the micro-dynamic behavior of granular material on an individual particle scale and the motion of every particle can be traced. Since DEM was first considered, many modified discrete element models have been developed. However, most existing DEM simulation are just for spherical particles or a specific type of non-spherical particles. In this work, a combination of several types of discrete element models including spherical models, super-ellipsoids, multi-sphere models (with no interaction and deformation between sub-spheres), and bonded-spheres models (with interaction and deformation between sub-spheres) were used to simulate the flow of granular materials with many different shapes. Several simulation cases including the flow of particles in tumblers, blenders, and other devices were performed to validate the modeling approaches. The simulation results show that it is practical and efficient to combine different discrete element models together for simulating both spherical and non-spherical particles in bulk solids and granular systems.

515703 Material Optimization of Perovskite Films using High Throughput Synthesis and Multi-Dimensional Analysis

Abstract:

Over the past years, Perovskites received special attention in the field of renewable energy as they are one of the prime candidates for energy storage and conversion materials. Perovskites act as the active light harvesting layer of a solar cell and provide excellent performance.[1] A common synthesis strategy is through the liquid phase under inert atmosphere.[2,3] However, this comes with the challenge of a huge parameter space that in principle would have to be first synthesized, then characterized and finally optimized based on the therefrom deduced knowledge.

To overcome these limitations, we follow a fully automated approach, which allows automated high throughput (HT) synthesis and HT characterization of perovskites with high accuracy. As an efficient route to explore the parameter space, we use design of experiments (DoE) to find optimized settings for process parameters such as synthesis and annealing time and temperature, but also material aspects like the molar ratio of the reactants.

For all experiments, a Chemspeed© technologies (CST) swing XL platform with a liquid handling tool and a heatable shaking rack was used. The robotic platform is enclosed within an Mbraun© glovebox with inert atmosphere (Nitrogen). After precursor preparation, deposition was realized by a liquid handling tool on glass substrates inserted in a customized aluminum 24 well plate, which is then transferred onto a hot plate for film formation.

In a first step, after annealing at different temperatures, a characterization cascade was developed and applied that contains the subsequent elimination of samples that exhibit undesired properties and side products. Fast absorption analysis, photoluminescence (PL) measurements and finally X-ray diffraction (XRD) were performed to identify optimum chemistry and process parameters. Although this is a very reliable procedure, it is still comparatively demanding in terms of the time effort. Therefore a more efficient, purely optical approach by which the experimental parameter space is only explored in terms of rapid emission analysis was developed in a second step. Noteworthy, both strategies led to the same optimum conditions, however, at significantly reduced time effort.

In conclusion, our work shows how by the combination of automation, DoE and efficient characterization, optimum conditions for new materials like perovskites are rapidly identified. This opens the door towards faster material development which is important whenever new perovskite structures, e.g. lead free materials for industrial applications, are strived for.

Literature

1. Green, M.A.; Ho-Bailie, A.; Snaith, H.J., "The emergence of perovskite solar cells", Nature Photonics 2014, 8, (7), 506-514.
2. Baikie, T.; Fang, Y.; Kadro, J.M.; Schreyer, M.; Wei, F.; Mhaisalkar, S.G.; Graetzel, M.; White, T.J., "Synthesis and crystal chemistry of the hybrid perovskite (CH 3NH3)PbI3 for solid-state sensitised solar cell applications", J. Mater. Chem. A 2013,1 , 5628-5641.
3. Yan, K.; Long, M.; Zhang, T.; Wei, Z.; Chen, H.; Yang, S.; Xu, J., "Hybrid Halide Perovskite Solar Cell Precursors: Colloidal Chemistry and Coordination Engineering behind Device Processing for High Efficiency", J. Am. Chem. Soc. 2015, 137, (13), 4460–4468.

509509 Solid Liquid Separation Via Particle Flow Instability

Abstract:

A new separation strategy using novel particle-flow-instability physics is successfully developed for a *difficult-to-separate suspension* in which there is some combination of a small density difference between solid and liquid, high viscosity, and small-sized particles. The method we propose here requires no dilution for medium, high or extremely high viscosity slurries and can produce effective solidâ€"liquid separation with low capital and operational inputs. The results have relevance for many biotechnological and/or chemical processes.

509543 Milling and Grindability Assessment of Pharmaceutical Materials

Abstract:

Milling is one of the most important unit operations in the pharmaceutical industries. Most of the Active Pharmaceutical Ingredients (APIs) and excipients are produced by crystallisation, and grown to a sufficiently large particle size suitable for isolation and drying. Therefore most of the materials have to be milled to desirable particle sizes. Milling is influenced by a combined effect of milling dynamics (input power, mill geometry, etc) and material properties. Considerable research has been carried out on analysing the particle behaviour in various mills. Nevertheless, the prediction of the milling rate and milled product properties (e.g. product size distribution) is still very challenging.

In this work, the milling behaviour of several pharmaceutical materials in the PicoPlex mill (pin mill unit of the Picoline, Hosokawa Alpine) has been studied. A combination of experiments and Distinct Element Method (DEM) simulations is used to explore the dynamic behaviours of the materials in the mill. The results are then used to validate a novel approach of grindability assessment of pharmaceutical materials using the Scirocco disperser of the Malvern Mastersizer 2000 (Bonakdar et al. 2016). Based on this method the ease of grindability of the materials in the mill can be predicted.

Bonakdar, T., Ali, M., Dogbe, S., Ghadiri, M., and Tinke, A. (2016). A method for grindability testing using the scirocco disperser. *International Journal of Pharmaceutics,* 501: 65-74

507857 Bi-Directional Thermal Control of Twin Screw Granulation Process Via a Specialised Annular Heat Pipe

Abstract:

The continuous process of twin screw granulation forms a hotbed in pharmaceutical research as it is expected to succeed the current batch production of granules. In essence, the scale up of the material output is a function of production time, instead of equipment size (for batch granulators). This way, a â€œjust-in-time productionâ€ of the desired amount of material can be performed according to specific needs. In this work, a twin screw granulator was utilised in the granulation of both the placebo and active pharmaceutical formulations. Subsequently, the same equipment was used in granulating ceramic raw material powders for tile pressing. During processing, the spatial constraint in the barrel and screw compartments caused friction, resulting in excess heat generation. For the thermally-sensitive pharmaceutical formulations, this must be controlled to avoid denaturation of the granule products. A common solution is by using flowing cool water, but a good control has been proven to be difficult to achieve as different granulation zones generate different heat levels (kneading blocks > conveyer blocks). Heat pipe is a passive and highly effective thermal conductor and stabiliser. A carefully-designed annular heat pipe, filled with a small amount of methanol was successfully employed to stabilise the temperature in the barrel space, cooling the granulating material before it reached its critical temperature. For the ceramic powders, the methanol filling of the heat pipe was replaced by water and it was then heated up to provide an in-situ drying operation as the powders granulating. The in-situ drying in this instance, reduces energy consumption for the downstream drying processing as the product granules contain less moisture to be removed. The annular heat pipe, with its bi-directional operation of heating and cooling is a technically viable temperature controller for twin screw granulators. Compared to the established active cooling methods, the heat pipe provides a passive alternative, with only a discreet amount of fluid required to last its operational lifetime, thereby providing an economic thermal control solution.

Mixing Grains with Different Elongation in a Rotating Drums

Claudia Piacenza[1,2], *Marco Marconati*[1], *Colin Hare*[1], *Andrea C. Santomaso*[2], *Marco Ramaioli*[1,*]

[1] *Department of Chemical and Process Engineering, Faculty of Engineering and Physical Sciences, University of Surrey, Guildford GU2 7XH, UK*
[2] *APT Lab, Department of Industrial Engineering, University of Padova, Padova, Italy*

[*] *Email: m.ramaioli@surrey.ac.uk*

Abstract

Mixing grains with different properties is a remarkably challenging process, relevant to many industrial applications. Rotating drums have been used extensively as model systems to study granular media flow and mixing and segregation. Numerous studies considered the mixing of grains with different sizes, but only few studies considered shape and elongation, which have been already identified as important characteristics affecting packing and inducing segregation. In this contribution, the mixing of binary mixtures of grains having the same volume, but different elongations was studied experimentally. The mixing dynamics of a layered granular medium was characterized in a rotating drum, highlighting the impact of the grain shape, of the drum angular speed and of the drum filling ratio. A mixed or segregated state is reached very rapidly, but experiments are continued to verify that the state reached is a steady state. The experiments demonstrated that below a critical difference of elongation, the grains can be mixed effectively. Conversely, when grain elongations are very different, a central core is formed, rich in the more elongated grains. In this case only limited mixing can be achieved. These results can guide the formulation of mixtures of grains with different elongations.

1. Introduction

Mixing is a process operation aimed at obtaining a homogeneous system starting from two or more components. On the contrary, when there is a lack of homogeneity within the mixture segregation occurs. Segregation phenomena represent a relevant problem in the case of mixing solid particles having different physical properties such as particle size, density, shape or surface roughness. Differently from liquids and gases, solid particles are not thermal activated system [1] and so in order to obtain a certain degree of mixing they need to be mechanically activated. Typically three mixing mechanisms are largely recognized:

- dispersion, analogous to diffusion in fluids, it involves single particles that randomly move due to collision between each other;
- convection, it involves a portion of the bulk material, with a given direction and velocity;
- shear mixing: it is peculiar to granular solids and it is a combination of the two previous mechanisms.

Powder mixers are relatively simple machines of low capital cost. They are classified into two types: tumbling mixers and agitated mixers. Tumbling mixers are the simplest. A totally enclosed vessel is rotated about an axis causing the particles within the mixer to tumble over each other on the mixture surface. Tumbling vessels provide low-shear environments. Convective mixers instead are characterized by the action of impellers or paddles to move the powders around and generate a well-mixed product. Convective mixers provide more shear into the mixture and can be used with cohesive materials.

Rotating drums belong to the tumbling mixers. They are extensively used in the chemical and process industries. With respect to mixing processes, rotating drums are often used to study and understand mixing and segregation phenomena; as a result, many experimental and numerical studies were developed on granular behavior in rotating drums [1-3]. Different granular flow regimes can occur in the drum, each one with its specific flow behavior. In particular, different flow regimes can be assessed by using the dimensionless Froude number, $Fr = \omega^2 R/g$ [4], where ω [s^{-1}] is the drum rotation speed, R [m] the drum radius and g [m/s^2] the gravitational acceleration. With $10^{-4}<Fr<10^{-2}$ and filling level $\varphi>0.1$ rolling regime takes place. This motion is characterized by a uniform, steady flow of a particle layer on the surface (cascading or active layer) while the larger part of the bed (plug flow region) is moved upwards by solid rotation with the rotational speed of the wall. A uniform and good mixing is achievable with this type of motion, but when particle size distribution is very broad segregation can appear. The rolling bed is preferred, since is the simplest regime providing favorable conditions for heat transfer and high quality of the product even when mass flow rates are large.

When particles to be mixed have different properties such as different size, different density or different shape segregation typically occurs and particles tend to separate. [5, 6]. Difference in size is by far the most important cause of particle segregation. Particle size distribution influences also strongly bulk properties such as packing fraction and bulk modulus [7].

Different size segregation mechanisms can be identified:

- trajectory mechanism: according to the Stokes' law the velocity of fine particles is lower than that of big particles. This can cause segregation when small particles move through air;
- percolation mechanism: if a bulk material is disturbed, a rearrangement in the particle packing occurs and for fine particles it is easier to fall down in the inter-particle voids;
- inertial mechanism: in pouring and filling operations bigger particles tend to flow selectively to the bottom of the pile since they have a higher momentum;
- elutriation mechanism: in presence of air the very fine particles can remain in suspension while larger ones tends to settled. This can occur in charging storage vessels or hoppers.

A homogeneous binary granular mixture, with differences in size and density, in a two-dimensional rotating drum typically gives radial segregation. Considering a rotating drum operating in the rolling regime filled with particles of different sizes, bigger particles mainly tend to the edge of the system, while smaller particles accumulate to the centre of the drum. The phenomenon is fast and the rotational speed of the drum seems not to affect the phenomenon; the size ratio of the particles is, instead, the parameter that mainly affects size segregation phenomena. Most of the experimental investigations about segregation in the rotating drum involved spherical or nearly-spherical particles; the experimental studies of the flow of non-spherical particles are, in fact, rather limited [6,8]. The shape of the particles, however, strongly influences the motion in the system. In particular, the static and dynamic angles of repose of non-spherical particles are found to be very different from the ones measured for spherical particles suggesting that differences in particle mobility can affect the segregation process [9].

2. Materials and methods

2.1 Particles

Both spherical and non-spherical particles were used for the experiments. Their size (diameter d for spheres, major and minor axis for ellipsoids a and b respectively) , aspect ratio AR, projected area A_p, volume V, and true density ρ are reported in Table 1.

Table 1. Physical properties of the particles

Type	d [mm]	a [mm]	b [mm]	AR [-]	A_p [mm^2]	V [mm^3]	ρ [g/cm^3]
Spheres S	8	-	-	1	50.3	268	1.09
Ellipsoids E1	-	10.48	6.98	1.5	57.5	267	0.96
Ellipsoids E2	-	18.44	5.26	3.5	76.2	267	0.98

Spherical particles with size 8 mm were used for the experiments. The ellipsoids were 3D printed with *AR* equal to 1.5 and 3.5. They were printed with the Ultimaker 3D printer and made of polylactic acid (PLA). The software used to realize the 3D design of the particles used in this work is FreeCAD, which is an open-source parametric 3D modeller allowing to obtain STL files.

The true density of all the particles used for the experiments was measured by gas pycnometry (Multivolume Pycnometer 1305, Micromeritics). The measured density of the spheres was very close to the typical density of the nylon 66 (1.14 g/cm^3). The PLA ellipsoids had a density that was lower than the typical density of the PLA (1.24 g/cm^3). This is because the particles were printed with a set infill density of the 20%, so they where partially empty inside.

2.2 Tumbling mixer

The quasi-2D drum was made of two transparent polycarbonate disks with a diameter of 30 cm and a thickness of 6 mm; between them a wooden spacer was placed, with the same external diameter, an internal diameter of 25 cm and a thickness of about 9.1 mm. The spacer allowed to create an internal volume of finite thickness separating the two polycarbonate disks.

a)  b) 

Figure 1. a) Front view and b) side view of the rotating drum (the drum is on the left side of the rigid structure).

The internal lateral wall of the volume was covered with sandpaper in order to provide friction and avoid undesired slip of the particles. To keep the drum in vertical position and allow its rotation, three additional wooden spacers were used, at a distance of about 16.3 cm. This rigid structure could rotate at variable rotational speed thanks to a motor connected to a voltage supplier. Two rollers mounted on a rigid support made the rotation of the drum possible; a pulley transmitted the motion to the driving roller, while the second roller was passive. The rotational speed could be changed by varying the motor voltage. The structure could be removed from the rollers and by unscrewing the bolts, the drum could be easily opened so that to load/unload the particles.

2.3 Image acquisition

Frontal pictures of the drum were taken by a color camera and analyzed by the image processing software ImageJ [10]. To avoid the reflection of the camera and other objects on the front transparent disk, a squared LED panel was used which allowed to have a dominant diffused light illuminating the drum. The LED had a power of 8 W and a color temperature of 4000 K; it was placed behind the drum and opposite to the camera. A circular sheet of paper, which was attached to the drum between it and the LED, diffused the light properly. The camera was a Basler ace camera – model acA2000-165uc, with a maximum resolution of 2040 × 1086 pixels. The focus and the diaphragm of the camera could be adjusted manually.

2.4 Experimental plan

The attention was focused on mixtures of particles with different shapes, since segregation phenomena due to size had been widely investigated in the past. Three mixtures were considered, with two possible initial top-bottom segregated configurations. Filling level and rotational speed where changed also as detailed in Table 2.

Table 2. Experimental tests performed with differently shaped particles.

AR 1 [-]	*AR* 2 [-]	ω [rpm]	φ [-]	Init. Config.
1	1.5	4.22	25%	E1/S
1	1.5	4.22	25%	S/E1
1	1.5	4.22	47%	E1/S
1	1.5	4.22	47%	S/E1
1	1.5	11.51	25%	E1/S
1	1.5	11.51	47%	E1/S
1	3.5	4.22	25%	E2/S
1	3.5	4.22	25%	S/E2
1	3.5	4.22	47%	E2/S
1	3.5	11.51	25%	E2/S
1	3.5	11.51	25%	S/E2
1	3.5	11.51	47%	E2/S
1.5	3.5	4.22	25%	E2/E1
1.5	3.5	4.22	25%	E1/E2
1.5	3.5	4.22	47%	E2/E1
1.5	3.5	4.22	47%	E1/E2

*AR*1 and *AR*2 indicate the aspect ratios of the particles in the binary mixture. It is reminded that the spheres (AR=1) and the two types of ellipsoids (AR=1.5 and AR=3.5) had the same volume. The E1/S initial configuration indicates a segregated configuration with short ellipsoids above spheres (this was assumed as basis configuration); the S/E1 configuration is inverted with spheres above short ellipsoids. A similar convention is applied to the experiments considering mixtures of ellipsoids. All the experiments were performed with binary mixtures with a composition of 50% and 50% on a number (and hence also volume) basis. The filling level φ of the 25% was obtained with a total number of particles equal to 200; with 320 total particles the filling level of the drum was equal to 47%.

2.5 Image analysis procedure

Image analysis techniques were used to identify the particles on the basis of their different shape. The procedure was automated using customized ImageJ macros. Contrast and brightness of the images were first adjusted. Color images were then set to gray scale and segmented by applying a threshold. The operation was carried out twice to separate light and dark particles with different thresholds. Watershed segmentation was then used to separate touching objects. This last operation was more difficult for the particles at the border of the drum; in some case, in fact, the ellipsoids were erroneously identified as spheres. Also, the presence of the border of the drum complicated the watershed operation.

The centre of mass of each particle ($x_{m,i}$ and $y_{m,i}$) was estimated and the radial distance (d_i) from the centre of the drum (x_c, y_c) calculated as:

$$d_i=[(x_c-x_{m,i})^2+(y_c-y_{m,i})^2]^{0.5}$$

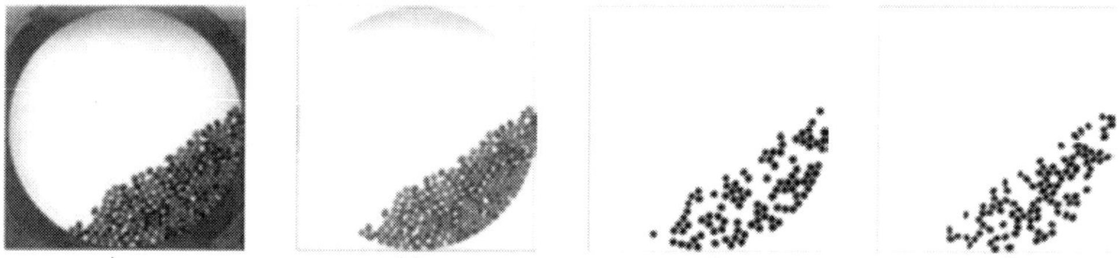

a) b) c) d)

Figure 2. Example of image processing steps. a) Original colour image; b) grayscale conversion and circle cut from the original picture; segmentation applied to c) light particles and d) dark particles.

The mean value of distance (d_m) for the spheres and for the ellipsoids was then calculated as:

$$d_m = \Sigma\, d_i/N$$

Only for very elongated ellipsoid (AR=3.5) the watershed operation failed resulting in over-segmentation errors. Single particles could not be identified unequivocally in this case; only one center of mass for all the elongated ellipsoid was estimated and the average distance (d_m) of the particles from the drum center calculated.

3. Results and discussion

Figure 3 shows two typical experiments, carried out at the lowest rotational speed and filling ratio, with ellipsoids (AR=1.5 at left and AR=3.5 at right) on top and spheres on bottom in the initial and final configuration. It is evident even by visual inspection that the final state presented some level of radial segregation and that a different final segregation intensity was reached in the two systems.

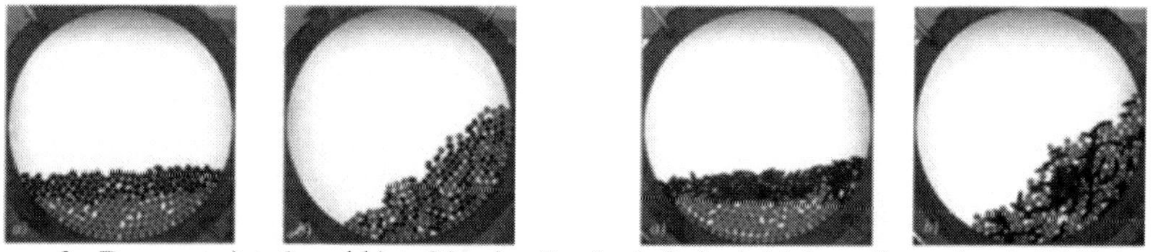

Figure 3. Drum at a) t=0 and b) t=30 min. On the left the initial configuration is E1/S (using ellipsoids with AR=1.5) and on the right the initial configuration is E2/S (AR=3.5). In both experiments ω=4.22 rpm and φ= 25%.

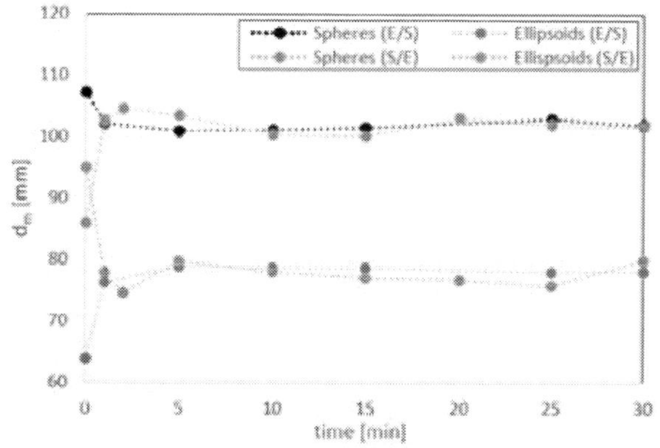

Figure 4. Evolution of the mean radial distance vs. time for the mixture of spheres and ellipsoids with AR=3.5. The initial configuration E2/S is compared to the initial configuration S/E2.

A more quantitative analysis was obtained by plotting the trend of the mean radial distance (d_m) of the two species at different time instants. As a typical example of the system evolution, the case of E2/S

is shown in Figure 4. The d_m reached a steady value after a short time. Dedicated experiments on a shorter time scale (2 minutes, not shown here) revealed that steady state value for d_m was reached after 30s for mixtures with ellipsoids E1 (*AR*=1.5) and after 60s for mixtures with E2 (*AR*=3.5).
At steady state, the radial distance was in both cases smaller for the ellipsoids, meaning that they mainly remained concentrated on top or at the center of the granular bed while the spheres mainly concentrated at the periphery of the drum (close to the drum wall). This tendency was much stronger for ellipsoids E2. So a slower segregation dynamics compensated by a larger segregation intensity was observed with the more elongated particles.
As Figure 4 shows, inverting the initial configuration from E/S to S/E did not affect the dynamics and the final result in all the experiments performed with these particles. Initial configuration resulted to have no effect on the segregation propensity of the mixtures analyzed.

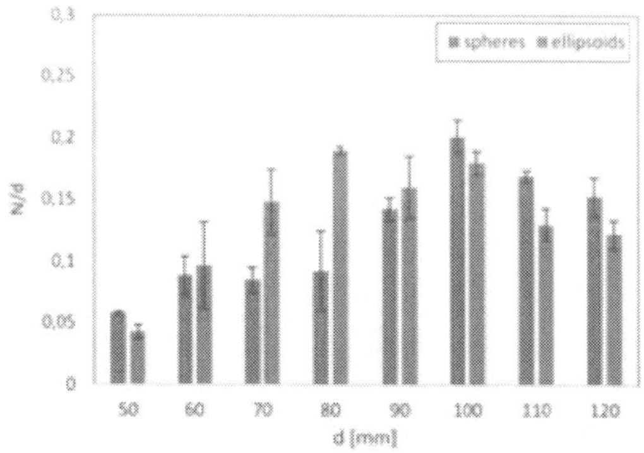

Figure 5. Distribution of the radial distance of spheres and ellipsoids with AR=1.5, obtained by averaging the results of three repetitions. Other conditions are t=30 min, ω=4.22 rpm, φ= 25%.

A more in-depth analysis can be carried out considering the distribution of the radial distance of the spheres and ellipsoids separately. The radial distance from the centre of the drum (*d*) was divided in classes of 10 mm and the number of each type of particles (*N*) in each class was computed. To obtain the distribution, *N* was normalized on the radial distance as *N/d* and plotted versus *d*. In the case of completely homogeneous mixtures and assuming that the medium takes the shape of a circular sector, *N/d* should be constant and equal for the two types of particles in each class. The peak of the distribution of the ellipsoids was found at lower distance than the peak of the distribution of the spheres. Moreover, at lower distance, the number of ellipsoids was higher than the number of spheres, especially for distances between 70 and 90 mm; at higher distance (closer to the drum wall), instead, the number of spheres was higher than the number of ellipsoids, thus confirming the previous analysis based on d_m.

Also, binary mixtures made of the two types of ellipsoids E1 and E2 (*AR*=1.5 and *AR*=3.5 respectively) were taken into account at the same filling levels of the previous experimental tests and at the lower rotational speed. Stable segregation was again observed with intermediate results with respect to the two previous cases (S+E1 and S+E2). So as a general comment on these experiments, stable segregation was always observed when mixing particles having same volume but different *AR*. The intensity of segregation increased with increasing the difference in *AR* between the particles. In particular particles with larger *AR* tended to accumulate at the center of the granular bed, while particle with lower *AR* accumulated at the periphery. Segregation occurred in the surface active layer where the more elongated particles, because of their reduced propensity to flow (particle rolling was prevented by their elongated shape), were systematically overtaken by the less elongated (and faster) particles. In this way a spontaneous separation of particles occurred in the active layer, with those having a lower *AR* arriving first at the drum wall, while those having an higher *AR* remaining trapped in the core.

3.1 Effect of filling level

The comparison of the behavior of the mixtures at the two filling levels can be carried out by normalizing the mean radial distance d_m of the two types of particle on the mean distance of all the particles d_c, calculated as $(d_{m,\ spheres}+d_{m,\ ellips.})/2$. The trends of d_m/d_c are shown in Figure 6

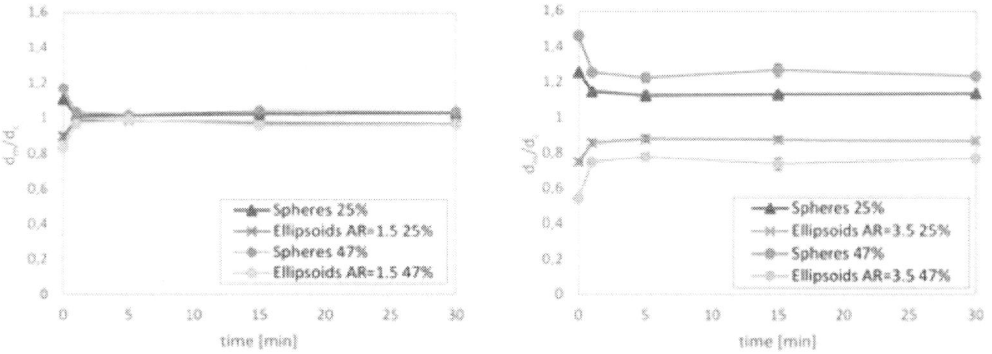

Figure 6. Trend of d_m/d_c vs. time for the mixture of spheres and ellipsoids with AR=1.5 (left) and AR=3.5 (right) at two different filling levels.

In the case of mixture of spheres and ellipsoids with AR=1.5, the curves at different filling levels overlap, showing that the behaviour of the mixture was not affected by the filling level of the drum. In the case of the other two mixtures (S+E2 and E1+E2, not shown), the curves did not overlap , especially for the case S+E2 shown in Figure 6 (right). An explanation can be found in terms of bed geometry. When increasing the filling level (below 50%) the length of the surface active layer increased as well. This gave more chances for particles to separate, in particular when the difference in AR was large enough (S+E2 and E1+E2). However for the system S+E1 where the difference in AR was the smallest, the difference in active layer length in the two cases was not sufficient to create observable differences in the segregation pattern.

3.2 Effect of rotational speed

In addition to experiments carried out at 4.2 rpm, tests were repeated at 11.5 rpm with spheres and ellipsoids (25% filling level). Figure 7 shows that rotational speed did not affect the segregation patterns, at least in the range of the analyzed speeds.

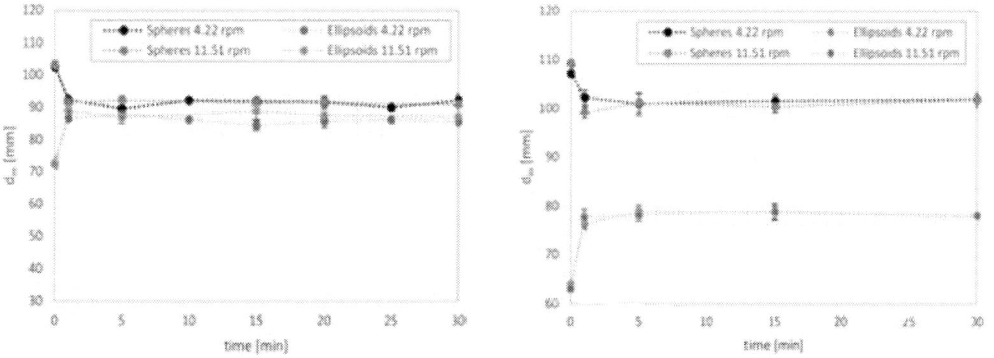

Figure 7. Trend of the mean radial distance vs. time for the mixture of spheres and ellipsoids with AR=1.5 (left) and AR=3.5 (right) at two different rotational speeds.

Conclusions

From the experimental tests described in the previous paragraphs, it was found that binary mixtures of particles with different shape but with the same volume do not mix uniformly and segregation was always observed. In particular, the particles with lower aspect ratio tended to stay closer to the drum walls, while the more elongated particles were found preferentially at the centre of the granular bed. The most significant case was represented by the mixture of spheres and ellipsoids with *AR*=3.5; for this mixture, in fact, the segregation was more evident and the non-spherical particles clearly accumulated in the centre of the granular bed. The same final state was obtained regardless of whether the longer particles were initially located to the top or the the bottom of the mixture. A proportionality between *AR* difference and segregation intensity was observed. Some variables such as rotational speed and initial bed configuration were found to not affect the segregation dynamics and intensity. Only filling level (in addition to *AR*) was found to affect segregation intensity. A mechanistic explanation was given in terms of differences in particle flowability occurring in the surface active layer (confirmed also by an analysis of the angles of repose in the drum). In this specific case therefore percolation and inertial effects (typical of size segregation) were not the main driving mechanisms for the observed segregation (the particle indeed had the same volume and similar mass so that sterical and inertial phenomena were minimized). Segregation mainly occurred because of differences in particle velocity in the active layer due to a reduced mobility of particles with larger *AR*.

References

1. GDR MiDi (2004), "On dense granular flows", The European Physical Journal E, 14, 4, 341–365.
2. Jarray A., Magnanimo V., Ramaioli M., Luding S. (2017) Scaling of wet granular flows in a rotating drum, EPJ Web of Conferences, 140,03078.
3. Volpato S., Canu, P., Santomaso A.C. (2017), "Simulation of free surface granular flows in tumblers", Advanced Powder Technology 28(3), 1028-1037.
4. Mellmann J. (2001), "The transverse motion of solids in rotating cylinders – forms of motion and transition behaviour", Powder Technology, 118, pp 251-270.
5. Windows-Yule C.R.K., Scheper B.J., van der Horn A.J., Hainsworth N., Saunders J., Parker D.J., Thornton A.R. (2016), "Understanding and exploiting competing segregation mechanisms in horizontally rotated granular media", New Journal of Physics, 18, 2, 023013, 2016.
6. Santomaso, A.C., Olivi, M., Canu, P. (2004), "Mechanisms of mixing of granular materials in drum mixers under rolling regime", Chemical Engineering Science, 59 (16), 3269-3280.
7. Kumar N., Magnanimo V., Ramaioli M., Luding S. (2015), "Tuning the bulk properties of bidisperse granular mixtures by small amount of fines", Powder Technology, 293, 94-112.
8. Ramaioli M., Pournin L., Liebling Th. M., (2005), "Numerical and experimental investigation of alignment and segregation of vibrated granular media composed of rods and spheres", Proceedings of the 5th International Conference on Micromechanics of Granular Media, 2, 1359-1363.
9. Santomaso, A.C., Canu, P. (2007), "Single particles properties vs. bulk flowability", Proceedings of 5th International Conference for Conveying and Handling of Particulate Solids.
10. Schneider, C. A., Rasband, W. S., Eliceiri, K. W. (2012), "NIH Image to ImageJ: 25 years of image analysis", Nature methods 9(7): 671-675.

507441 Generation of Particles with a Special Morphology By Desublimation of Copper Phthalocyanine

Abstract:

The morphology of submicron particles becomes more and more important concerning new functional materials. For example the special electric properties of particles can be used in electronic devices [1]. Moreover the electric, optical or catalytic properties of particles can be optimised by adjusting the morphology of the particles. This may imply a risk for the health but makes it possible to realize specific product characteristics. The morphology of the particles should therefore be considered e.g. for filter tests and for the developing and testing of sensor systems.

To explore the characterization of the particle morphology the DFG-AiF cluster project "multi-parametric characterization of particle-based functional materials using innovative online measurement systems" (MPaC) includes the development of online sensor systems for the size and structure of the particles as well as the development of generation methods for particles with a special morphology. The production of spherical particles can be achieved with different commercial aerosol generators. Several generation techniques for special particles are described in literature [2] [3] [4]. To test the sensor systems with particles of different shapes a new type of aerosol generator was developed. The aim for the new generator is to produce the particles in the gaseous phase and provide the particles directly for sensor or filter testing.

The developed generator uses the desublimation of copper phthalocyanine (organic pigment) to structure the surface of titanium dioxide. The bulk material copper phthalocyanine is dispersed into a nitrogen gas stream and transported to a tube furnace. Here the bulk material sublimates and the gas stream carries the material to a laval nozzle where the adiabatic expansion takes place. This leads to a rapid decrease of the temperature behind the laval nozzle and an oversaturation in the gaseous phase. In addition to the decrease of the temperature core particles of titanium dioxide were initiated behind the laval nozzle. The copper phthalocyanine desublimates on the core particles and causes a structured surface. The morphology of the generated particles is characterized by SEM. The particles show a punctual coating of the titanium dioxide particles with copper phthalocyanine. The size of the titanium dioxide particles is about 500 nm and the punctual coating of copper phthalocyanine measures a few nanometres.

The generation method and the design of the generator will be presented and the morphology of the generated particles will be described.

[1] K. Deppert, M. H. Magnusson, S. B. Carlsson, B. Gustafson, T. Junno, . L. Montelius, B. J. Ohlsson, L. Samuelson und L. E. Wernersson, „Future electronic devices by aerosol techniques," *Journal of Aerosol Science,* Nr. 31, 2000.

[2] J. Raula, A. Kuivanen, A. Lähde, H. Jiang, M. Antopolsky, J. Kansikas und E. . I. Kauppinen, „Synthesis of image-leucine nanoparticles via physical vapor deposition at varying saturation conditions," Journal of Aerosol Science, Nr. 38, 2007.

[3] M. Heurlin, M. H. Martin, D. Lindgren, M. Ek, R. L. Wallenberg, K. Deppert und L. Samuelson, „Continuous gas-phase synthesis of nanowires with tunable properties," Nature, Nr. 492, 2012.

[4] K. Nielsch, K. Deppert, M. H. Magnusson, G. Dumpich, E. Kruis und H. Fissan, „Formation of ultrafine particles from powders of compounds with incongruent sublimation behavior," Journal of Aerosol Science, Nr. 28, 1997.

516184 Rubbery Milling of Seed Endosperms for Improved Sustainability by Natural Functionality Preservation

Abstract:

Increasing demand for functionalised goods addressing individual needs requires a profound understanding of process-induced alterations. Development of models and process-structure-property relationships are essential to control the resulting quality and properties of processed food. A wide range of food systems contain solid or semi-solid particulates as major structure building blocks. Generation and tailoring of such particulates strongly depends on their size distribution, shape, morphology, phase behavior (amorphous/crystalline), interfacial properties and related inter-particle force interactions.

A novel processing technology was developed widening the frame of possible alterations, aiming towards tailoring such particle properties generating novel food structures or preserving techno-functional characteristics of the raw material. The technology employs targeted adjustments of the initial moisture content in combination with milling temperature. The resulting transition from the glassy to the rubbery state generates different material properties and processing behaviour. The technology is highly versatile, owing to the universality of moisture induced state transitions.

Seeds from locus bean gum or guar gum were wetted with excess water in order to adjust a rubbery state of the seed endosperm material. In this state milling within a novel type of impact mill with superimposed drying and sifting steps causes more gentle material disintegration compared to conventional dry milling. This preserves the high molecular weight fraction of e.g. galactomannans (micro-scale structure) from degrada-tion and allows at the same time the adjustment of a micro-porous morphology (macro-scale structure) improving wetting, water uptake and dispersing characteristics of the resulting thickener powders. Such structure preservation /modification increased the thickening characteristics extraordinarily as will be demonstrated.

From this prominent rubbery milling example we have entered in related systematic work addressing a wider range of seed endosperms with various â€œnature-synthesizedâ€ func-tionnalities to be further addressed.

/1/ Pollard, M., Eder, B., Fischer, P., and Windhab, E. J. (2010) "Characterization of galacto-mannans isolated from legume endosperms of Caesalpinioideae and Faboideae subfamilies by multidetection aqueous SEC." *Carbohydrate Polymers*, 79(1), 70-84.

/2/ Illmann, S., Stoob, M., Sieber, C., Konz, A.K. and Windhab, E.J. (2012) "New Thickening Quality of Galactomannan Polysaccharides by Tailored Impact Milling" *XVIth Int. Congress on Rheology*, Lisbon, Portugal, August 5-10

503422 Particle Technology Education at Purdue University

Abstract:

Degree programs and coursework at Purdue University focused on particle technology will be presented. Additional information may be found at: https://engineering.purdue.edu/CP3

506453 Teaching Particle Technology in Portugal â€“ University of Coimbra

Abstract:

Education in Particle Technology started in Portugal in the 80â€™s (last years of the decade). The Chemical Engineering degree of the University of Coimbra was the first engineering degree, in the country, to include this topic in the curriculum. Particle Technology was recognized by us, at that time, essential for the education of chemical engineers, in parallel to education in the traditional areas of Chemical Engineering which were still focused on monophasic systems.

Since then, Particle Technology has always been present in the Chemical Engineering curriculum of the University of Coimbra, in different configurations. During the first years particle characterization was combined with design of operations involving particulate systems (grinding and pneumatic and hydraulic transport). From 2005 to 2010, with the curriculum reformulation in relation with the Bologna Process, the two topics started to be addressed in different courses: one course on Materials Characterization and the process topic in a course entitled Structuring and Stabilization Processes, dealing with colloids and emulsions, grinding, agglomeration and spray drying and, for the first time, an integrated modelling strategy for particulate processes based on the population balance formulation was introduced in this course. The two courses were compulsory, at this time.

From the school year 2011/2012 till present, the configuration of this area in the Chemical Engineering degree of the University of Coimbra changed again, even if not deeply (see https://apps.uc.pt/courses/EN/course/361). The course on materials characterization, which includes general notions on particle characterization is kept as an autonomous course included in the 1st cycle of the Chemical Engineering degree of the University of Coimbra. A second course on design of processes involving particulate systems now exists in the 4th year of the MSc degree (1st year of the 2nd cycle), entitled Disperse Systems Technologies, which is an optional course, and which deals with all types of disperse systems, i.e gas, liquid or solids as disperse phase. This course starts with a review of the mechanisms/interactions occurring at the interface with the disperse phase, goes on, again, with a chapter on population balances modelling, and proceeds with strategies for disperse systems stabilization, linking this subject with formulation concepts, ending with the design of some processes involving particulate systems, such as colloids and emulsions, crystallization, grinding, agglomeration and spray drying. This course is also offered to other engineering degrees, such as Materials Engineering and Biomedical Engineering.

Also at the PhD level a course on Disperse Systems Technologies can also be offered during the formal lectures period of the PhD program on Chemical Engineering (or others), depending on the students background and interests.

Since the introduction of the Particle Technology topic in the curriculum of the Chemical Engineering degree of the University of Coimbra, the feedback from employers has been very positive, and the competences in this field acquired by our students have been quite valued by the job market. Till today, the Chemical Engineering degree of the University of Coimbra is still the one, in Portugal, with a stronger emphasis on this topic in the curriculum. In our curriculum, the topic of Particle Technology interacts directly with another course on Product Design, also in the 4th year of the MSc degree and, of course, also with the Project Design course (5th year of the MSc degree), besides other more traditional courses in a Chemical Engineering degree such as in Reaction Engineering and with courses related to environmental technologies.

At present, we have started developing multimedia tools for the teaching/learning of this topic, especially in the field of particle characterization, to be made available on-line, as is the case of the interactive video on nanoparticles characterization (http://labvirtual.eq.uc.pt/siteJoomla/index.php?option=com_content&task=view&id=277&Itemid=470, in Portuguese). Also, the European Federation of Chemical Engineering (EFCE) is considering investing in this type of tools for the field of particle technology, as the result of a collaboration between the Working Groups of Characterization of Particulate Systems and Education.

Acknowledgement

The authors would like to acknowledge the financial support of Pest/C/EQB/UI0102/2013 from FCT/MCTES (PIDDAC), co-financed by the European Regional Development Fund (ERDF) through the program COMPETE (POFC).

506629 Highlights of Particle Technology Teaching in Singapore

Abstract:

Our approach to education in particle technology is to incorporate elements of our research in this field, relevant industrial case studies and apply a Technology Enhanced Learning (TEL) approach for implementing pedagogical interventions to enhance the learning experiences of students.

For supercritical CO_2 processing, case studies in extraction and drying are used to teach different chapters in the curriculum such as:

- Particle characterization
  - The bulk density, size, shape of real industry samples are given to students for analysis using software such as ImageJ
- Flow through packed bed
  - The pressure drop and scale-up of packed bed design for supercritical CO_2 flow through packed bed are used as a worked example and assignment problem to help students appreciate the role of density and viscosity in the pressure drop in a packed bed
- Drying
  - Supercritical CO_2 as an alternative drying media as compared to hot air or superheated steam. Case studies and examples provided

For gasification and chemical looping combustion applications, case studies in the treatment of wastes and carbon capture are similarly used as part of the syllabus:

- Particle characterization
  - The effect of polydispersity on various fluidization characteristics (entrainment, transport dis-engaging height, clusters, segregation, etc)
  - The effect of agglomeration
- Reliability of correlations and simulations
  - Correlations in literature give predictions orders-of-magnitude apart
- Scale-up
  - Rule-of-thumbs in scaling up the various regimes

A design software has been developed within the paradigm of Technology Enhanced Learning (TEL) to facilitate learning via a repeated practice approach. The software generates detailed solution steps to typical engineering design problems encountered within this module. Students can compare the complete solutions generated with their own hand calculations, and thereby apply a repeated practice approach towards their learning of engineering design calculations. The software has been successful in enhancing the performance of students in the final examination.

507116 Particle Technology Course at the University of Salerno

Abstract:

The Course of Particle technology at the University of Salerno is a 6 EC, that is a 60 hours course including lectures, as well as classroom and laboratory practical. The course dedicated to Master student in Chemical Engineering. It a compulsory course within the â€œinnovative process and nanotechnologies curriculumâ€ can be optioned by any mechanical or chemical engineering student at the master level.

The course includes basic particle size and shape characterization concepts, fundamentals on interparticle forces, fundamentals of powder mechanics and silo discharge, fundamentals on-fluid particle interactions and fluidization.

The main learning outcome is the ability of the student to understand the phenomena occurring in processes involving particulate solids and to carry out basic design operation on storage and fluidized bed units.

507382 Professional Master of Engineering Degree in Particle Technology

Abstract:

Particles are ubiquitous in the process industries yet nearly absent in the undergraduate education of chemical engineers. This poster descibes a professional Master of Engineering degree program at University of Delaware (the MEPT) that closes that gap by exclusively focusing on particulate matter, its processing, and its incorporation into products. The program can be completed in a single year although we welcome already-employed students who wish to take it part time. Now in our fourth year, we will describe our program and our experiences including guiding distance learners successfully through the program.

In its 1-year realization, a core curriculum of 18 credits is augmented by 6 credits of electives which the student may use to tailor the program to their specific interests. The program is capped by a 10-week, 6-credit summer internship. The 12 core credits taken in the fall teach the fundamentals of particle transport, particle properties & characterization, particle formation and transformation kinetics and the mathematics of particulate systems. The 6 core credits taken in the spring teach the student to apply the fundamentals to both (a) equipment design (in a unit operations course) and (b) product and process design (in a product engineering course).

The presenters of this poster are the program directors, both with long industrial careers preceding this venture into education. In addition, we have developed a strong network of industrial partners who share a wealth of additional experience and expertise with our students both as adjunct professors and as internship mentors. We have successfully placed international students in internships at US companies via the curriculum practical training program (CPT) who have gone on to practice in industry after graduation under the optional practical training program (OPT).

507780 Recent Developments in Particle Technology at the Universidad Tecnica Federico Santa Maria

Abstract:

The Department of Mechanical Engineering of the Universidad Técnica Federico Santa María formed in 2003 a Research Center for Transporting Granular Materials (CITRAM) to support educational activities and experimental research, as an extension for those ME students taking the regular course on Bulk Solids Handling and those developing a thesis in this area.

Over the years, many ME students have studied, investigated and modeled different topics including: pneumatic conveying of solid particles of different nature, size and density; identification of flow patterns developed in the pipeline, minimum conveying velocity and pressure drop; abrasive wear using a novel rubber-wheel tester; segregation of ores in stockpiles and silos; flow of granular materials in hoppers and chutes; heat transfer of granular materials; among others. The aim of this poster is to present and share the results of this work.

508520 Using Perusall to Enhance Student Learning of Particle Technology at Graz University of Technology

Abstract:

Perusall is a digital platform that allows students to collaboratively annotate textbooks, papers, and other similar classroom reading assignments. In order to enhance student learning, Perusall was introduced to help teach Particle Technology at Graz University of Technology (TU Graz, Austria). At Graz University of Technology, two Particle Technology courses exist in sequence; the first course is focused on fundamentals topics, whereas the second course is focused more on applications and separations topics. Approximately 60-75 students are registered for each course. For both courses, Perusall was used to assign reading materials, which were primarily journal articles that complemented the course topics. Although these assignments were non-compulsory due to national requirements, students were asked to use Perusall to read and annotate the course readings. Student participation and input regarding this digital approach to learning was recorded. This poster will highlight how Perusall was implemented in both Particle Technology courses as a tool to enhance student learning.

508524 A Graduate Course in Fluidization and Gas-Solid Flow Systems

Abstract:

The objective of this course is: a) to provide students with a fundamental understanding and design of Fluidization and Fluid- Particle flow systems, b) to provide students with an understanding of Computational Fluid Dynamics (CFD) and the kinetic theory approach to fluid-particle flows c) to provide students with the knowledge required to solve flow, heat and mass transfer problems in processes which involve particles and d) to provide students with sufficient knowledge to read and understand technical articles in areas related to fluid-particle processes. The course is designed for Master and PhD students in Chemical, Mechanical and aerospace engineering. The course includes: Fluidization concepts and regimes including bubbling, turbulent and circulating beds; Particle characterization, diagnostics and size reduction and agglomeration/granulation; Concept and design of Standpipes, non- mechanical valves, gas distributors, cyclones and hoppers; CFD approach to fluid-particle flow systems and fluidization including kinetic theory and dispersed phase modeling; Computational transport phenomena of fluid-particle systems including method of moments; and examples from energy, chemical and pharmaceutical industries. In addition the students will visit PSRI to see large-scale fluidized beds in operation.

The students are required to finish a team project. The teams consist of 3-4 students working on a project which will count as their final exam. The projects will primarily involve background research (about 80%) but some of the investigation (about 20%) should involve taking an additional step from what is currently known. This additional step can take the form of a research proposal, new theory, new insight based on current information, new calculations, new simulation and/or new data. The results of the projects will be communicated to the class and to the Course Instructors via a written report of 10-20 pages, and an oral presentation involving the professors and all of the group members.

508742 Education of Fluid-Solid Multiphase Flow at Department of Mechanical Engineering, Osaka University

Abstract:

In mechanical engineering field, education curriculum in undergraduate course concentrates on the basic subjects of mechanical engineering such as mathematics, mechanics, material mechanics, fluid mechanics, thermodynamics, control engineering, design, etc. Concerning the course relating to powder technology, we have a class of "Fluid-Solid Multiphase Flow" in the curriculum of graduate school of engineering.

The contents of "fluid-solid two-phase flow" is as follows;

1 Introduction: examples of multiphase flow

2 Definitions and fundamental concepts

3 Particle response time and Stokes number

4 Fluid force acting on particle

5 Fluid force acting on particle in dense system

6 Particle-wall and particle-particle interaction

7 Heat and mass transfer

8 Continuous phase flow equations

9 Modeling of fluid-solid multiphase flow

10 Discrete particle modeling and simulation

11 Numerical simulation method for continuous phase flow

This course includes basic concepts of powder technology, dynamics of particles in fluid flow, derivation of basic equations for locally-volume-averaged fluid flow, discrete particle modeling including DEM (Discrete Element Method), and basics of numerical analysis and modeling.

509420 Bulk Solids Handling Education at the KSU Bulk Solids Innovation Center

Abstract:

Kansas State University Bulk Solids Innovation Center (BSIC) was started in 2015 with university and industry partnership and is located in Salina, Kansas. This center is dedicated to bulk solids handling research, education and outreach and equipped with world class pneumatic conveying system (both dense and dilute) full scale test lab and other material testing equipment. The vision of this center is to become valued resources for companies who produce or handle bulk solids during their operations. Currently, this center offers short courses in various subject lines related to bulk solids handling such as storage and flow, pneumatic conveying systems, discrete element method, dust explosion and control and mixing and segregation. Duration for these short courses varies from two to three and half days and offered as live course with class room teaching and practical training. The fundamental course on bulk solids handling is offered as distance education online format. This center is also in the process of starting two year online (distance education) master degree program named professional master of technology specialized in bulk solids handling and one year graduate certificate program (online) in bulk solids handling.

507117 Estimates of Powder Flow Properties at Low Consolidation from the Torque Measured in a Powder Rheometer Equipped with a Cylindrical Impeller

Abstract:

Small scale industrial application often used in pharmaceutical and other fine chemicals application involve the use of small intermediate bulk container and small hoppers as well as the use of powders in process involving low consolidation levels. The knowledge of flow properties of low consolidated powders is, therefore, necessary for the proper design, as well as for determining the suitability of newly formulated powder to existing processes lines and equipment. Conventional shear testing is well standardized and the measured powder properties are well defined. Unfortunately, however conventional shear testers are not always able to reproduce the correct powder history found in the application of interest. Therefore, new testing methods, more flexible in the powder preparation, are receiving growing interest.

The equipment used in this research project is the Powder Cell unit purposely developed by Anton Paar to fit on Anton Paar Rheometers and to carry out torque measurements on aerated and not aerated powders. In this instrument, the state of consolidation of the powder can be adjusted by changing the impeller depth and the aeration rate. A cylindrical impeller geometry has been tested on a number of free flowing as well as slightly cohesive powders. In order to be able to compare results with the properties measured with a conventional shear tester the impeller rotational speed was kept at very low levels. A proper powder conditioning procedure, involving powder fluidization and shear, was also set up in order to obtain reproducible results. A simple model was developed to relate the measured torque with the dynamic angle of internal friction in the powder measured with a conventional powder shear tester. The results indicated a good match between the dynamic angle of internal friction measured with the instruments and those measured with the conventional shear cell in the range of consolidation tested. In order to match the results, a pseudo hydrostatic state of stress had to be hypothesized within the measuring cell. In the same procedure the extrapolated torque at the start of the impeller and a zero consolidation was related to the powder cohesion. It was possible to conclude that in the tested configuration the instrument allows to successfully exploring powder behavior at low consolidation ranges which, in the appropriate conditions, well correlates with the powder flow properties measured with standardized equipment.

507137 Predicting Process Performance in Screw Feeders Using Powder Flow Measurements

Abstract:

Screw feeders are routinely used to ensure accurate and consistent feeding of powders between stages of a process. Despite widespread use, predicting feed rates typically relies on pre-existing performance information and extrapolation. This can result in the use of equipment that does not meet design requirements, often contributing to sub-specification operation and in extreme cases â€“ complete process failure.

Automated, multi-variate powder characterisation tools provide reliable and comprehensive measurement of a powderâ€™s response to process-relevant conditions. This data can then be correlated with process performance information to improve feeder efficiency and ensure high quality final products at the required rates.

Five powders were run through two different screw feeders, a DIWE-**GLD**-87 VR (a full flight single-screw feeder) and a DIWE-**GZD** (a flat-bottomed double-screw feeder) (Gericke AG, Switzerland). These materials were also tested using an FT4 Powder RheometerÂ® (Freeman Technology Ltd, UK) to evaluate their dynamic flow, bulk and shear properties. Multiple Linear Regression (MLR) analysis was then performed to evaluate the relationships between volumetric feed rate of the screw feeders and rheological properties of the powders.

MLR analysis demonstrated clear relationships between the GLD volumetric feed rate and two dynamic flow properties, Specific Energy (SE) and the Flow Rate Index (FRI). A correlation between GZD volumetric feed rate and Aerated Energy (AE) was also observed. The testing of two additional materials provided further support of these correlations and reinforced the models defined by the MLR analysis.

The two different screw feeders generate very different processing environments and therefore exhibited relationships with different rheological parameters. This emphasises the need for a multi-variate approach to powder characterisation.

This study shows the robust relationships that exist between powder flow properties and the volumetric flow rate delivered by different screw feeders and demonstrates how it is possible to generate a design and operating model that can be used to predict performance in order to optimise productivity.

509293 Using a Freeman FT4 Rheometer and Electrical Capacitance Tomography to Assess Powder Blending

Abstract:

Measurements of flowability for powder mixtures may be affected by segregation, and equally, segregated powders may be mixed during the measurement cycle. Experimental studies in powder mixing and segregation are hampered by a deficit in suitable, widely deployable measurement techniques. By using powders of fundamentally different flow and electrical (permittivity) properties this study assesses the influence of mixing or segregation on flowability results from a Freeman FT4 powder rheometer as well as the use of Electrical Capacitance Tomography (ECT) to measure powder mixing and mixedness.

Thus, the blending (or otherwise) of two powders of different physical properties (particle size, density, basic flowability) and contrasting electrical permittivity was studied in a Freeman FT4 rheometer. Two initial arrangements were used: one with the heavier, smaller powder on the top that would be expected to mix readily; and the other, the inverse, that would be expected to be resistant to the axial blending mechanism prevalent in the FT4. The torque and thus flow energy were tracked through 30 cycles of the FT4 impeller passing into and back out of the powder layer. In one case, the FT4 data were seen to evolve towards a steady state comparatively quickly, while, in the other, relatively little change in the basic flow energy was observed. The change in the first case is attributed to mixing of the powder, and the lack of change in the latter case to the absence of mixing.

Simultaneous ECT measurements using a two plane sensing arrangement were taken. The reconstructed tomograms and the basic average permittivity data show clearly the mixing in the first case and the absence of blending for the latter. This confirms not only the sensitivity of powder flowabilty measurements to segregation of component powders, but also the potential of ECT as a measurement tool for powder mixing and segregation studies.

507365 The Use of Modern Rheometers in Powder and Granular Media Measurements for Scientific and Industrial Purposes

Abstract:

Modern Rheometers are usually equipped with high precision optical encoders and air bearings and a very Precise DC Motors and fast normal force sensors. Furthermore flexible programmable software allows complex and conditional measurements to be programmed by the user. A vast array of peripherals allows to re-equip the baseline instrument to the specific need of the measurement in question. These features allow gathering of novel data not accessible to traditional Powder characterization equipment such as:

Using high torque resolution (+- 10 nNm) as a means of distinguishing between very similar powders

Oscillatory experiments such as frequency or Amplitude analysis allow the exact state of fluidization to be determined even for group c powders

Characterizing extremely free flowing granular media through the high torque resolution and relating this to the angle of internal friction / tensile strenght

Creep Tests through the use of the precise control of radial velocity of the measurement system

Control of consolidation from the fluidized state to high normal strain: allowing to control the of the state of consolidation and therefore being able to simulate process conditions better

Simzultion of low gravity condition through the use of Fluidizing air

Direct apparent viscosity measurements in fluidized beds and the direct measurement of the stokes number

Temperature and Moisture Control within the instrument

These options allow us to directly simulate process conditions to serve both as a calibration tool for calculative approaches as well as a tool for analytical. All of the measurements above will be presented and discussed both as an analytical tool by itself but also as a an industrial example.

Lastly the the use of these devices with inertial rheology is discussed

507727 Flow Assessments of Selective Laser Sintering Powders Via Fluidized Bed Rheology on a Rotational Rheometer

Abstract:

Selective Laser Sintering (SLS) is a rapidly expanding field of additive manufacturing technology where manufacturing is done by adding consecutive powder material layers followed by sintering of these layers. There are about a dozen of available additive manufacturing technologies but SLS is considered to be the most favorable for production of plastic parts. One big challenge which is limiting widespread use of this technology is the narrow availability of applicable polymers. Currently strict criteria are applied towards the sphericity of the particles as well as having an extremely narrow size distribution, this puts a considerable strain on the manufacturers of such powders, especially cost wise, and severely limits the widespread availability of exotic materials for high performance engineering purposes. The different properties of powders which influence SLS can be divided into intrinsic (thermal, optical, rheological) and extrinsic (powder flow and particle size). This talk will focus on extrinsic properties especially characterization of flow properties of SLS powders via fluidized bed rheology on a rotational air bearing rheometer. In this study several raw materials ranging from high density metal to low density polymer powders have been used. The work includes flow characterization of SLS powders with a suite of different experiments which include tensile strength, rotational cohesion strength, penetration, oscillation, shear rate sweep etc measurements on an air bearing rotational rheometer. The results are compared and discussed both from the standpoint of particulate mechanics as well as the perspective of the practical process engineer.

511400 A Surface Characterization Platform Approach to Study Food Powder Flowability

Abstract:

The surface properties of powders used in beverages have been investigated using IGC (Inverse Gas Chromatography) to assess surface energy/interactions and X-Ray Photoelectron Spectroscopy (XPS) to assess chemical surface composition. The influence of surface properties on powder flowability was studied. The physical attributes of the powders were also determined in the study, and flow properties were determined using a powder rheometer. Five coffee powder samples, generated by three different processes (spray-drying, freeze-drying and milling) and three cocoa powders were used for this study. The data show that for both types of powders â€" coffee and cocoa, differences in chemical composition at the surface are detectable by XPS. The results suggest that flow is inhibited in powders with higher levels of fat deposited at the surface. The results also indicate that while powder flow is influenced by physical attributes such as particle size distribution, which influence the flowability of powders, a major factor is that of the chemical composition at the surface, which in turn has a direct effect on interparticle cohesiveness. We present an assessment platform, geared toward understanding bulk powder behavior, based on the use of vapor sorption techniques, such as IGC and water sorption, in combination with surface structural characterization analytical tools (XPS) to optimize process and product development.

506169 Intrusive Probes in Riser Applications

Abstract:

Many of the probes used to understand hydrodynamics in circulating fluidized bed risers intrude into the environment they are measuring, although assumptions are typically asserted that the intrusive probes do not affect the data collected. This could be a poor assumption in some cases and conditions. We found that intrusive fiber optic probe measurements consistently mis-predicted the solids concentration compared to the non-intrusive pressure drop measurements outside the fully developed flow region of a riser containing fluid catalytic cracking catalyst or glass bead particles. The discrepancy was sensitive to superficial gas velocity, solid circulation rate, probe position and flow direction. Barracuda VRâ„¢ computational fluid dynamics simulations confirmed this, and indicated that particle momentum was lost at the leading edge of the probe and particles were spilling over to the probe tip. Accordingly, new probe designs were proposed to mitigate the intrusiveness of a fiber optical probe for more accurate characterization.

508342 Detection of Fluidization with Real-Time Magnetic Resonance Imaging

Abstract:

Unfluidized areas (dead zones) within gas-solid fluidized beds can deteriorate the mixing and heat transfer between the gas phase and the solid phase, and thus lower the performance, homogeneity and efficiency of the desired process. A thorough understanding of the conditions that lead to the formation of dead zones is therefore key in order to design, operate and scale fluidized bed reactors reliably. The experimental detection of dead zones inside three-dimensional (3D) fluidized beds can provide valuable information to validate theories and numerical models. However, due to the opacity of most granular materials, such measurements are challenging and require the use of intrusive probes[1,2] or non-intrusive tomographic techniques, e.g. x-ray tomography[3], electric capacitance tomography (ECT), positron emission particle tracking (PEPT)[4] or magnetic resonance imaging (MRI)[5]. While intrusive probes perturb the granular dynamics, the tomographic techniques suffer from limitations in either their ability to measure particle motion (x-ray CT, ECT, PEPT) or from poor temporal resolution (MRI)[6].

Here we present a MRI methodology that can discriminate between fluidized and unfluidized areas in 3D beds of considerable size (cylindrical bed; diameter 190 mm and height 300 mm) at a frame rate of 20 Hz. The high temporal resolution of this technique allows studying not only the presence of dead zones but also their formation and disintegration. The presented measurement technique bases on a recent development in real-time probing of granular dynamics with MRI which uses array detection of the magnetic resonance and engineered granular signal sources[7]. It exploits both spin saturation effects and diffusion attenuation to create a strong dependency of the signal on particle motion.

We anticipate the presented technique to be a starting point for more sophisticated studies on the formation of dead zones in fluidized beds, such as determining the influence of internals, the distributor or bed geometry. Moreover, we envisage that the technique is suited to provide fundamental physical insights on the nature of fluidization, its relations to jamming and the glass transition[8].

References

1. Geldart, D. & Kelsey, J. R. The use of capacitance probes in gas fluidised beds. *Powder Technol.* **6,** 45–50 (1972).

2. Werther, J. & Molerus, O. The local structure of gas fluidized beds -II. The spatial distribution of bubbles. *Int. J. Multiph. Flow* **1,** 123–138 (1973).

3. Mudde, R. F. Double X-ray Tomography of a Bubbling Fluidized Bed. *Ind. Eng. Chem. Res.* **49,** 5061–5065 (2010).

4. Wildman, R. D., Huntley, J. M., Hansen, J. P., Parker, D. J. & Allen, D. A. Single-particle motion in three-dimensional vibrofluidized granular beds. *Phys. Rev. E* **62,** 3826–3835 (2000).

5. Dennis, J. S. Magnetic Resonance Studies of Dead-Zones in Gas-Solid Fluidised Beds. (2013).

6. Pore, M. *et al.* Magnetic resonance studies of a gas-solids fluidised bed: Jet-jet and jet-wall interactions. *Particuology* **8,** 617–622 (2010).

7. Penn, A. *et al.* Real-time probing of granular dynamics with magnetic resonance. *Sci. Adv.* Accepted, (2017).

8. Goldman, D. I. & Swinney, H. L. Signatures of glass formation in a fluidized bed of hard spheres. *Phys. Rev. Lett.* **96,** 145702 (2006).

507536 Solids Mixing Studies in Fluidized Beds Using Fluorescent Tracer Technique

Abstract:

A fluorescent solid tracer technique has been used to study various properties such as solids mixing time, solids residence distribution time and the effects of internals on mixing in fluidized beds. Studying these parameters is of great importance in controlling the back-mixing of solids or verifying that solids segregate in a fluidized bed. The technique may also be used to compare the flow patterns of different solids, understand the hydrodynamics associated with different internals at different operating conditions, and to optimize the size and the shape of a reactor. In these sets of experiments, FCC catalyst was coated with a suitable UV dye in a 12-in (30-cm)-diameter spray column using a suitable solvent to dissolve the dye. The dyed catalyst was then injected into a 6-in (15-cm)-diameter Plexiglas^{TM} column fluidized at superficial gas velocities of 1 ft/s (0.31 m/s) and 0.25 ft/s (0.08 m/s) to analyze mixing of FCC equilibrium catalyst. Fluorescent probes, which were inserted at different axial and radial locations in the bed, captured the mixing phenomena of the catalyst in the bed. A photodiode converted an optical signal to an electrical signal and the data were recorded using a data acquisition system and was analyzed to study the mixing behavior of particulate solids in a small scale fluidized bed.

508410 *Stereoscopic PIV Study on Swirl Phenomena in Uniflow Cyclones*

Abstract:

Understanding the complex flow field in a uniflow cyclone is important for further design development. The uniflow cyclone is advantageous in certain applications due to the compactness e.g. in space-limited environments in engineering equipment and will be therefore investigated. This separator type uses centrifugal forces induced by a swirl vane inlet to separate particles from a continuous phase. The resulting flow field is characterized by a strong circumferential velocity component, which forces the fluid to rotate in downstream direction.

Stereo-PIV is used to evaluate the flow field along the cross section and to gain insights into the flow field characteristics [1]. 3C/2D-PIV is is applied to evaluate all three velocity components simultaneously in the measurement plane. Therefore the cameras and the laser optics are mounted on an 1D traversing mechanism for precise positioning of the optical PIV system in the measurement volume along the axis of the main flow. This study mainly focuses attention on detailed measurements on the flow field in the separation chamber of the uniflow cyclone to find the best geometric parameters by means of aerodynamic aspects [2,3]. For validation a set of dust-laden experiments are carried out to get information about separation efficiency and pressure drop (ISO 5011) [4]. As a result the effect on cyclone performance has been quantified with a consistent set of experimental data which shows the influence of vane angle and core diameter ratio variations.

[1] Pillei, M., Goller, R., Kofler, T., Wierschem, A., and Kraxner, M., "Experimental Investigation of the Velocity Distribution near the Swirl Generator of a Uniflow Cyclone for Performance Data Prediction," In *18th International Symposium on the Application of Laser and Imaging Techniques to Fluid Mechanics.*

[2] Chang, F. and Dhir, V. K., "Turbulent flow field in tangentially injected swirl flows in tubes," International Journal of Heat and Fluid Flow, vol. 15, no. 5, pp. 346–356, 1994

[3] Pillei, M.; Kofler, T.; Kraxner, M.; Fluid flow investigation in the separation chamber of a uniflow cyclone with 3C-PIV and CFD; Fachtagung "Lasermethoden in der Strömungsmesstechnik"; GALA Karlsruhe, 2014.

[4] ISO 5011. Inlet air cleaning equipment for internal combustion engines and compressors - Performance testing.

506444 Relative Importance of Cohesion and Internal Friction in Flowability of Pharmaceutical Powder

Abstract:

The flow function, measured by shear cell testers, has become the standard metrics for assessing powder flowability in the development and manufacturing of pharmaceutical solid dosage forms. Commercial shear cell testers calculate flow function by determining a linear yield locus through measurements of yield strengths of materials under different normal stresses. Cohesion and internal friction are two quantities defined in Mohr-Coulomb theory that describe the linear yield locus; hence they also determine flow function. The purpose of our study is to understand the relative contribution of cohesion and internal friction to flow function and to apply the findings to improve powder flowability during formulation development.

Through the analysis of 1130 ring shear tests of various pharmaceutical powders, we found that the flow function is strongly correlated with the ratio of pre-consolidation stress to cohesion. The powders included single-component or multi-component system of excipients and APIs that were prepared by simple blending or granulation. However, flow function had almost no correlation with internal friction angles.

We developed a model based on the Mohr-Coulomb theory to verify the experimental findings. The above observationâ €"flow function strongly correlates with the ratio of pre-consolidation stress to cohesion but not with internal frictionâ €"was a mathematical consequence ensued from Mohr-Coulomb failure analysis. The model suggested that if the powders exhibited a small difference between the angle of internal friction at steady-state flow and the angle of internal friction at incipient flow, the correlation remained valid. Most pharmaceutical powders, coincidentally, possess such characteristic. Therefore, in practice, cohesion has a strong correlation with flow function while internal friction has none.

We demonstrated an application by mixing colloidal silicon dioxide with Hypromellose powder to show that substantial increase in flow function was accompanied by large decrease of cohesion while the internal friction angle showed negligible change. In contrast, mixing colloidal silicon dioxide with anhydrous calcium phosphate powder has no effect in flow function when cohesion changed minimally. Practitioners in solid dosage drug development can apply this knowledge and focus on improving powder flowability by reducing powder cohesion instead of internal friction.

[1] This study has been published in Leung et al. 2017. *Journal of Pharmaceutical Sciences*. 106:1865-1873

507030 Evaluation of Mechanical Properties of Organic Crystals for the Prediction of Breakage during Isolation Processes

Abstract:

Crystal breakage during the isolation of active pharmaceutical ingredients (API) is an issue of great concern to the pharmaceutical industry. Conservation of the desired particle size distribution (PSD) throughout downstream processing is extremely important as changes are known to affect properties such as bioavailability, flowability and bulk density [1]. APIs are commonly produced by batch crystallisation as high aspect ratio crystals, usually needle-shaped, and they are separated from the mother liquor through pressure filtration. At industrial scale, isolation of the material is prone to significant changes in the PSD. This is typically attributed to agitation during drying [2]. In contrast, MacLeod and Muller [3] showed that the filtration process itself can have a profound impact on the PSD of needle-shaped crystals. The main objective of this work is to elucidate the phenomenon of crystal fracture under pressure filtration. To achieve this, mechanical properties, such as yield stress, elasticity and hardness, need to be determined at the single crystal level.

Namazu et al. [4] conducted 3-point bending test experiments using an inorganic single crystal (Si) fixed beam in order to evaluate the size effect on the mechanical properties of crystals. It was found that reduction of the crystal size leads to an increase in bending strength. Our group developed a single crystal 2-point bending test by fixing one edge of single microscopic needle-shaped crystals and applying a force at the other edge using Atomic Force Microscopy [5]. We employed this method to determine the micro-mechanical properties of β-L-glutamic acid (β-LGA), a needle-shaped organic crystal. Similar to [4], a high number of experiments demonstrated that the yield stress is a strong function of the crystal size. Using a derivative from Bernoulli's beam theory [3], it was found that the critical stress leading to β-LGA crystal snapping is equal to 13.8 MPa. As the crystals are by nature not identical, a Weibull analysis was also performed to model the distribution of the experimental observations. The Weibull's distribution function shape (k) and scale (λ) parameters were estimated and are equal to 2.07 and 1.896×10^7 respectively. The median of the Weibull distribution is equal to 15.9 MPa, meaning that 50% of the crystals would break under such a stress [6].

A solid mechanics model, implemented in COMSOL Multiphysics, was used to simulate the single crystal fixed beam bending. The model can very accurately predict the crystal deflection and stress profiles under the application of varying forces. The extension of such a model to carry out simulations at the filter bed scale, in conjunction with yield stress experimental findings, could be ultimately used as a predictive tool towards the evaluation of the propensity for particle breakage during API isolation at scale, and the improvement of the process design as well as operating conditions.

References

1. Kougoulos, E., C.E. Chadwick, and M.D. Ticehurst, *Impact of agitated drying on the powder properties of an active pharmaceutical ingredient.* Powder Technology, 2011. **210**(3): p. 308-314.

2. Lekhal, A., K.P. Girard, M.A. Brown, S. Kiang, J.G. Khinast, and B.J. Glasser, *The effect of agitated drying on the morphology of L-threonine (needle-like) crystals.* International Journal of Pharmaceutics, 2004. **270**(1-2): p. 263-277.

3. MacLeod, C.S. and F.L. Muller, *On the fracture of pharmaceutical needle-shaped crystals during pressure filtration: Case studies and mechanistic understanding.* Organic Process Research and Development, 2012. **16**(3): p. 425-434.

4. Namazu, T., Y. Isono, and T. Tanaka, *Evaluation of size effect on mechanical properties of single crystal silicon by nanoscale bending test using AFM.* Journal of Microelectromechanical Systems, 2000. **9**(4): p. 450-459.

5. Sohi, S.S., S.D. Conell, D. Harbottle, and F.L. Muller, *A method for the determination of mechanical properties of needle shaped crystals using Atomic Force Microscopy.* Master's Report, University of Leeds, 2016.

6. Hallac, F.S., I.S. Fragkopoulos, S.D. Connell, and F.L. Muller, *Micro-mechanical properties of needle-shaped organic crystals.* Organic Process Research and Development, 2017. *To be submitted.*

CHARACTERIZATION OF PARACETAMOL GRANULES FORMED BY BINDER DROPPING

Dr. Sheena M. Reeves, Prairie View A&M University, Prairie View, TX

Abstract

Particulate materials are widely use in pharmaceutical industry in drug formulations. Some formulations are made using the process of agglomeration or granulation. Most active pharmaceutical ingredients (APIs) are administered in a formulation containing excipients or binders to prevent segregation. Excipients are inert materials that are used to assist drug delivery; however, these additives sometimes cause mild adverse effects and hinder API dissolution. Glycerides and waxes are added to most APIs to increase stability. Drug bioavailability and solubility is a constant challenge with pharmaceuticals. A variety of excipients exist and serve various purposes including added stability, segregation prevention, and controlled dissolution. Binding agents under consideration are classified as semi-synthetic or synthetic and include a comparison of polyvinylpyrrolidone (PVP) and polyethylene glycol (PEG). Although it is considered biologically inert and safe, PEG is known to cause allergic reactions to some individuals. Since PEG is a common binder, a replacement binder should have similar characteristics and capabilities. The purpose of this research is to investigate 1) how binding agents influence the ability of a paracetamol granule to dissolve in a buffer solution (pH 7.0) and 2) the ability of the granule to maintain cohesiveness during compression testing.

Wet granulation was chosen as the method of granule formulation and was accomplished by dropping solutions onto a powder bed. Initial tests in this project involved developing paracetamol granules of similar size, shape, and weight for consistency and uniformity. Upon drying, images of the granules were analyzed using Image-Pro Premier software and included measurements for perimeter, aspect ratio, area, diameter, and roundness. To create a granule containing between 275 – 300 mg, five drops of binder were used. To test for granule stability, compression tests were conducted using a conical indenter and a flat surface. Granules produced from synthetic binder solutions proved more cohesive and stable. Five binder drops produced granules without excess agglomeration around the edges with an average weight around 300 mg. Dissolution rates were determined using an Agilent Bio-Dis Apparatus 3 and UV-Vis spectrometry. The concentration of the binder, the type of binder, and the dip speed of the apparatus were varied. Pure paracetamol powder or granules formed with deionized water, a very weak binder, served as a baseline for analysis. Further investigation of the granule cohesiveness was conducted using scanning electron microscopy (SEM).

507694 Investigation of Single-Droplet Drying Kinetics in an Acoustic Levitator, Micro-Sphere Morphology and Mechanical Properties of Pharmaceutical Excipients

Abstract:

In the last years there has been significant interest in advanced continuous manufacturing of pharmaceuticals. These manufacturing routes have significant advantages, including a smaller plant footprint (smaller equipment and facilities), reduced manufacturing time, elimination of unnecessary storage by streamlined production from start to finish, and lower manufacturing costs. Moreover, quality is ensured in real time, leading to a better overall efficiency of the manufacturing process.

One of the challenges in continuous production is the link between primary (API synthesis, crystallization, purification, filtration and drying) and secondary manufacturing (blending, granulation and tableting/capsule filling). Particle properties of crystalline APIs are tailored in primary manufacturing by wet chemical processes. To enable further processing of the crystalline API in secondary manufacturing, a drying step is necessary to remove the residual solvents.

One route to transform a fluid feed stream into dry powders is spray drying. A major benefit of spray drying over other technologies is that drying is conducted close to the wet bulb temperature of the slurry, and therefore it is suitable for heat-sensitive products. As long as the surface of a droplet/particle is covered with a liquid layer, drying occurs at the wet bulb temperature in the first drying stage. This liquid layer is removed by evaporation of the liquid phase of the slurry into the drying air until solidification occurs. Afterwards, drying progresses in the second drying stage at temperatures converging to the dry bulb temperature until the designed residual moisture content is reached.

In suspensions, solids are dissolved in the carrier liquid and suspended as particles. The drying of suspension slurries results in precipitation of dissolved material forming bridges between the primary particles. Therefore, the final microsphere properties are influenced by the amount of dissolved solids, since this determines the size and amount of bridges between primary particles. There is a clear correlation between more dissolved material in the slurry liquid and the formation of stronger bridges between primary particles. This alters significantly the final microsphere properties, as well as the corresponding processability.

In this work we focus on the correlation between slurry composition and the resulting material properties after drying. Drying experiments were conducted by acoustic levitation of a model compound in binary liquid mixture droplets. The acoustic levitation setup enabled microscopic observations of the drying behavior of single droplets. Variations in the suspension composition revealed the impact on the drying kinetics as well as the mechanical properties and the morphology of the final particles (microspheres). Different microsphere morphologies were observed, from very loose agglomerates to densely packed microspheres with crust formation. Electron micrographs of the microsphere morphology of different suspension compositions were in good agreement with the hardness measured by a compression strength test.

507256 Optimising High Shear Wet Granulation with the Combined Use of in-Line and at-Line Testing

Abstract:

High Shear Wet Granulation (HSWG) is used in many industries to produce homogeneous granules that deliver improved performance in various processes compared to the powdered components. Typically, only the properties of dried granules and/or final products are evaluated to understand the influence of changes in process design and formulation. However, in-line and real-time measurements can enable more proactive and flexible control of HSWG without disrupting the process.

A Drag Force Flow (DFF) sensor (Lenterra Inc, USA) is a thin, hollow cylindrical pin containing two optical strain gages which detect deflections caused by the force of flow. These sensors can provide highly sensitive in-line measurements of flow forces within granulators. Previous research has also demonstrated that these measurements correlate well with granule densification and resulting tablet properties.

In this study, three formulations were granulated with 40% wt/wt water in a GEA PharmaConnect™ High Shear Wet Granulator. The relationship between the in-line response of a Drag Force Flow (DFF) sensor and at-line flow property measurements of wet granules measured using an FT4 Powder Rheometer® (Freeman Technology Ltd, UK) was investigated.

By evaluating the real-time DFF sensor output, expressed as a Force Pulse Magnitude (FPM), it was possible to clearly identify the different phases of the granulation process, including water addition and end-point. The Basic Flowability Energy (BFE) values generated from FT4 dynamic flow testing displayed a similar trend to the FPM data, displaying an increase and subsequent decrease as a function of time after the addition of water. A relationship between BFE and binder level was also observed with higher levels of binder resulting in higher BFE values, indicating stronger, denser and larger granules.

Wet mass consistency measured in-line by the DFF sensor is a good indicator of the rheological properties of the granulate and correlates well with at-line BFE measurements that are known to influence Critical Quality Attributes (CQAs) of a tablet. The data provided by both the DFF sensor and FT4 can be correlated to granulation end-point and help define a design space of parameters that correlate to optimal process configuration.

507832 Determine the Flow Properties of Bulk Materials to Prevent Flow Problems

Abstract:

In many mining and industrial processes where large quantities of crushed ores and non-degrading bulk materials are handled and stored, silos, hoppers, stockpiles, feeders and chutes are widely used because they are economical and reliable (if properly designed and operated). However, they are not free of trouble and may experience flow problems such as arching, ratholing, erratic flow, limited storage capacity, limited discharge flow rate, caking, segregation and/or flooding.

Testing provides the best method to characterize a bulk material and its handling properties for proper equipment and system selection, design and operation. Hence, we strongly recommend performing laboratory tests with representative samples of the material (or materials) to be handled in a silo, hopper, stockpile, feeder and/or chute including cohesive strength, wall friction, compressibility, permeability, and other parameters such as bulk density, internal friction, chute, repose and draw down angles of the material, tested at similar conditions to those to exist in the plant including different moisture levels, consolidating pressure, storage time at rest, etc.

The purpose of this article is to highlight a proven, scientific method that can be utilized to ensure reliable storage, flow and discharge of bulk solids in silos, hoppers, stockpiles, feeders and chutes based on the Jenike flow-of-solids theory and laboratory testing. Knowledge of the flow properties of the material handled provides a design basis to ensure mass flow, avoid arching and prevent the formation of â€œratholesâ€ even if handling wet cohesive materials. The effect of an increase in water content of the ore is also discussed in details with experimental results.

508541 Discrete Element Method (DEM) Investigation of Particle Properties on Bulk Powder Flowability:Â Toward a Particle Calibration Library for Pharmaceutical Materials

Abstract:

Application of the discrete element method (DEM) to industrial processes can provide insight into the performance of powder processes. For improved confidence in simulation results, the simulation input parameters, namely the particle contact parameters, must be known with some certainty. For some processes, e.g. tablet coating operations, the particle size is such that the contact parameters can be directly measured and the number of particles is small enough to permit reasonable simulation times. In most industrial processes, however, the particle size may be too small to make meaningful direct measurement of the contact parameters and the number of particles far exceeds what is computationally tractable in an industrially-relevant time scale. One common approach to address this shortcoming is to make simplifications to the particle shapes, size, and size distribution such that a smaller number of particles are simulated. In doing so, the link between contact parameters for the real and model particles is dubious if not non-existent. A particle calibration approach where contact parameters for the model particles are selected such that the simulated bulk powder behavior matches that of the real material may be a useful approach to simulate large-scale processes with improved confidence.

In this work, the shear flow of particles in a periodic section of an annular ring shear tester is simulated to examine the effects of various particle properties on the bulk flow behavior. The impact of particle shape along with coefficients of restitution, sliding and rolling friction, and cohesion are examined through thousands of cloud-based DEM simulations. The simulated bulk flow behavior spans the range of experimental results determined for a wide array of typical pharmaceutical materials. Thus, this extensive simulation dataset provides a prerequisite to a potential DEM particle calibration utilizing a lookup table for contact parameter values. Simulated and experimental shear tester results will be presented with a focus on the link between the DEM input parameters and macro-scale bulk flow measurements. Further, an outlook for extension to calibration of DEM particle models will be discussed.

508638 The Effects of Particle Size Distribution and Moisture on the Flowability of Fibrous and Woody Biomass

Abstract:

One of the critical issues in the biomass industry is the poor flowability of the material. Several studies have been emerging in this area to contribute towards designing better equipment and handling the material to mitigate this problem. An understanding of the influence of physical and surface properties of the granular material that affect the flowability can contribute towards the designing of efficient feeders and hoppers. In this work, we compare the permeability, compressibility, shear properties and bulk density of fibrous and woody biomass for different particle size distributions, using the Freeman FT4 Powder Rheometer. With biomass being an organic matter, an exchange of moisture with the surroundings occurs. Thus, we investigate the flowability of the material in the presence of low moisture levels, in addition to the dry material.

508538 Discrete Element Method (DEM) Modeling to Study the Effects of Particle Size and Shape on Flowability: Toward More Realistic Representations of Actual Powders

Abstract:

The flowability of a powder is strongly dependent on the shape and size distributions of the constituent grains, yet most computational approaches for modeling powders make gross oversimplifications of both the shape and size distribution, i.e. approximating a powder with monodisperse spheres. In this work, the discrete element method (DEM) has been used to examine the effect of the mean particle size, particle size distribution, and particle shape on the powder flowability, using the unconfined yield strength (UYS) as a measure of flowability.

The particle size distribution, surface energy, and particle shape – imaged by microtomography –of D-Mannitol powder are experimentally characterized and used as inputs to a DEM model of the SSSpinTester. The SSSpinTester device is used to measure the powder's UYS at different levels of powder consolidation; the consolidation stresses are prescribed via the rotational speed of the powder test cell in the SSSpinTester.

It is computationally impractical to implement the actual particle size distribution in DEM simulations: billions of particles would be required for modeling even a gram of real material. Therefore, certain simplifications are often made when simulating real particles—such simplifications include scaling of particle sizes and truncation of fines to varying extents. In the present work, the impact of varying levels of particle-size scaling and extent of truncation of fines are systematically investigated. The influence of particle shape, varied by increasing the number of glued spheres used to represent complex, non-spherical shapes, is also studied.

The effects of varying particle size, investigated using both monodisperse and polydisperse spheres, reveal that as particles size decreases, the UYS values increase indicating reduced powder flowability. This increase in UYS value is attributed to the increase in the ratio of particle (contact) surface area to volume in the bulk, which causes increased surface (cohesive/frictional) interactions. The effects of varying particle shape, investigated using polydisperse non-spherical particles, also leads to an increased UYS values as the shape complexity (i.e., number of glued spheres) is increased, yet again indicating reduced powder flow. This apparent decrease in flowability is due to increased interlocking among non-spherical particles, which does not occur when spherical particles are used. The present results reveal that both particle size and shape play an important role in determining the unconfined yield strength; a practical implication being that the DEM representation of particles affects the flowability which, in turn, affects the flow predictions for industrial systems such as hoppers, blenders, etc.

507971 Linking Particle Properties to Powder Performance in the Process Plant

Abstract:

The Virtual Formulation Laboratory (VFL) proposes a practical tool of simulating interactions between components, resulting in information on the flow behaviour of particulate materials. Product desired performance within certain formulation specifications requires developing products that present right particle properties during the manufacturing processes and also when consumers use them. The manufacturing processes for these products present challenges that lead to delays in the path to productivity and increased development costs.

Building a knowledge database from existing components and fine-tuning the relationships based on physical properties (related to morphology and/or intrinsic properties) should allow prediction of behaviour in manufacturing scenarios and guide technology choice at an early stage.

This study investigates the relationship between particle-scale interactions dominated by the cohesive van der Waals force and the flow function. We developed a tool for prediction and optimisation of manufacturability of advanced solids-based formulations which will be used by industries allowing to quickly and accurately measure the flow properties of their materials for a range of purposes, from bench marking and quality control to assisting reformulation. In the framework of the collaborative project VFL, a large set of physical experiments using different devices commonly used for powder characterization are carried out with the objective of evaluate reproducibility. In these set of experiments we look at the particle size influence for evaluation of bulk behaviour considering microscopic particle properties and cohesion strength.

508625 Measurement and Imaging of Gas-Solid Systems Using Electrical Capacitance Volume Tomography

Abstract:

Electrical Capacitance Volume Tomography (ECVT) is a non-invasive imaging technology recently developed for multi-phase flow applications. It is based on distributing flexible capacitance plates on the peripheral of a flow column and collecting real-time measurements of inter-electrode capacitances. Capacitance measurements here are directly related to dielectric constant distribution, a physical property that is also related to material distribution in the imaging domain. Reconstruction algorithms are employed to map volume images of dielectric distribution in the imaging domain, which is in turn related to phase distribution. ECVT is suitable for imaging interacting materials of different dielectric constants, typical in multi-phase flow systems.

ECVT is being used extensively for measuring flow variables in various gas-solid flow systems. Recent application of ECVT include flows in risers and exit regions of circulating fluidized beds, gas-solid bubble columns, trickle beds, moving beds, and slurry bubble columns. ECVT is also used to validate flow models and CFD simulations. The technology is uniquely qualified for imaging phase concentrations in various energy applications as it exhibits favorable features of compact size, low profile sensors, high imaging speed, and flexibility to fit around columns of various shapes and sizes. ECVT is also safer than other commonly used imaging modalities as it operates in the range of low frequencies (~1 MHz) and does not radiate radioactive energy.

High temperature and pressure gas-solid reactors are widely used in industry. Deep understanding of flow behavior in such reactors is limited due to the complex hydrodynamics that exist in the bed. In this effort, ECVT is used to study flow behavior in various gas-solid systems and at different conditions. Case studies that involve bubbling beds, moving beds, and slugging at temperatures up to 900C will be presented. Real-time images and quantitative data acquired in those experiments provide a very unique insight into gas-solid systems at harsh conditions.

508341 Material Distribution of Gas–Solid Fluidized Beds Measured By Electrical Capacitance Tomography – a Big-Data Perspective

Abstract:

Material Distribution of Gas-Solid Fluidized Beds Measured by Electrical Capacitance Tomography - a Big-Data Perspective

Gas-solid fluidized beds are used in many industrial processes, such as coal gasification, power generation, granulation, and polymerization. To measure hydrodynamic characteristics of gas-solid fluidized beds, numerous intrusive and non-intrusive experimental techniques have been developed. The intrusive methods like optical probe, capacitance probe, and pressure measurement are easy to implement but only capable of providing local information concerning the fluid flow. While the non-intrusive techniques such as tomography can be used to visualize the entire flow field without causing any disturbance to the flow. Compared to other industrial process tomography techniques, Electrical Capacitance Tomography (ECT) shows advantages in terms of fast imaging speed, no radiation, robustness, and low cost. Moreover, considering the non-conductive nature of materials in gas-solid fluidized beds, ECT is a suitable measuring technique for hydrodynamic investigation of these reactors.[1]

In ECT, the sensing electronics measure variations in capacitance between pairs of electrodes, which are placed around the periphery of a pipe or vessel under investigation. These measurements are then used to reconstruct the cross-sectional permittivity distribution as a presentation of the material distribution inside the sensing area via a specific image reconstruction algorithm. However, two major difficulties are associated with the reconstruction process. First, the number of measured independent capacitance data is far less than the number of unknown image pixels, and therefore the problem is severely under-determined. Secondly, due to the ill-posed and ill-conditioned property of the reconstruction process, the reconstructed results are sensitive to raw capacitance measurement noise. To address these problems, different reconstruction techniques, including single-step and iterative algorithms, have been proposed and have been reviewed by Yang and Peng[2] and Cui et al.[3]

Although big progress has been made in the image reconstruction with ECT in the last two decades to improve the image quality, i.e., the accuracy of material distribution measurements, the correlation coefficient between the true image and reconstructed image by a certain technique is usually lower than 0.9. This apparently hiders ECT as an excellent technique for gas-solid fluidized beds measurements, especially when a direct comparison between ECT measurements and Computational Fluid Dynamics (CFD) simulations is needed, in which high-accuracy images are necessary.

In the last several years, big-data analysis has gained considerable attention and momentum in both academia and industry, and also see some applications in the field of chemical engineering.[4] The dynamic evaluation method, which we have proposed recently,[5] provides a way for image reconstruction with ECT using a big-data perspective, and is potential to improve the image quality greatly.

To meet different needs, the big-data image reconstruction method is categorized to two schemes: an on-line scheme and an off-line scheme. The procedure of the on-line scheme is as follows. First, CFD simulations of a gas-solid fluidized bed, which is the same as that used in experiments, are performed. Next, a generic sensitivity matrix is obtained from a probability distribution analysis carried out on the material distributions extracted from the CFD simulation results. Finally, material distributions in experiments can be obtained using the modified Tikhonov regularization method equipped with the generic sensitivity matrix. While the off-line scheme is due to fingerprint matching the normalized capacitance between CFD simulations and experiments. As an illustration, Figure 1 shows some typical results obtained by different image reconstruction algorithms and the second-scheme big-data method. As can be seen, the correlation coefficients obtained by the existing image reconstruction algorithms are all lower than 0.9, also some details especially on the boundary between bubble and emulsion phases are blurred. While the best image can be obtained by the big-data method with the correlation coefficient as high as 0.95.

Figure 1. Material distribution reconstructed by different image reconstruction algorithms and the second-scheme big-data method.

Literature Cited

1. Rasel RK, Zuccarelli CE, Marashdeh QM, Fan L-S, Teixeira FL. Towards multiphase flow decomposition based on electrical capacitance tomography sensors. *IEEE Sens J.* 2017;in press.

2. Yang WQ, Peng LH. Image reconstruction algorithms for electrical capacitance tomography. *Meas Sci Technol.* 2003;14:R1-R13.

3. Cui ZQ, Wang Q, Xue Q, Fan WR, Zhang LL, Cao Z, Sun BY, Wang HX, Yang WQ. A review on image reconstruction algorithms for electrical capacitance/resistance tomography. *Sens Rev.* 2016;36:429-445.

4. Qin SJ. Process data analytics in the era of big data. *AIChE J.* 2014;60:3092-3100.

5. Guo Q, Meng SH, Wang DH, Zhao YF, Ye M, Yang WQ, Liu ZM. Investigation of gas-solid bubbling fluidized beds using ECT with a modified Tikhonov regularization technique. *AIChE J.* 2017;in press.

507964 Heat Transfer in a Pressurized Fluidized Bed with Continuous Addition of Fines

Abstract:

Oxygen-fired pressurized fluidized bed combustor technology with downstream carbon capture and sequestration is considered a key approach to clean coal combustion. In such technology, the fluidized bed combustor operates at elevated pressures and houses an in-bed heat exchanger tube bundle. The goal of this work was to investigate the impact of pressure and presence of fine particles (i.e., surrogate for pulverised fuel) on the overall tube-to-bed heat transfer coefficient. Experiments were conducted in a pilot-scale fluidized bed with an inner diameter of 0.15 m under cold flow conditions. A tube bundle consisted of five horizontal staggered rows was completely submerged in the bed (simulating the combustor in-bed heat exchanger tubes). One of the tubes was replaced by a 900 W heating cartridge housed in a hollowed copper rod. Five thermocouples distributed at 45° intervals along the copper rod circumference measured the surface temperature and ensured that local effects were included. Through supplying DC power and monitoring bed and copper rod surface temperature, the overall convective heat transfer coefficient was determined. The bed material was large glass beads of 1.0 mm in diameter while the fines were glass beads of 60 µm in diameter and thus susceptible to entrainment. The fine particles were continuously fed to the fluidized bed over a designed period of time, and then captured downstream by a filter system. Fluidization was conducted at 101, 600 and 1200 kPa with excess gas velocities (U_g - U_{mf}) of 0.21, 0.29 and 0.51 m/s. Fine particle feed rates were 0, 9.5 and 14.4 kg/h. The results showed that average heat transfer coefficients matched the correlation developed by Molerus et al. (1995) within a 5% difference across all conditions when fines were not present. At atmospheric conditions, where slug flow regime occurred, the maximum heat transfer coefficient was at the bottom of the rod, while it moved to the side of the rod at high pressures where the bubbling regime occurred. Overall, the heat transfer coefficient doubled when pressure was increased from 101 to 1200 kPa. The addition of fine particles decreased the average overall heat transfer coefficient by 3-9%, but there was no effect on the angular profile across the tube surface. The fines likely reduced the turbulence adjacent to the tube surface negatively impacting heat transfer via gas convection, which is the primary mechanism.

References

Molerus, O., and Mattmann, W. (1992). Heat transfer mechanisms in gas fluidized beds. Part 3: Heat transfer in circulating fluidzed beds. Chemical Engineering & Technology, 15 (5), 291-294.

508614 Investigation of the Effect of Fluidization on Alkane Catalytic Cracking in Fluidized Bed Reactor By High Temperature ECT Measurement System

Abstract:

Ethylene and propylene are important chemical raw materials. Alkane cracking is one of the major rountes for olefins production. At present, thermal cracking of alkane has been widely used in commercial olefins plant, which normally requires reaction temperature higher than 800 °C, and is considered as one of the most energy consuming processes in chemical industrial sector. Catalytic cracking technology can effectively reduce the alkane reaction temperature and on-purpose produce light olefins. Because of the rapid deactivation of catalyst caused by coke deposition, fluidized bed has been regarded as the well suitable reactor for catalytic cracking. However, the catalyst fluidization in fluidized bed reactor could significantly alter the reaction performance. Own to the lack of efficient high temperature measurement technology for fluidized bed, the influence of fluidization on the alkane conversion and product selectivit under real reaction conditions has not been understood yet.

In our laboratory a high temperature ECT system has been developed for fluidized bed measurment. The high temperature ECT system includes a specially designed high temperature electrode (working temperature above 600°C) and the corresponding algorithm. By use of the high temperature ECT measurement system, we monitored the catalytic cracking reaction of alkanes and investigated the influence of fluidization conditions on the reaction performance.

507700 Morphological Stability of Microencapsulated Lipophilic Compounds By AFM Imaging

Abstract:

Micronutrients and nutraceuticals such as vitamins, carotenoids and polyunsaturated fatty acids (PUFAs) are essential for human and animal health and well-being. Fat-soluble vitamins, like Vitamin A, or PUFAs are often chemically instable (*e.g.* to oxidation, heat) and they cannot easily be added to a hydrophilic food environment like a beverage. Thus, such vitamins are mostly sold in microencapsulated product formulations to protect the active ingredient from the surrounding environment. The morphological stability of microencapsulated lipophilic compounds directly influences the long-term storage stability and performance in final applications. Thus, during the development of novel product formulations, the stability of the encapsulated active ingredient is typically monitored in time-consuming studies, storing the product formulations at different environmental conditions and assessing the concentration of the active ingredient at several time points. Here, a facile method to investigate the internal morphology and structural stability of microencapsulated vitamins has been successfully developed. The internal particle morphology of three commercial dry Vitamin A acetate formulations differing in their matrix composition (hydrocolloid/polysaccharide and plasticizer) was systematically characterized by atomic force microscopy (AFM). The evolution of the particle internal morphology after *per-se* storage at 25 and 40 °C for ten months was investigated. Furthermore, an accelerated stress test by *in-situ* heating and AFM imaging was developed. Surface roughness parameters from the AFM images were used to semi-quantitatively estimate the evolution of the particle internal microstructure, and thus the aging and destabilization of the lipophilic vitamin formulations. The influence of humidity level on the thermal morphological restructuring of one of the commercial samples was also investigated. This study has shown that such AFM investigations of microparticles are a novel analytical tool to rapidly estimate the physical stability of the lipophilic vitamin dried emulsions.

Mougin, K., A. Bruntz, D. Severin and A. Teleki, "Morphological stability of microencapsulated vitamin formulations by AFM imaging," Food Structure 9, 1-12 (2016).

508208 Multi-Sensor Measurements of Quantitative Particle Size and Shape Information in Crystal Slurries

Abstract:

Introduction

Particle size and shape are critical quality attributes for active pharmaceutical ingredients (API) as they have direct impact on downstream processing, as well as on the performance behaviour of the finished product. Obtaining a reliable and robust quantitative information of these particle attributes in real-time remains a great challenge across the multiple manufacturing steps and cannot be achieved by just using a single sensor.

In-line particle sizing technologies, such as Focus Beam Reflectance Measurement (FBRM) and Particle Vision and Measurement (PVM), measure the optical response of single particle (single scattering) to a single wavelength light source and measurable size ranges are limited by the optical resolution of the light source and the optics. The increase of turbidity in the medium leads to an increase of scattering effects and absorption contributions. When photons encounter several particles, they undergo multiple scattering, changing the direction during each scattering event. In these cases, it is very challenging to obtain reliable results from either FBRM or PVM. Spatially and angularly-resolved diffuse reflectance measurement (SAR-DRM) is a spectroscopic-based approach for in-line monitoring.[1] SAR-DRM collects multi-wavelength (UV-visible-NIR) diffuse reflectance spectra from optical fibres of multi-angle multi-space arrangements. Absorption of light occurs at specific wavelengths for different particles, which can be used to identify samples composition; and different particle sizes and shape will have a different scattering pattern. SAR-DRM spectrum from each angular and spatial arrangement yields differences responses to the changes in the scattering and absorption properties of the sample. The technology requires multiple scattering on the incident light, therefore, it could complement other single particle based measurement technologies for in-line process monitoring.

In this work, crystal slurries are quantitatively characterised by using an in-line multi-sensor approach. Multivariate regression analysis is employed to establish calibration models for estimating particle size and solid loading. The knowledge acquired is then applied to a wet milling process, which is commonly used in pharmaceutical industries to improve the solubility and bioavailability of poorly water-soluble drugs by causing the breakdown of coarse particles into finer ones through mechanical forces.

Methodology

1. Characterisation of particle attributes, evaluation of the optical probes and modelling strategies

The investigation was carried out on crystal slurries of metformin HCL of different particle size ranges (<53, 53-90, 90-125, 150-180, 180-250 and 250-355 μm) and solid loadings (from 1 up to 40 wt.%). The samples were monitored using FBRM, PVM and SAR-DRM. Mathematical algorithms were applied to FBRM and PVM data to extract particle size distribution (PSD) and aspect ratio [2, 3]. Prior to each experiment, samples were analysed using off-line technologies such as laser diffraction and imaging.

Characterisation of the particles attributes served as an input to validate SAR-DRM sensitivity, accuracy and capability to track the differences in size and solid loading in crystal slurries. Multivariate analysis was applied to establish a performance matrix for all probes and combined methods in order to acquire a reliable and robust model to estimate particle size and concentration.

2. Application of the strategy to a wet milling process

A wet mill was coupled to the 250ml vessel via a recycle-loop configuration in gently suspended conditions in order to perform measurement and extract samples for off-line measurements. Breakage of metformin crystals through this wet milling method produced a range of end sizes from different starting particle sizes and concentrations. The significant change in particle attributes undergoing this system was then continuously monitored by FBRM, PVM and SAR-DRM. The same multivariate analysis was applied to assess the robustness of the model to this complex system.

Conclusions

The results show that FBRM and PVM are valuable tools to monitor particle size and shape distributions in low dense mediums. Increasing the solid loading, regardless of the particle size used, particles start overlapping and pictures lose their resolution which leads to ambiguous results in both FBRM and PVM. Larger particles are more likely to be overlapping or touching the picture frame and consequently, more likely will be discarded by the PVM imaging algorithm. In case of FBRM, large particles generate chords split, shifting the CLD towards smaller lengths.

SAR-DRM is capable of handling high solid loadings and relatively large particle sizes and is sensitivity to both particle size and solid concentrations, performing better than FBRM and PVM for such suspensions. SAR-DRM can be a complementary technique to the current in-line particle analysis to achieve robust process monitoring and enable improved control & optimisation of crystallisation processes.

References:

[1] Thennadil. Suresh and Yi-Chieh. Chen, Measurement apparatus and method, U.S. Patent EP2783199A1 (2014).

*[2] Agimelen. Okpeafoh, Hamilton. Peter, Haley. Ian, Nordon. Alison, Vasile. Massimiliano, Sefcik. Jan, Mulholland. Anthony, Estimation of particle size distribution and aspect ratio of non-spherical particles from chord length distribution, Chemical Engineering Science, **123**, 629-640 (2015).*

*[3] Agimelen. Okpeafoh, Jawor-Baczynska. Anna, McGinty. John, Dziewierz. Jerzy, Tachtatzis. Christos, Cleary. Alison, Haley. Ian, Michie. Craig, Andonovic. Ivan, Sefcik. Jan, Mulholland. Anthony, Integration of in situ imaging and chord length distribution measurements for estimation of particle size and shape, Chemical Engineering Science, **144**, 87-100 (2016).*

508226 Monitoring API Crystal Breakage in Wet Milling Using Inline Imaging and Chord Length Distribution Measurements

Abstract:

Introduction

Particle size and shape are critical attributes that affect the performance of particulate products. The monitoring and control of particle size and shape is an ongoing challenge in the field of particle technology. Current practices for determining particle size and shape mostly rely on offline analysis, in which the measurement is carried out after the process has finished. In addition to the offline analysis being carried out at the end of the process, the particle size and shape may be altered by the filtration and drying process required to extract particles from a liquid medium. Hence, the particle size and shape determined at the end of the process may not be representative of the size and shape of the particles during the process.

The Particle Size Distribution (PSD) and shape (quantified by the aspect ratio - ratio of particle width to length) of the particles in a suspension can be monitored in situ using data from suitable instruments. The Focus Beam Reflectance Measurement (FBRM) and Particle Vision and Measurement (PVM) sensors are inline probes that can be used for this purpose. The FBRM sensor measures a chord length distribution (CLD) which depends on the PSD and aspect ratio of suspended particles. The PSD and aspect ratio of the particles can subsequently be estimated from the CLD data by solving an inverse problem [1]. The PVM sensor captures images of particles in a suspension from which the PSD and aspect ratio of the particles can be estimated using a suitable image processing algorithm [2].

In this work, we employ previously developed algorithms [1,2] for solving the inverse problem associated with CLD data and analyse images from the PVM sensor to monitor the PSD and aspect ratio of particles during wet milling of organic crystals. Wet milling is an emerging application carried out in industry to reduce the size of particles to a suitable range. Organic crystals typically undergo breakage from wet milling. However, various systems and process conditions produce different effects which pose different challenges to the CLD inversion as well as the image analysis. We study the strengths and limitations of the CLD inversion and image processing under various conditions and find that they complement each other in obtaining quantitative information about particle size and shape.

Materials and Methods

A 1L stirred tank vessel charged with different particle suspensions was connected to an IKA MagicLab Wet Mill (a rotor-stator mill) in a recycle-loop configuration. The stirred tank was equipped with temperature controller and the FBRM and PVM probes were used for data capture. The suspension was passed through the wet mill at a series of increasing rotor speeds with real time monitoring. Benzoic acid, paracetamol and metformin hydrochloride were used in three separate experiments. In each process, the CLD data at every stage was captured with the FBRM probe while images of the particles were acquired using the PVM probe. Before the start of each process, samples of each material were analysed using the Malvern Morphologi G3 instrument (offline particle imaging) for particle size and aspect ratio estimation. Similarly, samples of the filtered and dried milled particles were analysed for particle size and aspect ratio estimation at the end of each wet milling process using the Morphologi G3 instrument.

Results

Obtained results from the employed method has shown the relative sensitivities of the two modalities to changes in particle size and shape due to the wet milling. The effect of bubbles produced during the process on the two modalities has also been studied. When the particles are sufficiently large and within the resolution of the PVM instrument camera, and not too large so as to into the image frame, the inline imaging method gives PSD estimates which are closer to online estimates of similar materials. This is especially true for systems composed of long needle-like particles such as benzoic acid and metformin HCL mixed with shorter or more rounded particles. This is because the CLD method needs to find a compromise PSD which the measured CLD at an appropriate aspect ratio. In systems like these, the aspect ratio distribution estimated by the imaging method tends to be dominated by the smaller particles, while the mean aspect ratio estimated by the CLD method tends to be dominated by the long needle-like or rod-like particles. However, the inline imaging method is limited to particles of sizes above about 30Î¼m, whereas the CLD method can go to smaller sizes. In addition, the PSD estimated by the inline imaging method becomes less representative as the sizes of the particles approach the size of the image frame. The accuracy of the PSD estimate is also affected by the proportion of objects that are captured outside the focal plane of the camera. Similarly, the PSD by the CLD method becomes less representative as the length of needle-like or rod-like particles approach the diameter of the circular trajectory of the FBRM laser spot making the chord length probability estimate 22 less accurate. Hence, in systems composed of a mixture of needle-like or rod-like particles and more rounded particles of various sizes, a combination of both the CLD and inline imaging methods (probably in a multi-objective approach) should give more robust estimates of the PSD and the aspect ratio. This will be particularly important to real-time monitoring and control of crystallisation processes where various process conditions could lead to the production of particles of various sizes and shapes and effects such as bubbles which could be very challenging to capture by a single sensor method.

Conclusions

We have employed wet milling processes of slurries of different crystalline materials to assess the strengths and limitations of two different inline particle monitoring modalities namely CLD and PVM imaging. The materials were carefully chosen as they produce crystals of different morphologies and mechanical strengths. The results from both CLD inversion and PVM image analysis demonstrate the ability of our software tools to quantitatively monitor particulate processes with respect to particle size and shape distributions. The results from the three case studies have identified a number of strengths and limitations of each method depending on the process conditions and nature of particles in each system. The CLD inversion algorithm can be applied to dense particle suspensions whose images may be difficult to analyse. However, the CLD data is affected by effects such as bubbles or particle transparency. The image processing algorithm allows the distribution of aspect ratios of particles to be estimated, but it is limited by the resolution of inline images

References:

*[1] O. S. Agimelen et al., Estimation of particle size distribution and aspect ratio of non-spherical particles from chord length distribution, Chemical Engineering Science **123**, 629 (2015).*

*[2] O. S. Agimelen et al., Integration of in situ imaging and chord length distribution measurements for estimation of particle size and shape, Chemical Engineering Science **144**, 87 (2016).*

508703 Investigating Surface Energy Heterogeneity of Processed Lactose Powders

Abstract:

Particle surface properties are known to be affected by the processing routes and processing conditions used for generating particles. Inverse Gas Chromatography (IGC), particularly Finite-Dilution Inverse Gas Chromatography (FD-IGC), has been shown to be a versatile technique for surface energy characterisation of solids. In this study, lactose powders processed by both milling and sieving the similar crystalline Î±-lactose monohydrate starting material were used. The sieved samples used in this study were divided into top, middle and bottom fractions. FD-IGC was used to examine the surface energy heterogeneity of these processed crystalline Î±-lactose monohydrate powders. The dispersive component of surface energy at different surface coverages of lactose was determined in this work. The results indicated that all lactose samples investigated showed some degree of heterogeneity. Unmilled, coarse lactose powders showed lower dispersive surface energy values and fine powders, produced by milling, were found to have the highest dispersive surface energy values. The surface energy distribution obtained from the deconvolution of the surface energy data obtained from FD-IGC using a Boltzmann distribution model and simulation-fitting approach suggested possibly three different surface energy regions on the processed lactose samples. The surface energy region with highest dispersive surface energy ranges from 44-49 mJ/m^2 followed by a 38-45 mJ/m^2 region and a region with lowest dispersive surface energy of 26-31 mJ/m^2. For sieved samples, the surface energy distributions for the top and medium sieved fractions suggested that the surface energy of large, unbroken Î±-lactose monohydrate crystals consists of roughly equal amounts of higher and lower energy surfaces whereas, the bottom fraction with smaller particles had greater proportion of high energy sites and less of the lower energy sites than the other two sieved fractions. Overall, FD-IGC for energetic heterogeneity determination helped in characterising the changes in surface properties of powders produced by different processing operations. Moreover, the experimental analysis combined with modelling approach and surface energy deconvolution were able to quantifiably determine different surface energy regions on the processed lactose samples.

508743 Nonlinear Radial Stress Response during Uniaxial Die Decompression

Abstract:

A nonlinear curve is generally observed at a plot of the radial stress versus axial stress during the entire loading/unloading procedure of a uniaxial die compression. At initial stages of unloading, the relationship is generally accepted to be linear. This observation serves as the basis of the argument that the compact exhibits linear elastic behavior during this period, and elastic constants such as Young's modulus and Poisson's ratio can be easily calculated. As such, the calculated elastic constants are further assumed to stand for the overall elastic constants of the powder compact. Here we propose, however, that the presumed linear relationship during the initial unloading is instead a transitional moment during which the die wall friction force changes direction. This is observed through analysis of compression in which axial force transmission and the coefficient of friction (CoF) are evaluated continuously during both the loading and unloading phases of compaction. Additionally, the effect of compact thickness on these parameters is studied, and observed to have a significant effect on force transmission but no impact on CoF values. This result supports the assertion that friction forces change direction when the compression cycle transitions from loading to unloading, which provides an alternative framework in which to understand the apparent profile linearity. The phenomenon is observed in materials exhibiting a broad array of mechanical properties. Using this analysis, we propose an alternative approach to determining fundamental elastic constants, in which an assumption of linear elastic strain was made to the overall (instead of a small portion) elastic recovery of a compact during unloading. With this new approach the elastic constants such as Young's modulus and Poisson's ratio can be calculated which is different from published data.

509921 Elastic Waves in Particulate Stiff Granular Soft Material Mixtures: Experiments and Simulations

Abstract:

The elastic response of disordered mixtures of granular and soft matter is investigated by means of wave-propagation, both experimentally and numerically. This allows inferring fundamental properties of granular and soft disordered materials such as elastic moduli and dissipation mechanisms. Mixtures are prepared with different volumes of soft matter mixed with hard matter to identify the transition from a rigid to a soft granular skeleton. We compare physical experiments in a triaxial cell equipped with piezoelectric wave transducers and Discrete Element Method simulations (DEM). DEM simulations offer deeper insights into the micromechanics in and at the transition between the stiff- and soft-dominated regimes. Interestingly, we find that the behavior is highly non-linear and also non-monotonic with increasing the percentage of soft content. The presence of soft particles alters the formation of force chains. The qualitative agreement between experiment and simulation is well captured despite of complex interaction between soft and stiff particles that render several ongoing challenges like, e.g., the presence multi-body-interactions instead of pairwise contacts.

509651 Effect of Size and Mechanical Properties of Particles on the Microstructure and Tensile Strength of Compacted Powders

Abstract:

Compaction is one of the most common industrial processes used to achieve densification of powders and to develop tensile strength or resistance to crushing. Several factors, including powder properties, applied compaction pressure, and the shape of tooling affect the compaction process itself and the final outcome. Even though several prior studies have addressed this topic in detail, the compaction process is still not well understood, particularly in terms of developing a mechanistic relationship between particle scale (micro scale) properties and the powder compact (macro scale) properties such as relative density and tensile strength. In order to understand the effect of particle properties, such as particle size, plasticity, elasticity, and surface bonding energy, on the powder compact, we ran several experiments and discrete particle simulations. We studied lactose and microcrystalline cellulose powders, which are one of the most commonly used excipients in pharmaceutical industry, and also metallic powders. The exact particle size distributions of these powders are represented in the simulations, and the mechanical properties of the particles are calibrated using the powder compaction experiment results. Then, we validated our simulation results based on experimental indirect tensile strength test. We found that the simulation results are quantitatively similar to experimental results. Once the simulations are validated, we used the simulations to investigate the microstructure evolution during the compaction process, particularly that of the size and the mechanical properties of the particles. The evolution of the microstructure is studied by analyzing the evolution of the magnitude, orientaiton, and spatial and temporal distribution of the contact forces and the bonding forces between the particles. In this presentation, we will discuss the experimental and simulation methods and results in detail.

508041 Random Packings of Polydisperse Adhesive Microspheres with Gaussian Size Distribution

Abstract:

We investigate random loose packings of polydisperse spherical micro-particles by means of a specially developed adhesive DEM approach. The results show that the differences of the packing properties (packing fraction and mean coordination number) between polydisperse packings and monodisperse cases with the same mean particle radius are well described by a quadratic function of the normalized standard deviation of the particle size (*Ïf*). Based on the previous literature (Liu, Li, Chen, 2016), we can uniquely predict the packing fraction of polydisperse adhesive packings via *Ïf* and a dimensionless adhesion parameter (*Ad*). Furthermore, by looking into the microstructure of the polydisperse packings, we found that the typical functions such as radial distribution function and angular distance distribution function, which are widely used to characterize the monodisperse packing structures, are not applicable. The local characteristic peaks are erased due to the size polydispersity.

508168 Comprehensive Characterization of Stressing in Mills By Particle Probes

Abstract:

In industry mills are the working horse for particle size reduction. However, a clear picture and a detailed understanding of the mechanisms involved in stressing and particle breakage are far from being complete. During comminution distributions of stresses are applied. Both, the type of stressing (one sided or two sided, loading by compression or impact) and the acting stress energies are distributed. So called mill functions link the acting stresses and their frequency to the used apparatus and its operational conditions. So far the aforesaid functions are widely unknown. Within this contribution a recently proposed novel approach based on stressing of fully characterized single particle probes is applied to wet comminution in stirred media mills [1]. Our approach relies on the fact that metallic and oxide probe particles deformed by single particle compression in a scanning electron microscope [2] exhibit the same morphology than particles recovered from stirred media milling, i.e. stressing in the technical apparatus can be mimicked by uniaxial compression [3]. By finite element modelling of the uniaxial compression and image analysis of geometry changes for individual particles (c.f. ratio of waist diameter to initial particle size) the energy absorption of the particle probes can be described. To access the stress number the formation of contacts upon stressing is treated by formal kinetics. Thus, the stressing history in terms of stress energy and stress number can be determined experimentally and stress energy and the stress number distributions in the mill become accessible. For the lab-scale stirred media mill operated at exemplary process conditions it will be shown that the involved stress energy and stress number distributions are broad and depend on grinding bead size, particle size and material. Furthermore, it will be shown that the grinding kinetics in the submicron size range is predominantly determined by particle capture. Our methodology is not limited to wet comminution and applicability to other types of mills will be shown. In summary, the proposed methodology will be useful for the understanding of stressing in technical mills, the development of models, the design of mills and ultimately to the optimization of comminution processes.

[1] A. Strobel, S. Romeis, S. Wittpahl, P. Herre, J. Schmidt, W. Peukert, Characterization of stressing conditions in mills – A comprehensive research strategy based on well-characterized model particles, Powder Technol. 305 (2017) 652–661.

[2] S. Romeis, J. Paul, M. Ziener, W. Peukert, A novel apparatus for in situ compression of submicron structures and particles in a high resolution SEM, Rev Sci Instrum 83 (2012).

[3] J. Paul, S. Romeis, M. Mačković, V. Marthala, P. Herre, T. Przybilla, M. Hartmann et al., In situ cracking of silica beads in the SEM and TEM - effect of particle size on structure-property correlations, Powder Technol. 270 (2015) 337–347.

CHARACTERIZATION AND OPTIMISATION OF AUTOCLAVE PROCESS PARAMETERS OF PARTICULATE REINFORCED POLYMER MATRIX COMPOSITES FOR AEROSPACE AND MARINE APPLICATIONS

Adeodu O. Adefemi[1], Anyaeche O. Christopher[2], Oluwole O. Oluleke[3]

[1]Department of Mechanical and Mechatronics Engineering, Afe Babalola University, Ado-Ekiti, Nigeria.
[2]Department of Industrial and Production Engineering, University of Ibadan, Ibadan, Nigeria
[3]Department of Mechanical Engineering, University of Ibadan, Ibadan, Nigeria

Corresponding Author
[1]adeoduac@abuad.edu.ng, [2]osita.anyaeche@ui.edu.ng, [3]oluwoleo2@asme.org

ABSTRACT

Polymer matrix composites (aluminium-polyester and carbon black-polyester) are used in diverse applications due to their easy adaptability in providing tailored mechanical properties usually achieved by curing. Their production process in autoclave curing results in long processing cycle, which invariably affects the quality negatively. There is dearth of information on determining optimum process parameters for curing in autoclave oven. The objective of this study was to determine optimum process parameters for the production of aluminium-polyester and carbon black-polyester composites in autoclave oven.

Process parameters for polymer matrix composites include cure temperature, degree of cure and cure cycle according to ISO 4049 standard. Three sets of samples of polyester reinforced with aluminium and carbon black powders separately in 30.0, 20.0 and 10.0% reinforcement composition, were weighed, mixed, cast into moulds and cured in autoclave oven according to ASTM D790 standard. Temperature and degree of cure of each sample were observed inside the oven by using thermocouple and differential scanning calorimetry, respectively, at an interval of five minutes until full cure was achieved. The samples were examined for tensile properties and voids using universal testing machine and optical microscope according to ASTM D638 standard. An optimisation problem for the autoclave process was formulated for the minimisation of cure time as objective function subject to maximum and minimum allowable temperatures, as constraints. The primal problem was solved using finite difference method and optimised by genetic algorithm. The tensile strengths and micrographs of aluminium-polyester and carbon black-polyester composites from both optimised and conventional control processes were compared using t-test at $\alpha_{0.05}$.

Autoclave oven achieved cure temperatures of 90 and 75 °C, degree of cure of 85.0 and 80.0% in 270 and 170 minutes for aluminium-polyester and carbon black-polyester, respectively. Tensile strength of 30.0, 20.0 and 10.0% aluminium-polyester cured in autoclave had 0.0092, 0.0068 and 0.0048 MPa, respectively, while 30.0, 20.0 and 10.0% carbon black-polyester composite had 0.0032, 0.0028 and 0.0020 MPa, respectively. Micrographs of aluminium-polyester samples cured in autoclave showed mean void size of 206 µm, while micrographs of carbon black-polyester samples showed mean void size of 337 µm. Autoclave optimum process parameters were able to achieve cure temperatures of 90 and 75 °C, degree of cure of 99.0 and 100.0% and also minimised the cure cycle to 30 and 23 minutes for aluminium-polyester and carbon black-polyester composites, respectively, compared to 230 and 170 minutes, respectively in the conventional control experiment. The 30.0, 20.0 and 10.0% optimised aluminium-polyester composites had 0.012, 0.011 and 0.0096 MPa, while 30.0, 20.0 and 10.0% optimised carbon black-polyester composites had

0.010, 0.0076 and 0.0068 MPa tensile strength, respectively. The micrographs of the optimised aluminium-polyester samples showed mean void size of 25.2 μm, while carbon black-polyester samples showed mean void size of 23.6 μm. There was an improvement in the microstructures of optimised aluminium-polyester and carbon black-polyester by 88.0 and 93.0%, respectively, compared to the conventional control samples. The variation between optimised and conventional control samples were significant for both composites.

The optimum process parameters for aluminium-polyester and carbon black-polyester composites were established. These optimum parameters are useful for curing composites with desirable microstructures and tensile strength properties

Keywords: Autoclave oven, Cure cycle, Microstructural, Polymer matrix composites, Tensile strength

Word count: 518

1. INTRODUCTION

Curing is described as a cross linking process of making molten fibre or particulate reinforced polymer matrix composites into solid final product [1]. The modeling and optimisation of the process have been largely applied in different industrial sectors as a means of gaining more knowledge in order to improve the quality of the product [1]. Shoddy manufacture of composite products can be traced to different reasons; degeneration of resin as a result of exothermic effect, dry spot and formation of voids due to trapped volatile gases or partial filling [2] [3] [4]. In view of these, investigation and setting of production process parameters in order to predict the implication of design variables on the final product, maximize the rate of production and minimize the cost of production cannot be under estimated. In addition, trustable process optimisation reduces waste of resources and promotes environmental friendly process. According to [1] [5] effective production process is said to be achieved within short cycle. This is practically impossible with autoclave curing system due to mode of heat transfer and the system set up. When composites are cure via electromagnetic means, small amount of the heat generated are conducted away, thereby resulting to thermal degeneration of the product or composite burn out [1] [3] [5]. The objectives of setting optimum production process parameters in batch processing are to minimize the cure cycle and establish optimum cure and temperature profiles with maximum and minimum temperature to avoid degeneration of the resin and to initiate cure, respectively, are set as constraints [4].

Optimisation of batch curing to minimize processing time has been studied by some researchers in the recent times. According to [6], look ahead strategy was used to minimize the cure cycle of dense polyester reinforced with carbon fibre. This technique examined a commutation between two techniques, a simple gradient based descent method and heuristic search, for a near optimum solution. [1] Exploited SQP algorithm based on thermo-chemical analysis to compare the optimisation of epoxy cure profile between thick and thin laminates composites of complexity in determine the best cure temperature. The size of the part influences the quality of the final product. Udaykumar and shy [7] also compares SQP algorithm with heuristic method to determine cure temperature profile for polyester-glass fibre system involving five stages among which are heating and cooling zones. Optimal cure profile for epoxy resin reinforced with carbon fibre laminates with 5 and 10cm thickness

were also established using GA, taking into consideration, maximum temperature and maximum heating rate allowed as constraints. Optimal cure and pressure profiles were equally found using SQP algorithm for a thick graphite-epoxy laminate. The temperature profile was first solved and subsequently establishing the optimum pressure profile in order to keep the void content under acceptable limit [8]. The work was compared with that established by Criscioli [9], there was a reduction in pressure profile by 45%.
There is dearth of information on setting optimum process parameters for curing in autoclave oven using genetic algorithm (GA). The aim of this study was to set optimum process parameters for the production of aluminum-polyester and carbon black-polyester composites in autoclave oven. The objectives of the study are to:

1. Study the major production process parameters of the polymer matrix composites in terms of cure temperature (T_c), actual cure cycle (t_c) and ultimate degree of cure (a_c) for autoclave curing technique
2. Examine the effects of the curing technique and bulk thickness on tensile strength and microstructural properties of the composites produced.
3. Determine the optimum production process parameters in terms of optimum temperature evolution and optimum degree of cure that minimizes the cure cycle
4. Examine the effect of the optimum production process parameters on the tensile strength and microstructural properties of the composites produced

2.0 PROCESS MODEL

The autoclave composite manufacturing process model is made up of filler-resin mixture consolidated uniformly in a mould to the required laminate thickness. This model is subdivided to thermal model, which is required to solve energy equation for temperature distribution through the laminate. Also, cure kinetic model known as cross linking chemical reaction of the curing process.
The assumptions of the process model are: heat is conducted uniformly across the length and breadth of the laminate and the thickness of the laminate is insignificance to other dimensions. Therefore, the process is predominantly one dimension, in the sense that the temperature varies with time (t) and location in x direction [1].
The energy equation for one dimensional heat of conduction through the laminate is represented by equation 2.1

$$\rho C_p \frac{\partial T}{\partial t} = \frac{\partial}{\partial x}(K \frac{\partial T}{\partial x}) + \rho H_r \frac{\partial \alpha}{\partial t} \qquad 2.1$$

The parameters are defined as follow: ρ is the density of the laminate, C_p is the specific heat capacity, T is the temperature distribution across the laminate, k is known to be the thermal conductivity, and Hr is the heat of reaction per unit mass of the laminate.
The kinetic model is represented by an auto catalytic model used by [10] [11] [12] given in equation 2.2

$$\frac{\partial \alpha}{\partial t} = g(\alpha, T) = A \exp(-\frac{\Delta E}{RT})\alpha^m (1-\alpha)^n \qquad 2.2$$

$\frac{\delta \alpha}{\delta t}$ is the rate of cure of the composite, which is dependent on the rate of heat transfer from the source ($\frac{\delta T}{\delta t}$) and the heat of reaction within the composite. The heat of reaction is a function of the composite material in consideration.

Where A is Arrhenius constant, ΔE is the activation energy and R is the universal gas constant. The parameters m and n are reaction order constants.

Model equations 2.1 and 2.2 are subject to initial condition given by

$$T = T_o \text{ and } \alpha = 0.1 \text{ at } t = 0 \text{ (note: } T_o = 25^0\text{C)} \qquad 2.3$$

This is the pre-cured state of the sample. The value of alpha is the minimum cure at pre-cured temperature

There are two possible boundary conditions associated with equation 2.1. The first boundary condition is a convective heat transfer between the source and the surfaces of the laminate given by

$$k\frac{\partial T}{\partial x} = -h(T - T_{\infty}) \text{ at } x = H \qquad 2.4$$

Where h is the heat transfer coefficient at the surfaces, T_{∞} is the temperature of the oven which may be function of time, according to some pre-determined cure cycle, and H is the thickness of the laminate. The second possible boundary condition is when the system is isolated at the surface given by

$$\frac{\partial T}{\partial x} = 0 \qquad 2.5$$

The following assumptions are made in order to build up the thermal models for the cure in autoclave [5]:

3.0 METHODOLOGY

Equipment and Material Selection

Process parameters as identified by ISO 4049 standards are cure temperature, cure cycle and degree of cure. The following material and equipment were used to examine the process parameters. The materials were selected were tailored to the applications in the field of aerospace and marine engineering. These materials are the major constituents in the production of parts in the industries. Also due to easy adaptability to desired mechanical properties. The percentage weight ratio of reinforcement to matrix was chosen based on strength to weight ratio, thermal conductivity, machinability and other mechanical factors as applicable in production in aerospace and marine industries.

Raw Material Selected

Aluminum and carbon black powder, Unsaturated Polyester Resins, Methyl ethyl ketone peroxide as catalyst and cobalt accelerator

Composite Manufacturing Process

This is the initial stage of the manufacturing process to set out the samples to be used for the study. This includes all necessary processes that lead to the production of the final part. The following are the set of sequential steps that was used for the lay-up polymer matrix composites according to ASTM D790 standard. The fabricated moulds were properly prepared and clean. Correct amount of particulates reinforcement were weighed into containers on the basis of 30, 20, and 10% reinforcement weight fraction. Correct amount of resin and hardener were weighed in ratio 4:1 and mixed with respective weight fraction of the reinforcement to add up to 500grams of total mixture. Resin and particulate fillers were evenly and slowly mixed with the aid of mechanical blender inside a fume cabinet (for safety) in order to ensure a homogeneous composite mixture. Each separate mixture were poured into moulds and transferred into autoclave oven. The curing of composite involves two different stages. The initial curing of the samples was done under ambient temperature for 24 hours. Posts curing with autoclave helps composites reach its full physical characteristics. As the curing progresses, the temperature of the samples was taken at five minute interval with thermocouple attached to the ovens and the degree of cure was determined by means of differential scanning Calorimetry until full cure was reached. The samples were examined for tensile properties and voids using universal testing machine and optical microscope according to ASTM D638 standard. An optimisation problem for the autoclave process was formulated for the minimisation of cure time as objective function subject to maximum and minimum allowable temperatures, as constraints. The primal problem was solved using finite difference method and optimised by genetic algorithm. The tensile strengths and micrographs of aluminium-polyester and carbon black-polyester composites from both optimised and conventional control processes were compared using t-test at $\alpha_{0.05}$.

Formulation of the objective function for cure optimisation

The objective function is a system of mixed ordinary and partial differential equations of the form:

System A1: 30% Aluminum-Polyester Composite

$Min:$ 3.1

$$t_c = f(T,\alpha) = \left\{(1191.5*4.77)\frac{\partial T}{\partial t} = 0.051(\frac{\partial^2 T}{\partial x^2}) + (1191.5*285.7*\frac{\partial \alpha}{\partial t})\right\}$$

$$\frac{\partial \alpha}{\partial t} = \left\{2991.4*Exp(\frac{-4500}{8.3*120})*\alpha^{0.34}(1-\alpha)^{1.66}\right\}$$

System A2: 20% Aluminum-Polyester Composite

$Min:$

$$t_c = f(T,\alpha) = \left\{(1261*6.1)\frac{\partial T}{\partial t} = 0.164(\frac{\partial^2 T}{\partial x^2}) + (1261*285.7*\frac{\partial \alpha}{\partial t})\right\}$$

$$\frac{\partial \alpha}{\partial t} = \left\{2991.4*Exp(\frac{-4500}{8.3*120})*\alpha^{0.34}(1-\alpha)^{1.66}\right\}$$ 3.2

System A3: 10% Aluminum-Polyester Composite

$Min:$ 3.3

$$t_c = f(T,\alpha) = \left\{(1330.5*10.52)\frac{\partial T}{\partial t} = 0.186(\frac{\partial^2 T}{\partial x^2}) + (1330.5*285.7*\frac{\partial \alpha}{\partial t})\right\}$$

$$\frac{\partial \alpha}{\partial t} = \left\{2991.4*Exp(\frac{-4500}{8.3*120})*\alpha^{0.34}(1-\alpha)^{1.66}\right\}$$

$Subject:$

$Max[T(x)] \geq 90$

$Min[T(x)] \geq 25$

$Min[\alpha(x)] \geq 0.8$ 3.4

System B1: 30% Carbon black-Polyester Composite

$$Min:$$

$$t_c = f(T,\alpha) = \left\{(1520*5.05)\frac{\partial T}{\partial t} = 0.081(\frac{\partial^2 T}{\partial x^2}) + (1520*359.8*\frac{\partial \alpha}{\partial t})\right\}$$

$$\frac{\partial \alpha}{\partial t} = \left\{2950*Exp(\frac{-42000}{8.3*120})*\alpha^{0.3}(1-\alpha)^{1.6}\right\} \qquad 3.5$$

System B2: 20% Carbon black-Polyester Composite

$$Min:$$

$$t_c = f(T,\alpha) = \left\{(1480*6.5)\frac{\partial T}{\partial t} = 0.167(\frac{\partial^2 T}{\partial x^2}) + (1480*359.8*\frac{\partial \alpha}{\partial t})\right\}$$

$$\frac{\partial \alpha}{\partial t} = \left\{2950*Exp(\frac{-42000}{8.3*120})*\alpha^{0.3}(1-\alpha)^{1.6}\right\} \qquad 3.6$$

System B3: 10% Carbon black-Polyester Composite

$$Min:$$

$$t_c = f(T,\alpha) = \left\{(1440*11.33)\frac{\partial T}{\partial t} = 0.186(\frac{\partial^2 T}{\partial x^2}) + (1440*359.8*\frac{\partial \alpha}{\partial t})\right\}$$

$$\frac{\partial \alpha}{\partial t} = \left\{2950*Exp(\frac{-42000}{8.3*120})*\alpha^{0.3}(1-\alpha)^{1.6}\right\} \qquad 3.7$$

$$Subject:$$

$$Max[T(x)] \geq 80$$

$$Min[T(x)] \geq 25$$

$$Min[\alpha(x)] \geq 0.75 \qquad 3.8$$

Where t_c is the curing time, which is a function of temperature (T) and degree of cure (α). T_c is the cure temperature, T_o is the minimum initial temperature to initiate cure reaction and a_c is the ultimate degree of cure.

The mixed PDE/ ODE problems were solved using method of lines. This converts mixed PDE/ ODE to a system of ODE's by discretizing using finite difference method. The systems of ODE's was solved using Matlab called ODE 15S. Optimum solution of the autoclave process was finally done with genetic algorithm.

Systems parameters used as input data for the solution of primal problems for Autoclave Heating

Table 3.1 System Parameters for ODE Model

Parameter	Unit	Aluminum-Polyester Composite	Carbon black-Polyester Composite
Heat of reaction (ΔH_r)	J/g	285.7	359.8
Activation Energy (E_a)	kJ/mol	42.5	42
Arrhenius Constant (Ar)	s^{-1}	2991.4	2950
Kinetic Constant (k)	s^{-1}	2950	
Reaction Order (n)		1.66	1.6
Reaction Order (m)		0.34	0.3
Gas Constant ®	kJ/kmol.k	8.3	8.3
Temperature of the reaction (T_{input})	0c	90	80

Table 3.2 System Parameters for PDE Model by% wt. of Reinforcement

Parameter	Unit	30% Al.Cpst	20% Al.Cpst	10% Al.Cpst	30% CB.Cpst	20% CB.Cpst	10% CB.Cpst
Density (ℓ)	kg/m^3	1191.5	1261	1330.5	1520	1480	1440
Specific Heat Capacity (C_p)	kJ/kg.k	4.77	6.1	10.52	5.05	6.5	11.33
Thermal Conductivity (K)	W/m.k	0.051	0.164	0.186	0.081	0.167	0.186

Table 3.3 GA Parameters used for Cure Optimisation

GA Input	Values
Number of Generation	50
Individuals per Population	50
Individual per Reproduction	40
Elite Individual	4
Size of Pareto	30
Mutation Probability	0.005
Cross-over Probability	0.5

Source: Giacomo (2014)

4.0 RESULTS AND DISCUSSION OF RESULT

Temperature and Degree of Cure Profiles for aluminum-polyester and Carbon black-Polyester composites

The results were presented in the figures below. The experimental data were the same as presented in Adeodu et al. [5].

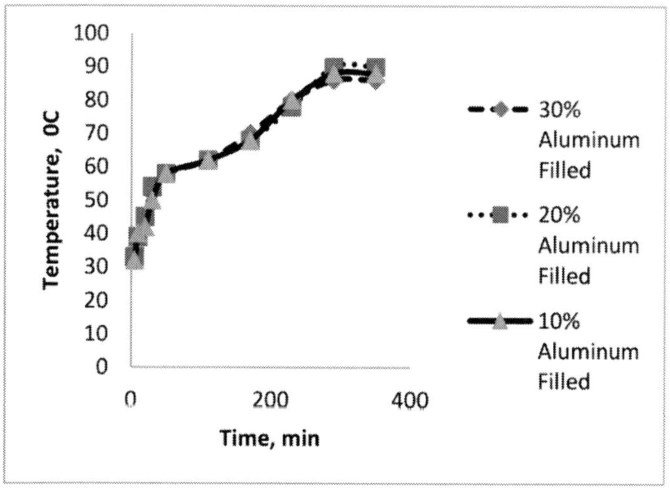

Figure 4.1 Temperature Profile for Aluminum Composites

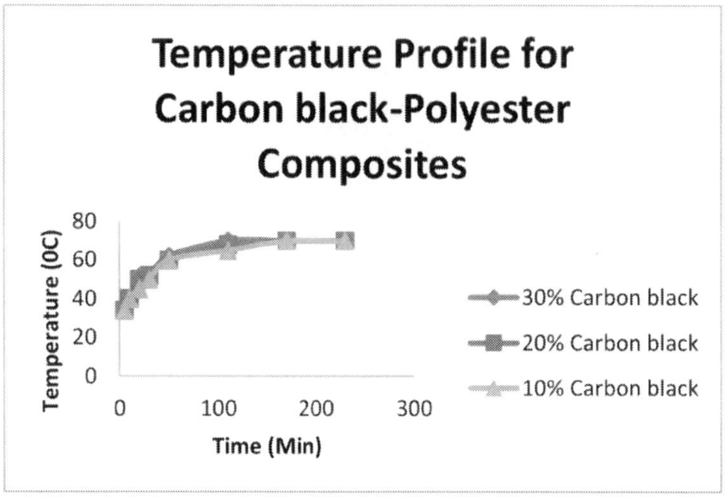

Figure 4.2 Temperature Profile for Carbon black Composites

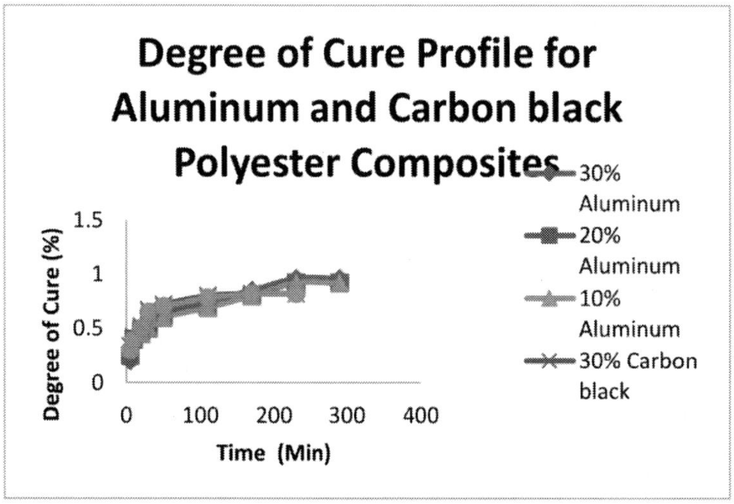

Figure 4.3 Degree of Cure Profile for Aluminum and Carbon black-Polyester Composites

Figures 4.1 and 4.2 show the temperature profile for aluminum-polyester and carbon black-polyester composites respectievly. These shoe temperature responses of the composites to time as the curing progresses.

In figures 4.1 and 4.2, the responses show similar increasing trends as the curing progresses in respective of the percentage reinforcement composition. it was observed in figure 3.6 that aluminum-polyester composite achieved a cure temperature of 80^0C at 230min cure cycle while carbon black-polyester composite achieved cure temperature of 70^0C at 170 min. the cure cycle for carbon black-polyester is shorter than that of aluminum-polyester composite due to the fact that carbon black is more exothermic in reaction than that of aluminum-polyester composite. Therefore, the rate of heat transfer from the autoclave couple with the heat of reaction from the cross-linking process was able to speed up the curing process in carbon black-polyester.

Figure 4.3 shows the cure characteristic of the composites. These are the responses of the composites to degree of cure with respect to time. Aluminum-polyester composites show an average degree of cure of 0.94 = 94% at cure temperature of 80^{0}C, while carbon black-polyester composites show an average degree of cure of 0.82 = 82% at cure temperature of 70^{0}C. according to Philip [2] and Min Li et al., [1], there is direct relationship between temperature evolution and degree of cure.

AutoclaveTensile Profile

The experimental Tensile Strength data were the same as presented in Adeodu et al., 2015

Figure 4.4 shows autoclave tensile strength profiles for aluminum-polyester and carbon black-polyester composites respectively. The two composites show similar increasing tensile strength trends as the percentage reinforcement composition increases. Aluminum-polyester shows higher tensile strength than carbon black-polyester composites. The reason for these is that the higher the degree of cure of polymer composites, the better the mechanical property in term of strength.

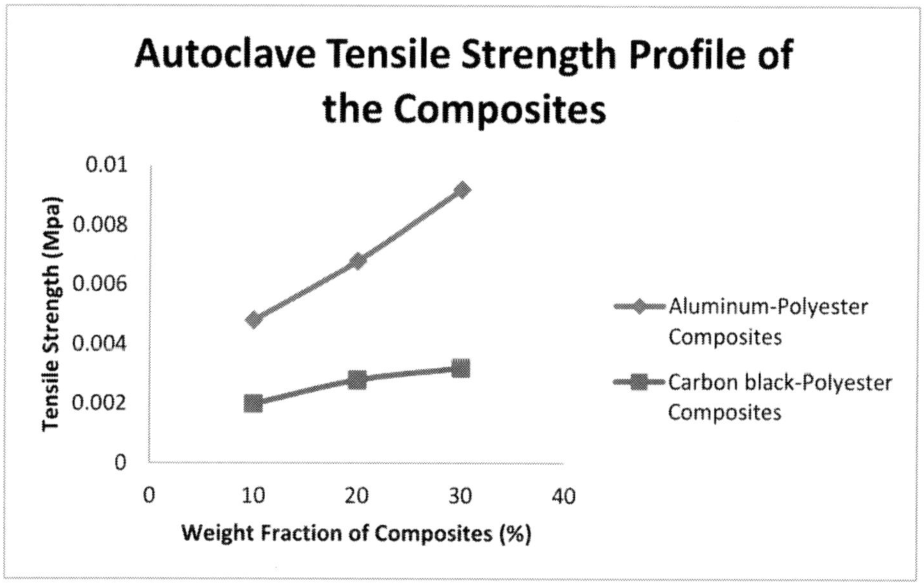

Figure 4.4 Autoclave Tensile Strength Curve

Autoclave Microstructural Analysis

Figures 4.5 and 4.6 show the micrographs of aluminum-polyester and carbon black-polyester composites respectively cured in autoclave. From the image analysis, 30% carbon black-polyester composite shows more voids compared to aluminum-polyester composite. Carbon black-polyester composite had 35 voids/ sq μm with mean void size of 191 μm while aluminum-polyester composite had 30 voids/sq μm with mean void size of 156μm. Voids are mostly caused by trapped volatiles during curing. According to Philips [2], higher temperature from the part surface during autoclave heating tends to progress curing faster at the surface of the part, sometimes forming a hard impermeable skin on the part surface which can trap volatile given off during cross linking reactions

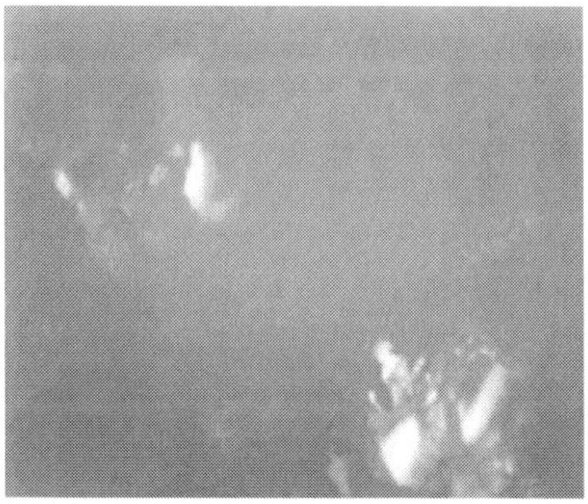

Figure 4.5 Micrograph of Carbon black-Polyester Composite (Mag 300X)

Figure 4.6 Micrograph of Aluminium-Polyester Composite (Mag 300X)

Cure Optimisation Results and Discussion

Figures 4.7 – 4.12 illustrate the optimisation outcomes for the last three generations for the varying weight fractions (30%, 20% and 10%) of aluminum-polyester composite and carbon black-polyester composite for autoclave curing respectively. Tables 4.1 and 4.2 also present the cure results as applied to the two systems for the last three generations of convergence.

Figures 4.7(a) and (b) show the temperature evolution and degree of cure profiles along the thickness of the laminate. From the kinetics of the reaction process, aside the heat transfer into the laminate from the autoclave, there is also heat of reaction due to curing of the laminate. The heat is liberated due to the exothermic nature of the system. So each curve in figures 4.8(a) and 4.9(a) show extent of heat evolutions and degree of cure profiles as the curing of the laminate progresses along the thicknesses (normalized lengths) of the laminates for various percentage weight fractions of the reinforcements 20 and 10 % respectively.

According to figures 4.7(a) and (b), 30% aluminum-polyester composite, when the curing process was about 0.6 along the thickness (normalized), the temperature evolution was 485K and the degree of cure was slightly above 0.8. As the curing progresses, more heat is liberated due to more cross-linking of the composite, resulting to higher heat evolution. At the final stage for 30% aluminum-polyester composite, the composite is assumed to be fully cured with ultimate cure value of 0.99 with temperature evolution of 512K at a final processing time of 0.667hr (40 minutes).

Figures 4.8- 4.9 show the temperature evolution and degree of cure profiles for both 20% aluminum-polyester and 10% aluminum-polyester composites respectively. The profiles show similar but slightly different trends to that of 30% aluminum-polyester composite. For

20% aluminum-polyester composite, the temperature evolution at final processing time of 0.556hr (33 minutes, 33sec.) is 505K at ultimate cure of 0.994. While for 10% aluminum-polyester composite, the temperature evolution at final processing time of 0.561hr (approx. 33 minutes) is also 505K at ultimate cure of 0.997.

It was generally observed that there was an increasing trend in the final processing times of the aluminum-polyester composites as the percentage composition of the reinforcement increases from 10% - 30%. This is supported by [1] and [5]. The matrix is majorly responsible for curing reaction. Therefore, higher composition of the matrix increases the rate of cross linking, thus speed up the processing time. This is finally impacted on the ultimate degree of cure of the composites as there was an observation of decreasing ultimate cure from 10% - 30% aluminum-polyester composites

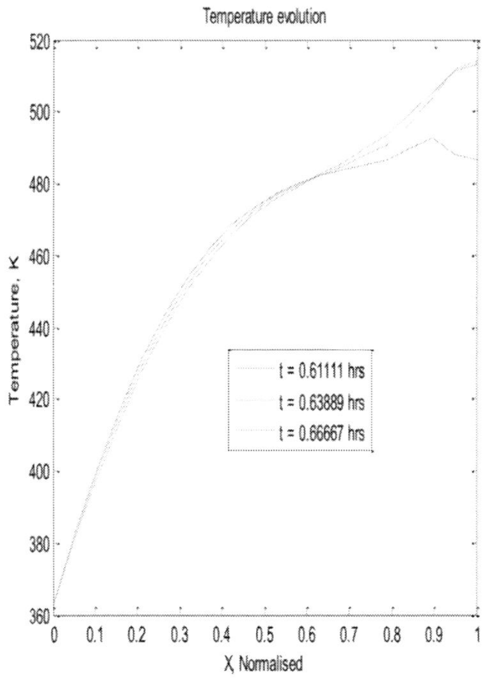

Figure 4.7(a) Temperature Evolution Profile for 30% wt. Aluminum-Polyester Composite

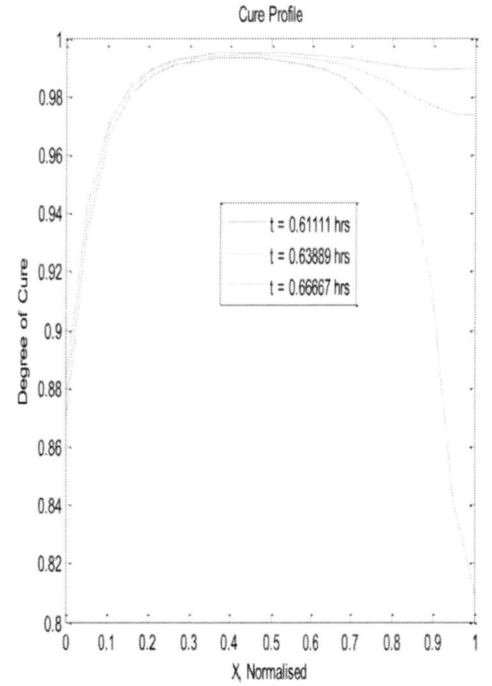

Figure 4.7(b) Degree of Cure Profile for 30% wt. Aluminum-Polyester Composite

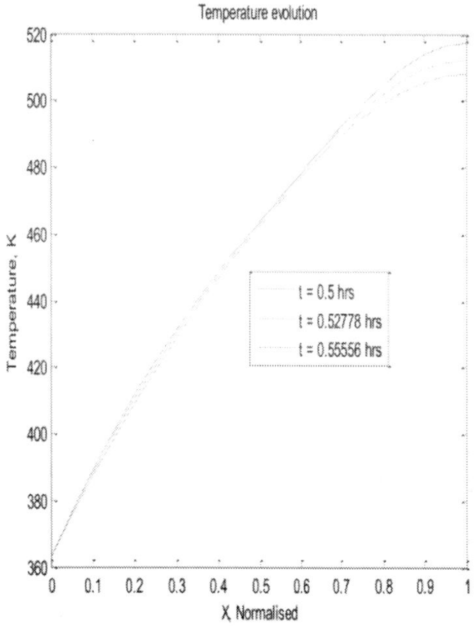

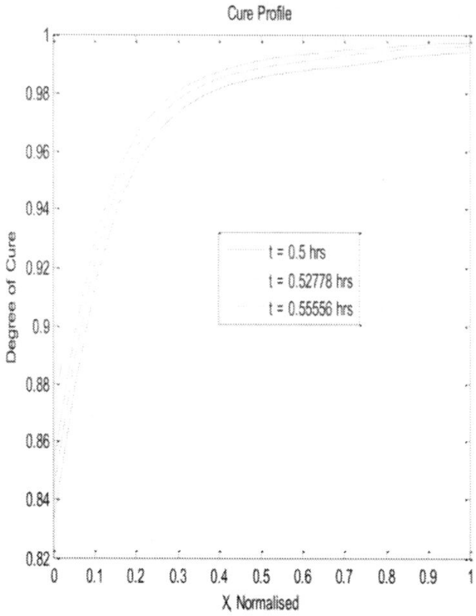

Figure 4.8(a) Temperature Evolution Profile for 20% wt. Aluminum-Polyester Composite

Figure 4.8(b) Degree of Cure Profile for 20% wt. Aluminum-Polyester Composite

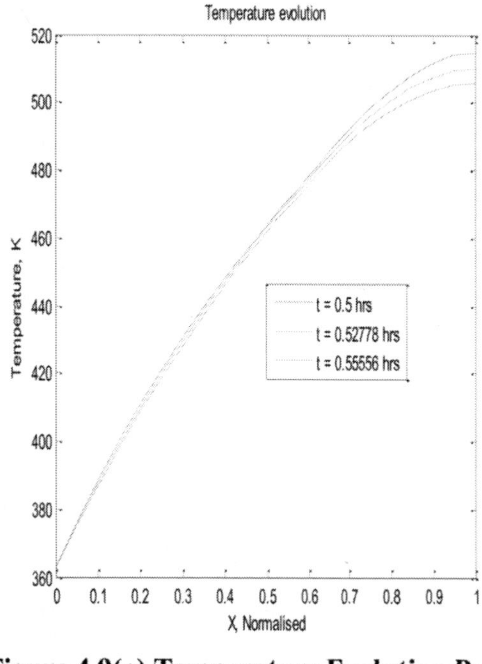

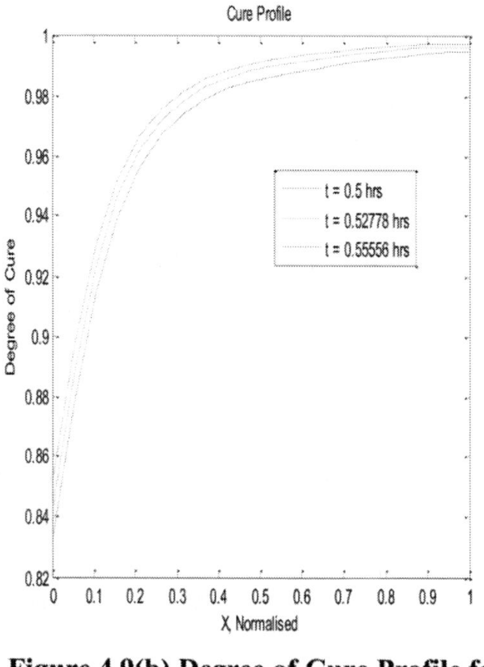

Figure 4.9(a) Temperature Evolution Profile for 10% wt. Aluminum-Polyester Composite

Figure 4.9(b) Degree of Cure Profile for 10% wt. Aluminum-Polyester Composite

Table 4.1 Summary of Temperature Evolution and Degree of Cure Profile for Aluminum-Polyester Composite

System	Processing Time (Hr)			Temperature Evolution (K)			Degree of Cure		
	30% wt.	20% wt.	10% wt.	30% wt.	20% wt.	10% wt.	30% wt.	20% wt.	10% wt.
Aluminum-Polyester Composite	0.611	0.5	0.5	485	515	515		0.992	0.995
	0.639	0.528	0.531	512	510	510	0.97	0.993	0.997
	0.667	0.556	0.561	512	505	505	0.99	0.994	0.997
Mean	0.639	0.528	0.531	503	510	510	0.98	0.99	1.00
Standard Deviation	0.028	0.028	0.031	15.59	5	5	0.014	0.001	0.001

Figures 4.10 - 4.12 show likewise the temperature evolution and degree of cure profiles for carbon black-polyester composite. The curves in the figures 4.10(a), 4.11(a) and 4.12(a) show the progression of heat evolution as the curing progresses along the normalized thickness of the laminate for varying percentage weight fractions of the reinforcement 30, 20 and 10 % respectively.

Figures 4.10(a) and (b) shows the temperature evolution and degree of cure profiles during the curing of 30% carbon black-polyester composite. In figure 4.10(a), there was a sharp decline in temperature evolution at around 0.75 along the normalized thickness of the laminate which corresponds to the temperature of 510K. This is likely to correspond to the region of glass transition temperature of the polymer, where the polymer suddenly changes its state from glass like (hard and brittle) to molten or rubber like material.

As the curing progresses, the cross linking is initiated with more heat evolution. The cross linking finally reached the peak at 585K. The composite is assumed to be fully cured at this stage with ultimate cure value of 0.999 at 0.389hr (23 minutes, 34 seconds) processing time.

Figures 4.11 – 4.12 show the temperature evolutions and degree of cure profiles for both 20% carbon black-polyester and 10% carbon black-polyester composites respectively. The profiles show much more similarity in trends which were more different from that of 30% carbon black-polyester composite. Both composites indicated very close proximities in the temperature evolution trend with final temperature evolution of 595K and at processing time of 0.388hr (23 minutes, 28seconds). The ultimate cure is approximately 1(full cross linking) at the period.

The closeness in the proximities of the parameters (temperature evolution and processing time) despite varying the percentage weight fractions of the reinforcements (aluminum and carbon black) may be due to degree of homogeneity composites as determined by [13]

Comparing the variations in the production process parameters in both aluminum-polyester composites and carbon black-polyester composites, carbon black-polyester composite shows higher differences in temperature evolution trend than that of aluminum-polyester composite with respect to percentage weight fraction of reinforcement (30%, 20% and 10%). The temperature evolution for carbon black-polyester composites ranges from 585K-595K, while that of aluminum-polyester composites ranges from 505K-512K. This shows that carbon black-polyester composite is more exothermic than aluminum-polyester composite as reflected in the mathematical modeling of the process and experimentation conducted [5]. The rate of heat evolution shows the extent of chemical reaction taking place (cross linking), which invariably contributed to shorter processing time of carbon black-polyester composite (0.388-0.391hr) as compared to aluminum-polyester composite processing time ranging from 0.561-0.667h

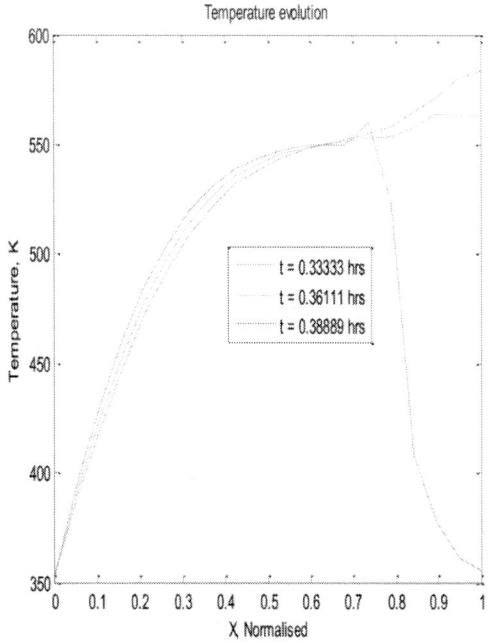

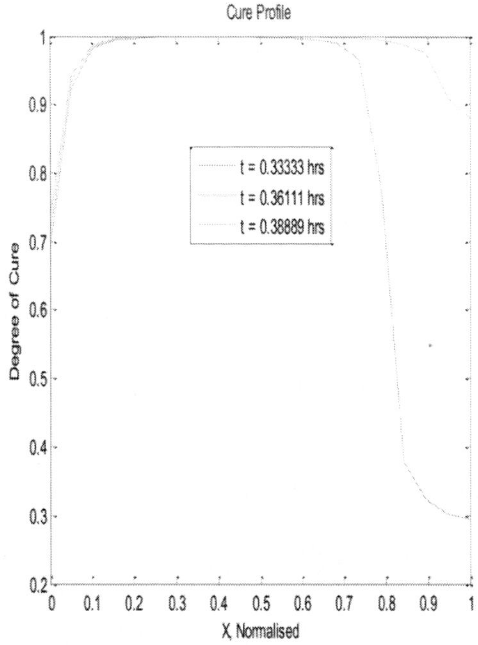

Figure 4.10(a) Temperature Evolution Profile For 30% wt. Carbon black-Polyester Composite.

Figure 4.10(b) Degree of Cure Profile for 30% wt. Carbon black-Polyester Composite

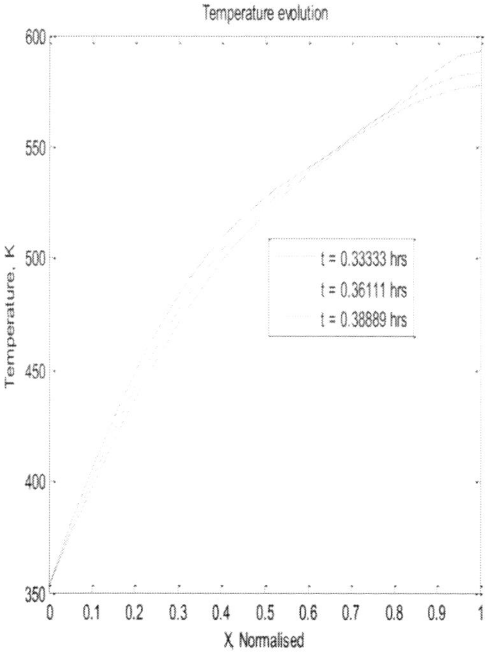

Figure 4.11(a) Temperature Evolution Profile for 20% wt. Carbon black-Polyester Composite.

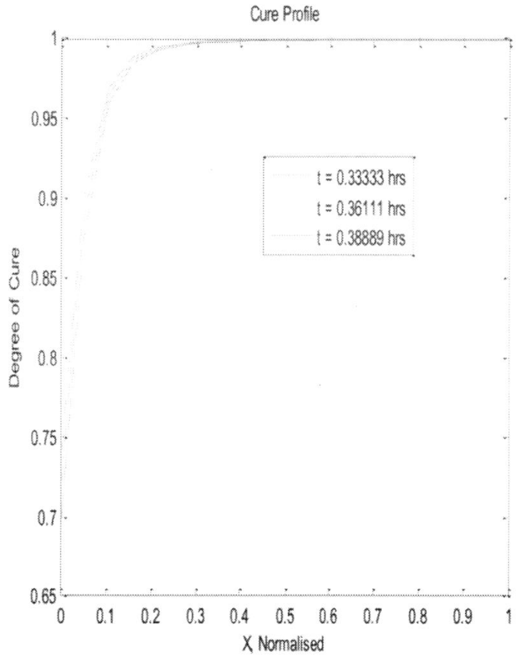

Figure 4.11(b) Degree of Cure Profile for 20% wt. Carbon black-Polyester Composite

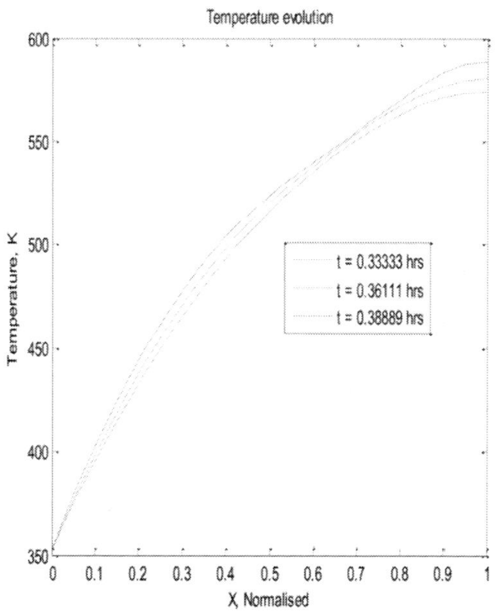

Figure 4.12(a) Temperature Evolution Profile for 10% wt. Carbon black-Polyester Composite.

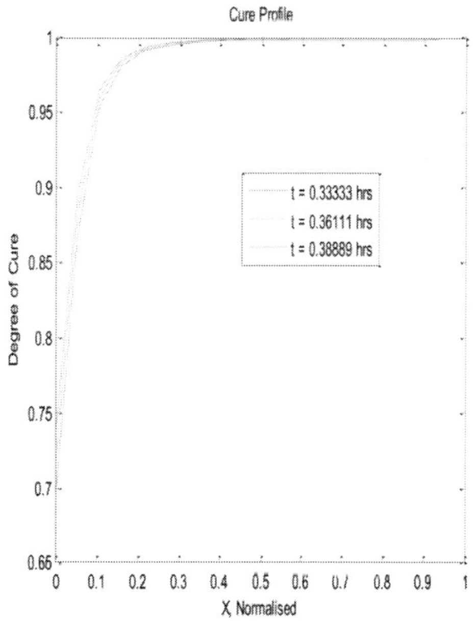

Figure 4.12(b) Degree of Cure Profile for 10% wt. Carbon black-Polyester Composite

Table 4.2 Summary of Temperature Evolution and Degree of Cure Profile for Carbon black-Polyester Composite

System	Processing Time (Hr)			Temperature Evolution (K)			Degree of Cure		
	30% wt.	20% wt.	10% wt.	30% wt.	20% wt.	10% wt.	30% wt.	20% wt.	10% wt.
Carbon black - Polyester Composite	0.333	0.333	0.333	560	595	590	-	1.00	0.99
	0.361	0.361	0.361	562	585	585	0.89	1.00	1.00
	0.389	0.388	0.391	580	575	575	0.99	1.00	1.00
Mean	0.361	0.361	0.362	567	585	583	0.94	1.00	1.00
Standard Deviation	0.028	0.028	0.029	11	10	8	0.07	0.00	0.01

Table 4.3 Optimum Temperature Evolution and Degree of Cure of Composites

System	Processing Time (Hr)			Temperature Evolution (K)			Degree of Cure		
	30% wt.	20% wt.	10% wt.	30% wt.	20% wt.	10% wt.	30% wt.	20% wt.	10% wt.
Aluminum-Polyester Composite	0.639	0.528	0.531	503	510	510	0.98	0.99	0.995
Carbon black - Polyester Composite	0.361	0.361	0.362	567	585	583	0.94	1.00	1.00

Table 4.4(a) Temperature Profile Result of Optimised Aluminum-Polyester Composite

30% Aluminum – Polyester Composite at 503K		20% Aluminum – Polyester Composite at 510K		10% Aluminum – Polyester Composite at 510K	
Time (min)	Temperature (^{0}C)	Time (min)	Temperature (^{0}C)	Time (min)	Temperature (^{0}C)
5	30	5	30	5	32
10	55	10	50	10	40
15	70	15	68	15	84
20	90	20	90	20	90
25	92	25	90	25	92
30	92	30	90	30	92

Table 4.4(b) Temperature Profile Result of Optimised Carbon black-Polyester Composite

30% Carbon black – Polyester Composite at 567K		20% Carbon black – Polyester Composite at 585K		10% Carbon black– Polyester Composite at 583K	
Time (min)	Temperature (^{0}C)	Time (min)	Temperature (^{0}C)	Time (min)	Temperature (^{0}C)
5	45	5	45	5	50
10	59	10	60	10	63
15	80	15	80	15	80
20	82	20	80	20	80

Discussion of Results of Cure Temperature Profile

Table 4.4(a) shows the cure temperature profile result of autoclave for varying weight fraction of optimised aluminium-polyester composite. This shows the time response of the varying weight fractions of aluminum-polyester composite to their respective optimum temperature evolutions. 30 % aluminum-polyester composite was cured at 503 K, while 20 % and 10 % aluminum-polyester composites were both cured at 510 K.

Also table 4.4(b) shows the cure temperature profile result of autoclave system for varying weight fractions of carbon black-polyester composites. It shows the time responses of the varying weight fractions of the composites to their respective optimum temperature evolutions. 30 % carbon black-polyester composite was cured at 567 K, 20 % carbon black-polyester composite was cured at 585 K, while 10% carbon black-polyester composite was cured at 583 K.

Each of the composites was cured separately and simultaneously one after the other, with the operating temperature of the oven set at each optimum temperature evolutions of each composites. The temperature readings were taken at every interval of 5 minutes to determine the ramp rate. These processes were carried out to ascertain that the cure temperatures of the composites were achieved at a reduced time corresponding to the minimized time without compromising the initial quality that has been established.

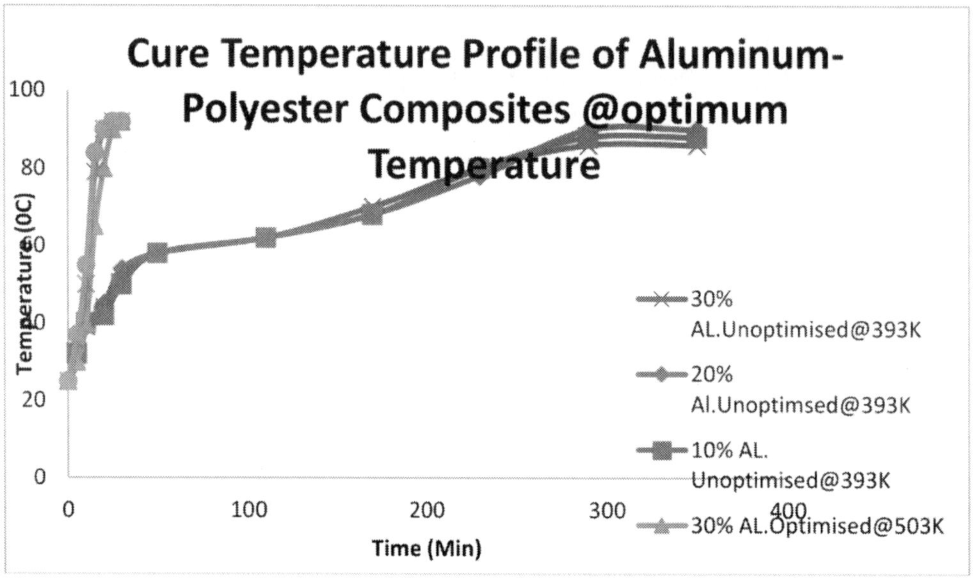

Figure 4.13(a) Optimum Cure Temperature Profile for Aluminum-Polyester Composites

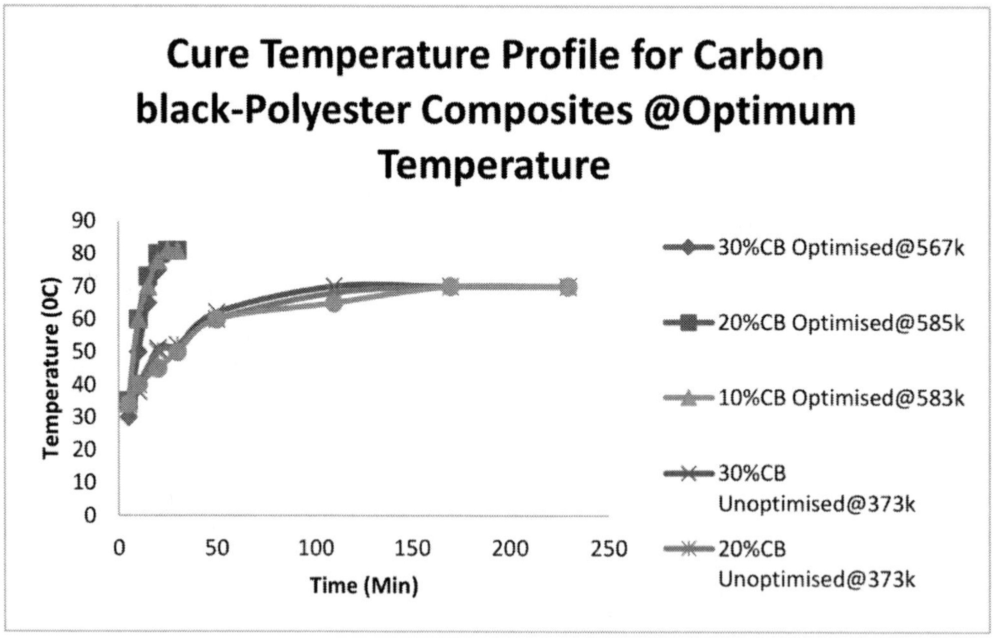

Figure 4.13(b) Optimum Cure Temperature Profile for Carbon black-Polyester Composites

Figures 4.13(a) and (b) show the cure temperature curves for varying weight fractions of both aluminum-polyester and carbon black-polyester composites at their respective optimum temperature evolutions.

The cure plots actually show that aluminium-polyester composites attained cure temperature of 90 °C within 20-25 minutes, while carbon black-polyester composites attained the cure temperature of 80 °C within 15-20 minutes. The results obtained for both aluminium-polyester and carbon black-polyester composites correspond to the results obtained in the computer simulated optimisation process in which aluminum-polyester cured within 30-36 minutes and carbon black-polyester cured within 20 minutes.

The summary of the processing time (cure cycle time) of experimental validation and the process optimisation scheme for both composites were presented in table 4.5

Table 4.5 Summary of Composites cure cycle

Composite System	Computer Simulated Conventional Autoclave Optimisation Scheme Cycle time (min)	Experimental Validated Conventional Autoclave Optimisation Scheme Cycle time (min)	Experimental Validated Conventional Autoclave Unoptimised Scheme Cycle time (min)
Aluminum-Polyester Composite	30-36	20-25	230
Carbon black-Polyester Composite	20	15-20	170

Results and Discussion of Tensile and Micro-structural Properties

Table 4.6(a) Maximum Force and Tensile Strength of Optimised Aluminum Composite Curing

Percentage weight fraction of samples	Cross-sectional area of samples (mm^2)	Maximum force (KN)	Tensile strength (MPa)
30% Aluminum-Polyester composite at 503k	2500	30	0.012
20% Aluminum-Polyester Composite at 510K	2500	28	0.011
10% Aluminum-Polyester Composite at 510K	2500	24	0.0096

Table 4.6(b) Maximum Force and Tensile Strength of Optimised Carbon black Composite Curing

Percentage weight fraction of samples	Cross-sectional area of samples (mm^2)	Maximum force (KN)	Tensile strength (MPa)
30% Carbon black-Polyester composite at 567k	2500	25	0.01
20% Carbon black-Polyester Composite at 585K	2500	19	0.0076
10% Carbon black-Polyester Composite at 583K	2500	17	0.0068

Tables 4.6(a) and (b) present maximum force and tensile strength for varying weight fractions of optimised conventional autoclave curing of the composites. The tables show similar trend of progression of direct relationship between maximum force and tensile strength ranging from 10% weight fraction to 30%

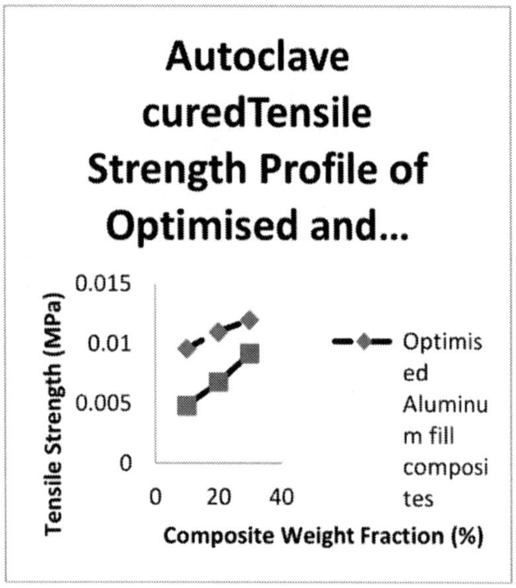

Figure 4.14(a) Tensile Strength curve of optimised and unoptimised aluminum-polyester composite.

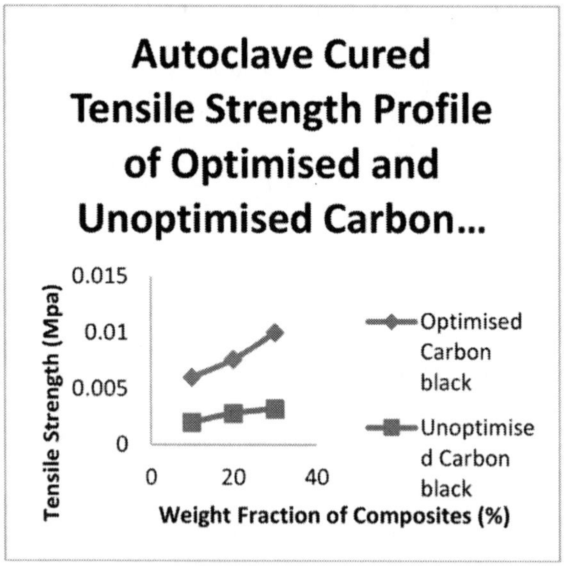

Figure 4.14(b) Tensile Strength curve of optimised and unoptimised Carbon black-polyester composite

Figures 4.14(a) and (b) compared the autoclave cured tensile strength profiles of optimised and unoptimised composites. There was observed an improvement in the tensile property of the optimised composites compared to the unoptimised composites. For instant, 30% optimised aluminium-polyester composite has 0.012 MPa while that of unoptimised 30% aluminium-polyester composite has 0.0092 MPa. Also, 20% optimised aluminium-polyester composite has 0.011MPa and the unoptimised counterpart has 0.0068MPa.

Similar observation was observed in the case of carbon black-polyester composites. 20 and 10 % optimised carbon black-polyester composites have tensile strengths of 0.0076 MPa and 0.0068 MPa respectively, while their unoptimised counterparts have 0.0028 MPa and 0.0020 MPa respectively. The results agree with the statements from the literatures by [5] [14] [15] [16] [17] [18] that bonding level at reinforcement-matrix interface increases as temperature increases. Therefore, curing of the composites at elevated temperatures increased the interfacial bonding of the optimised composites.

Comparison of Microstructure properties between Optimised Cured Composites and Unoptimised Cured Composites

Figures 4.15 and 4.18 show the micro-structural scans of conventional autoclave cured optimised and unoptimised composite samples. These figures were examined with Olympus BX41M Optical Microscope with magnification 300x. They show effect of elevated and normal cure temperature on the microstructure properties in term of voids in aluminium-polyester and carbon black-polyester composites. Voids are caused by trapped hydroscopic water and water vapour bubble within the parts during lamination [19] [20] [21]

Figure 4.15(a) Micrograph of Optimised 20% Carbon black-Polyester Composite

Figure 4.15(b) Micrograph of Optimised 30% Carbon black-Polyester Composite

Figure 4.16(a)Micrograph of Unoptimised 20% Carbon black-Polyester Composite

Figure 4.16(b)Micrograph of Unoptimised 30% Carbon black-Polyester Composite

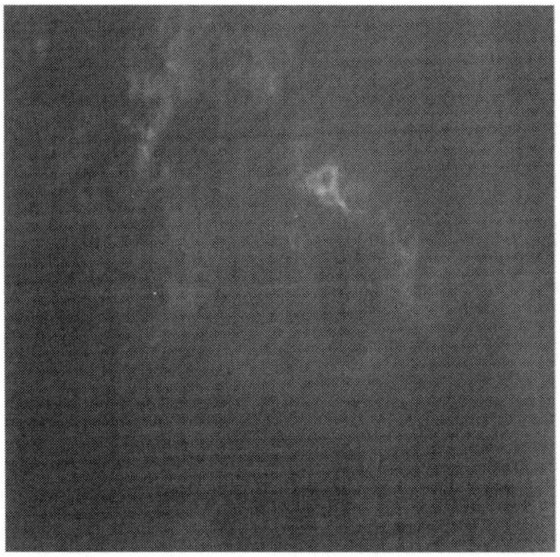

Figure 4.17(a) Micrograph of Optimised 20% Aluminum-Polyester Composite

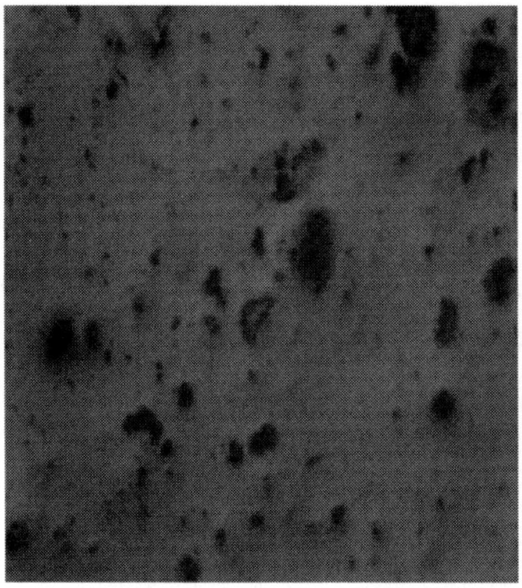

Figure 4.17(b) Micrograph of Optimised 30% Aluminum-Polyester Composite

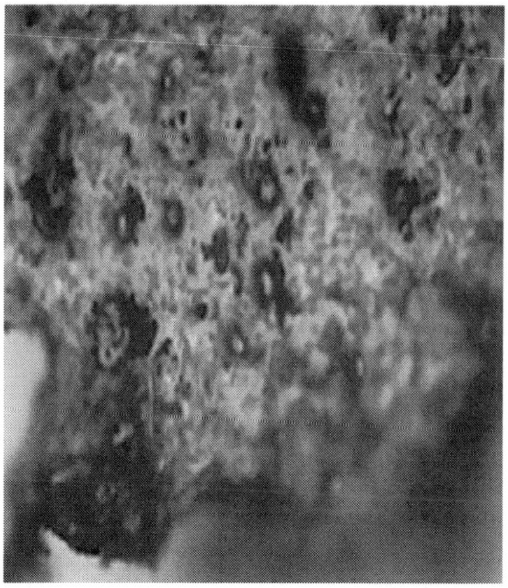

Figure 4.18(a) Micrograph of Unoptimised 20% Aluminum-Polyester Composite **Figure 4.18(a) Micrograph of Unoptimised 30% Aluminum-Polyester Composite**

In figures 4.15 and 4.16, it was observed that the unoptimised cured samples have more voids more than the optimsed cured sample. 20 and 30% unoptmised carbon black-polyester composite had 35 voids per sq μm with mean void size of 156 μm and 30 voids per sq μm with mean void size of 337 μm respectively, while 20 and 30% optmised carbon black-polyester composites had 5 voids per sq μm with mean void size of 52.3 μm and 3 voids per sq μm with mean void size of 23.6 μm respectively.

In figures 4.17 and 4.18, 20 and 30% unoptmised aluminum-polyester composites had 30 voids per sq μm with mean void size of 191 μm and 25 voids per sq μm with mean void size of 206 μm respectively, while 20 and 30% optmised aluminum-polyester composites had 1void per sq μm with mean void size of 12 μm and 5 voids per sq μm with mean void size of 25.2 μm respectively.This could be traced to an improvement made to the statement made by Philips [2], that some polymers undergo condensation type chemical reactions during curing which produce volatile reactants. The higher exterior temperature of conventional heating technique tends to advance curing at the surface of the parts, sometimes forming a hard impermeable skin on the part surface which can trap the volatiles given off during the condensation type reaction. Trapped volatiles usually result in excessive voids and delaminations, essentially rendering the finished part useless. Curing at an elevated temperature than normal shortens the cure cycle by speeding up the condensation reaction thereby reducing the amount of volatile given off. Therefore, there is an improvement in the quality of the optimised cured samples compared to unoptimised cured samples.

Statistical Analysis between Optimised and Unoptimised Process

The tensile strength and micrographs of aluminium-polyester and carbon black-polyester composites from the optimised and unoptimised processes were compared using T-test at α=.05. The T-test $P(T \leq t)$ for aluminium-polyester and carbon black-polyester are 0.0109 and 0.0318 respectively with α value of .05. These show that levels of variation between optimised and unoptimised samples are significant.

5.1 CONCLUSIONS

1. Autoclave optimisation of aluminum-polyester composite achieved optimum cure of 99% at minimized cure cycle of 33 minutes compared to the experimental investigation with ultimate cure of 85% at cure cycle of 230 minutes
2. Autoclave optimisation of carbon black-polyester composite achieved optimum cure of approximately 100% at minimized cure cycle time of 23 minutes as compared to the experimental investigation with ultimate cure of 80% at cure cycle of 170 minutes
3. Autoclave optimisation improved the tensile strength of aluminium-polyester and carbon black-polyester composites by 61 and 68 % respectively, compared to the unoptimised autoclave curing
4. There was an improvement in the microstructures of optimised aluminium-polyester and carbon black-polyester by 88 and 93 % respectively, compared to the unoptimised samples
5. Application of Genetic Algorithm as tool for the optimisation of production process was feasible as minimization of cure cycle for the polymer matrix composites were achieved subject to the set constraints

5.2 RECOMMENDATIONS

1. Further research on electromagnetic curing of composites to solve challenges of thermal runaway and composite burn-out is recommended
2. Cost implication on the optimisation of the production process for the polymer matrix composites is also recommended for further study

REFERENCES

1. Min Li, Qi Zhu, Philippe H. Geubelle and Charles L. Tucker, (2001). Optimal Curing for Thermoset Matrix Composites: Thermochemical Consideratons. Polymer Composites. Vol. 22, No. 1, pp 118 – 131
2. Philip C. Sturman and Rexford N. Y. 1999. Induction Heating of Polymer Matrix Composite Fibre Strands. SAMPE Journal, Vol. 26, No 4.
3. Ku, H. S., Siores E and Ball, J. A. R. 2011. Productivity Improvement Through the use of Industrial Microwave Technology, Proceeding of the International Conference on Composite Material Productivity Improvement, University of South Queensland, Australia
4. Giacomo S., 2014. Optimisation of the VARTM process. A Ph. D. Thesis, School of Aerospace, Transport and Manufacturing, Cranfield University, United Kingdom.
5. Adefemi O. Adeodu, Christopher O. Anyaeche, Oluleke O. Oluwole, Charles U. Omohimoria (2015). Effect of Microwave and Conventional Heating on the Cure Cycles of Particulate Reinforced Polymer Matrix Composites. International Journal of Materials Science and Applications. Vol. 4, No. 4, pp 229-240, doi: 10.11648/j.ijmsa 20150404.12
6. Pillai V., Beris A. N and Dhurjati P., (1997). Journal of Composites Material. 31, 22
7. Udaykumar, H. S. and Shyy, W., 1993. Optimal Cure Cycles for Thermoset Composites Manufacture. 29th National Heat Transfer Conference, Vol. 241, 8-11 August, Atlanta, US, pp. 23.
8. Rai, N. and Pitchumani, R., 1997. Optimal Cure Cycles for the Fabrication of Thermosetting-Matrix Composites. Polymer Composites, vol. 18, no. 4, pp. 566-581.
9. Ciriscioli, P. R., Springer, G. S. and Lee, W. I., 1991. Expert System for Autoclave Curing of Composites. Journal of Composite Materials, vol. 25, no. 12, pp. 1542-1587.

10. White S. R. and Hahn H. T., (1992). Journal of Composites Material. 26, 2423
11. Kim C., Teng H., Tucker C. L. and White S. R. (1995). Journal Composites Material. 29, 1222.
12. Pillai, V., Beris, A. N. and Dhurjati, P., 1996. Heuristics Guided Optimization of a Batch Autoclave Curing process. Computers and Chemical Engineering, vol. 20, no. 3, pp. 275-294.
13. Halpin, J. C., Kardos, J. L., and Dudukovic M. P., 1983. Pure & Appl. Chem. 55, 893-906.
14. Chawla Nikhilesh and Yu Linshen (2001). Mechanical Behavior of Particle Reinforced Metal Matrix Composites. Advanced Engineering Material Vol. 6, page 1438-1656
15. Shao-Yunfu, Xi-Qiao Feng, Bernd Lauke, Yiu-wing Mail (2008). Effect of Particle Size Particle-Matrix Interface Adhesion and Particle Loading on Mechanical Properties of Particulate-Polymer Composites. Science Direct. Composites: part B, 39 Page 933-961
16. Hassan S. B., Oghenevweta E. J., Aigbodion V. S. (2012). Potential of Maize Stalk Ash as Reinforcement in Polyester Composites. Journal of Minerals Materials Characterization and Engineering. Vol. 11, No 4, Page 445-459
17. Asueke F., Aigbodion V. S., Abdulwahab M., Fayomi O. S.I., Popoola A. P.I., Nwoyi C. I., Garba B. (2012). Effect of Bone Particle on the Properties and Microstructure of Polypropylene-Bone Ash Particulate Composites. Results in Physics 2 Page 135-141
18. Hossain M. R., Hossain M. F., Islam M. A. 2014. Effect of Wood Properties on the Behaviors of Wood Particle Reinforced Polymer Matrix Composites. J. Sci. Res 6(3) Page 431-443
19. Islam R. A and Chan Y. C. (2003). Effect of Microwave Preheating on the Bonding Performance of Flip chip on Flex Joint. Micro electronics Reliability. Vol. 44, pp 815-821
20. Nikki Sgriccia and Hawley M. C. (2007). Thermal Morphological and Electrical Characterization of Microwave Processed Natural Fibre Composites. Composites Science and Technology. Vol. 67, pp 1986-1991
21. Yizhuo Gu, Min Li, Zuoguang Zhang and Zhijie Sun (2012). Measuring Methods and Process Analysis of Void Formation Conditions for Resin Matrix Composites. Laboratory of Aerospace Materials and Performance. Beijing 100191, China

508233 Breakage of Frac and Tailing Sands in Centrifugal Pumps

Abstract:

A natural proppant for hydraulic fracturing, known as frac sand, is currently widely used as a substitute for manufactured proppants, which are considerably more expensive to produce. However, frac sand is significantly more costly than, for example, speciality or construction sand, as its preparation involves washing, classification, and dewatering to meet the required specifications. The mineral industry is well-versed in unit operations such as solids washing, size classification, and solids separation, all of which may produce unwanted solids tailings. Some of these tailings could be easily converted to frac sand. In this work, we compare particle size distributions (PSD) and the degree of particle breakage of frac and tailings sand - the latter is currently backfilled at a high cost in tailings ponds. Optical image analysis was used to measure PSDs of the two types of sand before and after slurry circulation by a centrifugal pump. The difference between the cumulative particle size distributions, before and after the centrifugal pump circulation, was used as a measure of particle strength. Accurate PSD measurements were further analysed to extract very detailed information about particle breakage behaviour. Critical effects of the slurry solids concentration and slurry residence time on breakage behaviour of sand were assessed.

Particle Measurement in High-Temperature Gas based on Light Scattering

Li-feng LU[1,3], Xiao-lin WU [1,3], Zhong-li JI [1,2], Ming-xing WANG[1,3]

[1] Department of Chemical Engineering, China University of Petroleum, Beijing, P. R. China

[2] Department of Mechanical and Transportation Engineering, China University of Petroleum, Beijing, P. R. China

[3] Beijing Key Laboratory of Process Fluid Filtration and Separation, China University of Petroleum, Beijing, P. R. China

Abstract: In order to achieve the particle measurement at 650 °C, a single particle measurement method and device suitable for high-temperature gas were developed. The influence of gas temperature on optical sensor of optical particle counter (OPC) was investigated in detail based on geometric optics theory and heat transfer theory. The thermal expansion of aerosol tube along the airflow direction was caused by the heat conduction between the tube and the gas, resulting in the position of the light apertures to shift to obscure the partial light beam. The invar material can be adopted to eliminate the above mentioned disadvantages. According to the principle of gas state equation, the optical parameters (refractive index) of the gas at different temperatures change with the change of its state parameters. Based on the geometrical optics theory and the geometric mathematical relationships, the theoretical and experimental dynamic models of the optical measuring volume with the change of the gas temperature were established. The established models were then used to analyze the influence of gas temperature on the optical measuring volume. The results showed that the influence of the temperature on the optical system is negligible. Meanwhile, particle detection experiments were conducted at different temperature to achieve a good performance of OPC.

Keywords: High-temperature gas; Optical measuring volume; Dynamic model

1 Introduction

With the continuous improvement of environmental protection and the safety requirements for industrial production, the technology of high-temperature purification is drawing more and more attention, especially in the fields of clean coal power generation, coal gasification, petrochemical industry and metallurgy [1]. For example, nuclear power plants often produce a large number of low radioactive solid wastes, and the combustible wastes usually account for a considerable proportion. The incineration and pyrolysis are effective treatment techniques for the combustible waste, during which the temperature of flue gas is 600~800 °C. With the aim to purify the flue gas, the German Karlsruhe Nuclear Research Institute, the French nuclear power station and the Japanese nuclear power station have installed the silicon carbide ceramic filter tube, aluminum silicate fiber and alumina ceramic filter tube at gas temperatures above 650 °C, respectively, and investigated their properties properties [2-4]. These filters have a direct influence on the purification quality of gas and the particle concentration in the downstream. In an industrial cracker system, the particle concentration is required to reach 0.014 mg/m^3 after the filters; the Integrated Gasification Combined Cycle (IGCC) requires the particle size to be smaller than 1 um with a particle concentration of less than 2.2 mg/m^3 and in the range of 1~10 um with less than 5.8 mg/m^3, and to remove the dust size of larger than 10 um [5]. In a Shell Coal Gasification Process (SCGP), the particle concentration is required to be less than 20 mg/m^3 [6]. In order to detect the concentration and size of the particles in pipes in real time for calculating the separation efficiency of evaluating the purification performance of the filtration system, the particle detection technology is needed to be improved [7].

The widely-used particle detection devices in industry are mainly based on the principle of light scattering [8-10], and the particle size and concentration are obtained through an inversion of the scattering light intensity and particle size, which are suitable for the conditions of below 200 °C. The method of first cooling and then detecting can be adopted for the particle detection under higher-temperature conditions, but droplet condensation and sublimation will inevitably occur and result in an incorrect particle samples.

With the aim to achieve the particle detection at 650 °C, an optimization was conducted on the existing particle detector (OPC) in this paper. And based on geometric optics theory [11] and heat transfer theory, a detailed research was performed on the influence of gas temperature on OPC optical sensor, including the dynamic change of the aerosol tube and the optical measuring volume (OMV). Finally, the experiments were conducted on the on-line particles measurement of different temperatures.

2 Experimental setup and method

Figure 1 gives the experimental setup for detecting particles in high-temperature gas, mainly including the heating unit, optical detection unit and gas purification and processing unit.

The heating unit consists of the vibration particle feeder and the air heater. The gas brings the particle out of the feeder, and is heated up to 1000 °C by the heater, which is monitored quickly by a PID temperature control instrument with easy operation.

The optical detection unit is used to achieve the experimental measurement of the characteristics of the OPC optical system and the on-line detection of particles, including the optical sensor and the OPC control host. The optical sensor is shown in Figure 2, which is composed of an aerosol tube and optical system. Aiming to avoid the influence of high-temperature gas on optical system, the aerosol tube (shown in Figure 3) has been developed, and is in a distance of 15 mm to the optical lens. In addition, the insulation coating with lower thermal conductivity (-80-1700 °C) is adopted to treat the outer wall of the aerosol tube. A K-type micro thermocouple was bent and installed into the light aperture to monitor the gas temperature at the center of the aerosol tube in real time without affecting the airflow and optical paths. Figure 4 shows the optical system of the sensor, which is composed of an incident light path and an observed scattering light path. Both the paths are arranged in 90° with the same optical lens and elements [12-14], and the intersected area is the center of the aerosol tube to form an OMV [15]. The intersected area is enough small and allows particles to pass through it one by one [16, 17]. The OPC host consists of the light source system (wavelength: 300-700 nm) [18], the photoelectric detection system and the signal analysis and processing system. In the particle detection, particle is illuminated by the incident light when passing through the OMV, and the scattering light is collected and transmitted to the photoelectric detector to achieve the conversion of a pulse voltage signal from an optical signal. Finally, the tested particle size is calculated based on the calibration curve of particle size and voltage amplitude [14, 19].

The gas purification and processing unit is composed of a vacuum pump, a flow controller, a dust collector and a heat exchanger. The high-temperature gas form the aerosol tube is completely cooled during the heat exchanger, the flow rate of which is adjusted by the flow controller. In order to avoid the damage of particles on the pump, the dust is captured and collected by a filter in the dust collector.

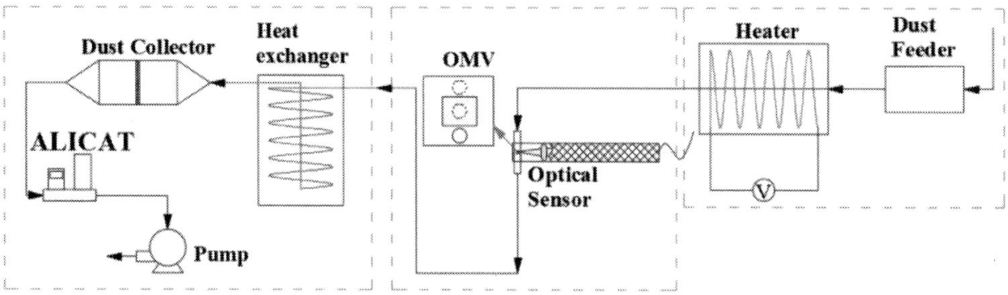

Figure 1. Experimental setup for detecting particles in high-temperature gas

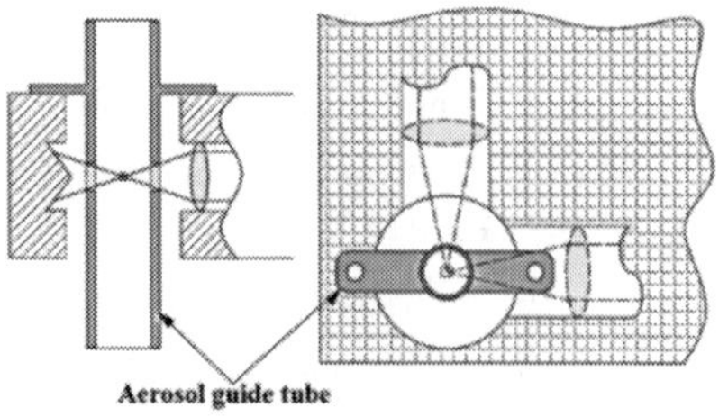

Figure 2. Optical sensor

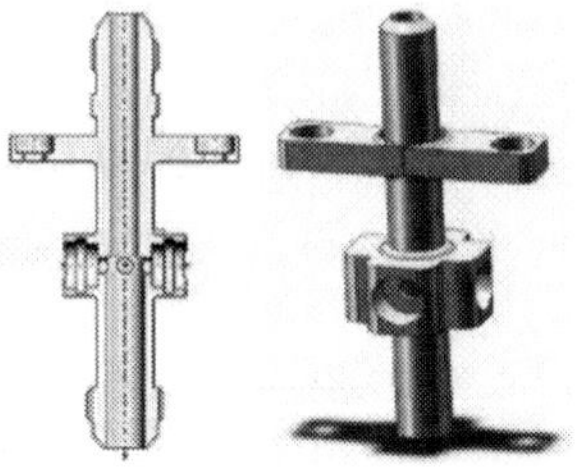

Figure 3. Aerosol tube for high-temperature gas

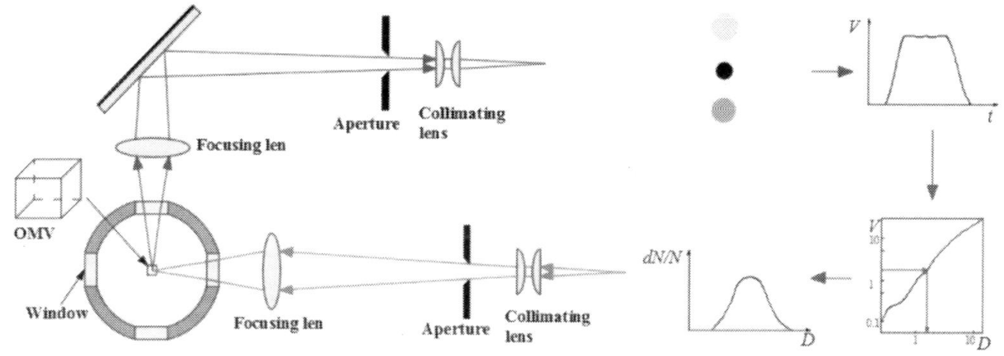

Figure 4. Optical system and principle of OPC

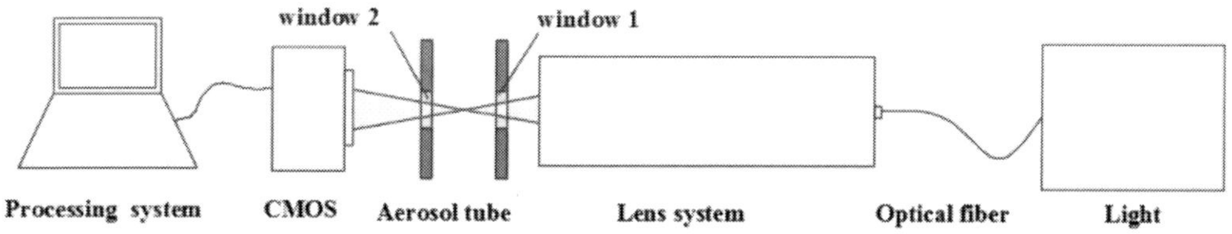

Figure 5. Evaluation and analysis device for light beam

3 Analysis and discussion

3.1 Influence of temperature on aerosol tube

The aerosol tube shown in Figure 3 has four apertures along the center axis, in which the glasses with a thickness of 2 mm were installed, as shown in Figure 4. When the gas flows in the tube, it can reduce the damage of high-temperature gas to the optical system elements. But the heat resistance of the tube has a direct influence on the imaging of the optical system at the center of the aerosol tube.

Figure 5 gives an evaluation and analysis device for light beam, including the light source system, the light path system, the aerosol tube and the light beam analyzer (CINOGY CMOS-1201, Germany), which can be equipped at the optical testing unit as shown Figure 1. For the gases at different temperatures, the light passes through the glasses in the aerosol tube and images on the CCD of CMOS-1201 to detect the spot of the light beam in real time.

After eliminating the background noise, the spot image at different temperatures can be acquired. When the aerosol tube is made of 304 stainless steel, the measurement results are shown in Figure 6. It can be found that the shape of the light spots is distorted and shows the imperfection of the imaging with an increase in gas temperature. That's due to thermal expansion along the airflow direction caused by the change of gas temperature, which results in the shift of light apertures position to obscure the partial light beam. A bad result must be acquired for smaller particle size and lower particle count efficiency because of the decreased incident light. Meanwhile, the shape of the OMV was changed to greatly reduce the performance of OPC detection by a failure of eliminating the error of edge detection in optical system.

With the aim to eliminate the above mentioned disadvantage, the invar material with a linear expansion coefficient of 1.6×10^{-6} m/°C was adopted for high-temperature experiment. The results are shown in Figure 7. At different temperatures, no deformation was found in the shape of light spots, indicating that the material can meet the testing requirement for high temperature up to 650 °C with a good heat resistance.

Table 1. Comparison of the application conditions of the two materials

Material	Light divergence angle /°	Gas temperature /°C
304 (Stainless steel)	<10	<200
4J36 (Invar)	<30	<700

The application conditions of the two materials are

summarized in Table 1. Compared with the stainless steel, the invar material is of great significance for the development of high-temperature detection device.

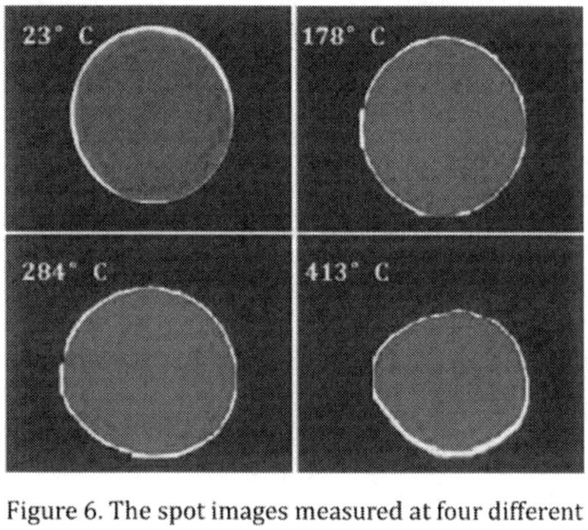

Figure 6. The spot images measured at four different temperatures of gas for 304 stainless steel.

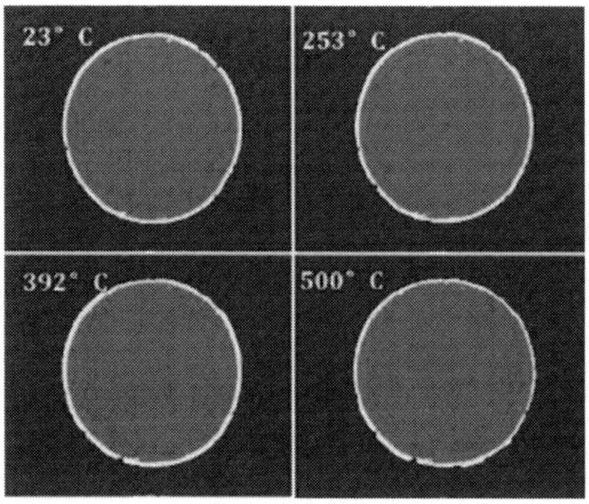

Figure 7. The spot images measured at four different temperature of gas for 4J36 (Invar).

3.2 Influence of temperature on optical system

In general the optical sensing system in the particle detection device is only applied at atmospheric temperature, but can also satisfy the temperature up to 150 °C. Under high-temperature condition, the gas volume is going to expand with the reduction in the gas density. The transmission characteristic of light beam could be changed, resulting from the decrease in the optical performance of the sensor. In order to investigate the influence of temperature on optical system, the specific causes are analyzed by theoretical method, and the extent of its influence is determined through experimental verification.

3.2.1 Theoretical method

By calculating the refractive index of the gas at different temperatures, the change of the OMV with the temperature can be obtained.

a) Calculation of gas parameters

The gas medium is air, so its refractive index can be calculated by the formula proposed by Rueger at the working temperature, given by:

$$n_{ph} = \frac{273.15}{1013.25} \cdot P \cdot N_{sph} \cdot \left(a_0 + \frac{a_1}{T} + \frac{a_2}{T^2} + \frac{a_3}{T^3} + \frac{a_4}{T^4}\right) \times 10^{-6} + 1 \quad (1)$$

Where P is the gas pressure; a_0, a_1, a_2, a_3 and a_4 are the correlation coefficient and constant; N_{sph} is the gas refractive index at atmospheric temperature, given by:

$$N_{sph} = (n_{sph} - 1) \times 10^6 = 287.6155 + \frac{1.62887}{\lambda^2} + \frac{0.01360}{\lambda^4} \quad (2)$$

The wavelength range is 400-700 nm. The gas refractive indexes at different temperatures in the range of 25-650 °C are calculated by Eq. 1 and 2. The results are shown in Figure 8. It can be found that the refractive index of air decreases with the increase in the gas temperature. And it can be inferred that the light divergence angle at high temperature is going to increase compared with that at atmospheric temperature. Consequently, the position of the beam focus is offset, leading to larger optical measuring volume, as shown in Figure 9.

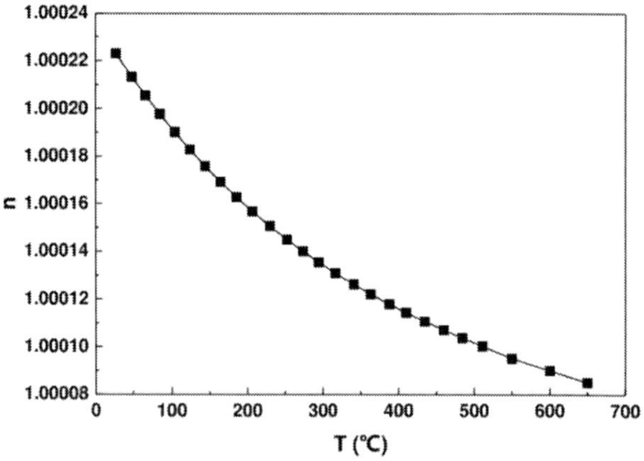

Figure 8. Gas refractive index at different temperature

In order to analyze the above unfavorable influence, a model of the OMV and gas temperature is needed to be built, mainly including the functions of the determination of the optical parameters and the calculation of the dynamic variable.

b) Calculation of optical parameters

On the basis of the Lorentz-Lorenz formula and actual gas state equation, the relationship between gas pressure and refractive index is given by:

$$\frac{n^2-1}{n_0^2-1}=\frac{\rho}{\rho_0}=\frac{PT_0Z_0}{P_0TZ} \quad (3)$$

Where T_0, P_0, ρ_0, Z_0 and n_0 are the gas temperature, pressure, density, compression factor and refractive index under atmospheric condition, respectively; T, P, ρ, Z and n are the gas temperature, pressure, density, compression factor and refractive index under high-temperature condition, respectively.

The light divergence angle is calculated by the refraction law shown below:

$$n_0' \sin\alpha = n \sin\beta \quad (4)$$

Where n_0' and α are the air refractive index and light divergence angle under atmospheric condition; β is the light divergence angle under high-temperature condition.

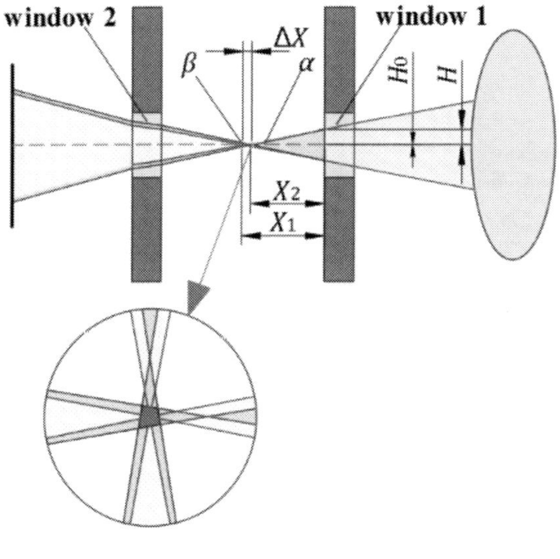

Figure 9. Schematic of the enlargement of the OMV

c) Calculation of dynamic variable

Based on the geometric optical theory and the mathematic relationship, the distance between the light focus position and the glass window 1 is triangular with the equivalent spot radius of the inner side of the glass window 1. According to X_1 and X_2 (the distance between the light focus position and the inner side of the glass window 1 under atmospheric condition and high-temperature condition), the theoretical model of the OMV with gas temperature was built. X_1 and X_2 are calculated by:

$$X1=\frac{H-H_0}{\tan\alpha} \quad (5)$$

$$X2=\frac{H-H_0}{\tan\beta} \quad (6)$$

The change of the light focus position (ΔX=$X1$-$X2$) can be calculated by Eq. 5 and Eq. 6.

The change of the equivalent spot diameter of the OMV can be calculated based on the triangle relation, as shown below:

$$\Delta D = 2\Delta X \tan\beta \quad (7)$$

Eq. 6 and Eq. 7 are the formulas of the theoretical dynamic model of the OMV with the change of the gas pressure.

Using the model, the optical parameters at different gas temperatures in the range of 25 ~ 700°C were calculated with three light divergence angles (4.02°, 9° and 13°, tested under atmospheric condition, respectively). The results are presented in Figure 10. It can be found in Figure 10 that ΔX increases with the gas temperature with approximate two patterns. In addition, ΔD changes by a smaller margin (less than 1 um). However, ΔD and ΔX satisfy a trigonometric relation. It can be also noted that the enlargement rate of the OMV volume (δ) has a small increase with the increase in gas temperature. Under the constant temperature, δ is going to increase for the larger light divergence angle, but it is still relatively smaller (<1%) compared with the OMV at atmospheric condition. Consequently, the influence of the gas temperature on the OMV can be ignored based on the above theoretical analysis.

3.2.2 Experimental method

In order to verify the reliability of theoretical analysis, an experimental dynamic model of the OMV with gas temperature variation was established based on the image analysis and the mathematical geometry relations, including the measurement of the change of light spot, the calculation of the geometric relation and the calculation of the relative variable.

By the refraction law, the changes of light spot on the inner and outer side of the glass window 2 under high-temperature condition (ΔD and $\Delta D'$) meet ΔD=$\Delta D'$. Taking the light focus position as the central axis, a triangle relation was formed by the distances between the light -

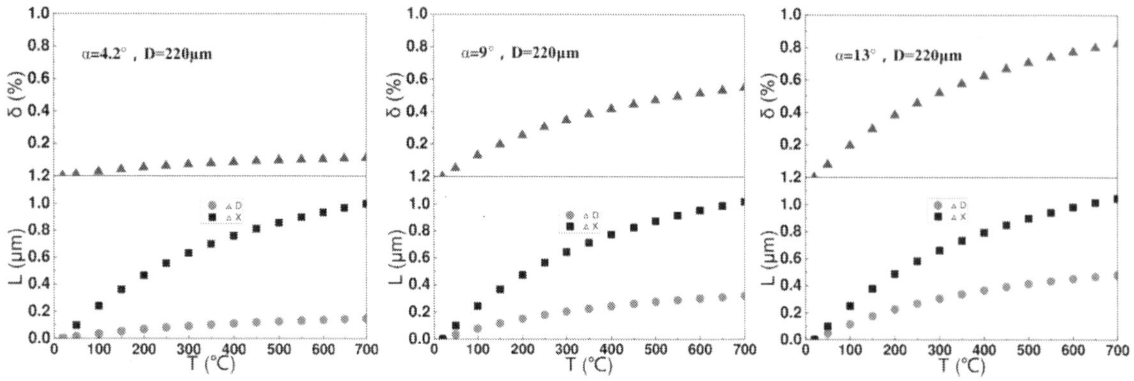

Figure 10. Results calculated based on theoretical dynamic mode with three light divergence angles (4.02°, 9° and 13°)

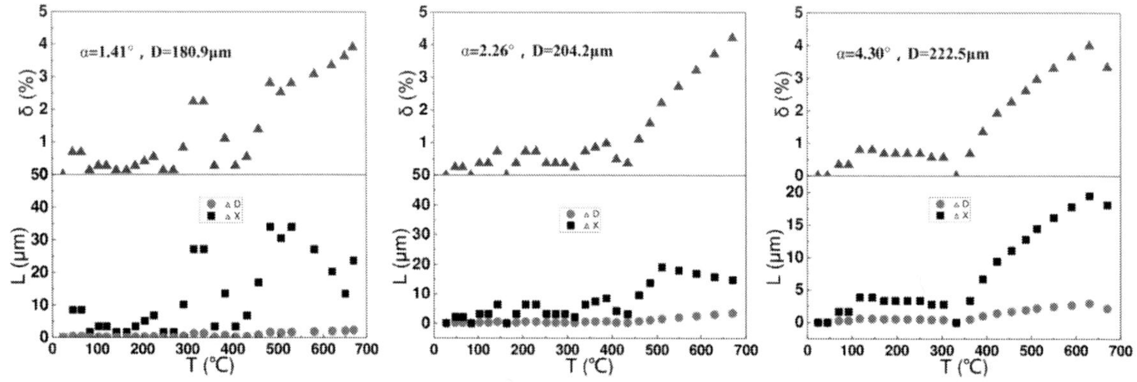

Figure 11. Results calculated based on experimental dynamic model

focus position and the inner side of the glass windows 1 and 2 (x and y), the difference of the light spot equivalent radius on the inner side of the glass windows 1 and 2 (H and h, $h = H - \frac{\Delta D}{2}$) and the equivalent radius of the OMV (H_0). Therefore, we have the following equations.

$$\frac{H - H_0}{h - H_0} = \frac{y}{x} \tag{8}$$

$$x + y = k \tag{9}$$

$$\beta = \arctan\left(\frac{h - H_0}{x}\right) = \arctan\left(\frac{H - H_0}{y}\right) \tag{10}$$

where β is the light divergence angle under high-temperature condition; k is the distance of the inner sides of the glass windows 1 and 2.

The changes of ΔX and ΔD under high-temperature condition to those under atmospheric condition are shown as follows:

$$\Delta X = \frac{y - x}{2}, \text{ or } \Delta X = y - \frac{k}{2}, \text{ or } \Delta X = \frac{k}{2} - x \tag{11}$$

$$\Delta D = 2\Delta X \tan \beta = \Delta D' \tag{12}$$

Eq. 11 to Eq. 12 are the formulas of the experimental dynamic model.

Using the above established experimental dynamic model, the ΔX, ΔD and δ of four cases in Table 2 were calculated, and the results are shown in Figure 11. It can be found that these three parameters increase obviously with the gas temperature increasing, but cannot meet the two function growth mode as the results of theoretical model. That may be caused by the reasons listed below: 1) the fluctuation of gas flow in the pipeline causes the unstable gas temperature field; 2) the beam quality analyzer is slightly influenced by other vibration sources; 3) the aerosol tube is affected by the gas flow fluctuation. But according to the experimental results, the maximum values of ΔX and ΔD are very small (less than 50 μm and 5 μm at the temperature below 700 °C, respectively) compared with the equivalent diameter of the OMV. Consequently, the ratio of the volume of the OMV is less than 5%. To sum up, the gas temperature has greatly weakened influence on the OMV, which can be ignored in actual technical development.

Table 2. Optical parameters of experimental model

number	light divergence angle /°	spot equivalent diameter /μm
1	1.41	180.9
3	1.94	285.1
4	2.26	204.2
5	4.30	222.5

3.2.3 Experiments

The performance of the optical sensor is investigated at different temperatures up to 650°C. The experimental results showed that the temperature of the optical sensor is around 50°C with the gas temperature in the range of 200-300 °C, and is going to increase about 10 °C at the gas temperature of 400-500 °C. The temperature is in the controllable range with a maximum value of 70 °C under the gas temperature of 650 °C. Therefore, the developed optical sensor can be used for the detection of particles in high-temperature conditions for a long time.

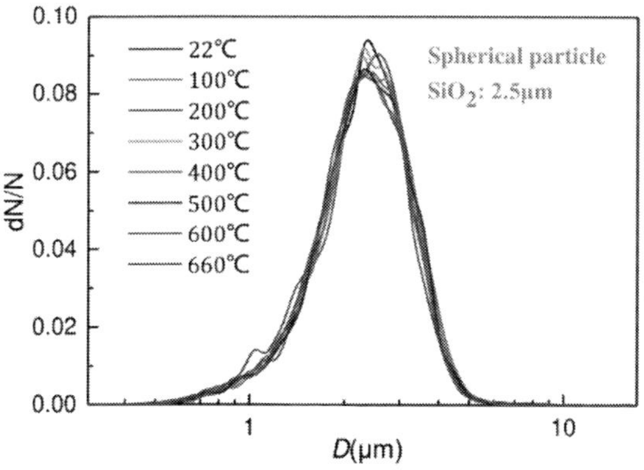

Figure 12. The results obtained under high-temperature condition

In the experiment of particle detection, the standard silica spherical particles with a strong temperature resistance ($D_{particle}$=2.5 μm) were tested at different gas temperatures. The testing time is 60 s. Figure 12 shows the test results. It can be found that all of the particle size distributions are in good agreement and have a good normal distribution, which meets the test results of single particle size, and the peak particle size is about 2.5 μm. Consequently, the optical sensor worked well for the particle detection under higher-temperature condition.

4 Conclusion

The emission of high-temperature flue gas causes serious air pollution and partial waste of available heat. In order to meet the relevant emission standards, additional purification units (high-temperature ceramic filters, etc.) are installed in the industrial process to collect the dusts in the flue gas. As the purification units are important for controlling the dust concentration, it is necessary to calculate the separation efficiency through a real-time measurement of the dust concentration in the flue gas in the upstream and downstream of the purification equipment.

In present work, a single particle measurement method and device suitable for high-temperature condition were developed. In order to complete the particles measurement at 650 °C, the beam images in optical sensing system under different temperature conditions were measured in real time by CMOS sensor. It was found that the beam images were deformed when the light divergence angle increased to more than 10°, and the deformation became more serious as the temperature increased. Depending on the principle of heat transfer, thermal expansion along the airflow direction caused by heat conduction between the duct and the gas is the main cause of this phenomenon. By changing the aerosol tube material t o one with an ultra-small linear expansion coefficient (Invar: 4J36), the optical system can be applied to the condition with temperature up to 700 °C, and in particular the light divergence angle extends to 30°.

According to the principle of gas state equation, the optical parameters (refractive index) of the gas at different temperatures change with the change in its state parameters. According to the results calculated by the formula of Ruegur's atmospheric refractive index, it was found that the refractive index of the gas decreases with the increase in temperature, which leads to the increase in the divergence angle of the beam in the aerosol tube, and shift of the position of the optical beam. Based on the geometrical optics theory and the geometric mathematical relationships, the theoretical dynamic model of the optical measuring volume with the change of the gas temperature was established. By calculating the amount of the beam focus offset and the volume of the optical measuring volume at different temperatures, it was found that the optical measuring volume increases with the temperature increasing. However, when the gas temperature was 650 °C, the increase rate of optical measuring volume was extremely small, so the influence of the temperature on the optical system can be negligible.

The performance of the whole detection system was tested by using 2.5 μm standard spherical silica particles. The satisfying measuring results were obtained and they could be applied to the actual on-site measurement at the temperature of 650 °C.

Acknowledgement

The authors thank the supports of "The National Key Research and Development Program of China (NO.2016YFB0601100)"; and would also like to thank the laboratory and staff of the Beijing Key Laboratory of Process Fluid Filtration and Separation, China University of Petroleum.

Reference

[1] J. Meyer, H. Umhauer, K. Reuter, A. Schiel, G. Kasper, M. Förster, Concentration and size measurements of fly ash particles from the clean gas side of a pressurised pulverised coal combustion test facility, Powder Technology, 180 (2008) 57-63.

[2] H. Leibold, F. Dirks, V. Rüdinger, Particulate emissions from a LLW incinerator and off-gas cleaning with a new type of ceramic candle filter, Waste Management, 9 (1989) 87-94.

[3] B.L. Perkins, Incineration facilities for treatment of radioactive wastes: a review, Incinerators, (1976).

[4] D. Fino, N. Russo, A. G. Saracco, V. Specchia, Multifunctional Filter for Treatment of the Flue Gases from Municipal Waste Incinerators, Industrial & Engineering Chemistry Research, 44 (2005) págs. 9542-9548.

[5] Scheibner, Wolters, Schumacher hot gas filter long-term operating experience in the nuon power buggenum IGCC power plant, (2002).

[6] J.M. Wang, Shell coal gasification technology and its application in China, Coal Quality Technology, (2006).

[7] Q. Xu, Z. Ji, L. Yang, Performance assessment of high temperature flue gas on-line particle size analyzer, Ciesc Journal, 63 (2012) 3506-3512.

[8] B.Y.H. LIU, V.A. MARPLE, K.T. WHITBY, N.J. BARSIC, Size distribution measurement of airborne coal dust by optical particle counters, Aihaj, 35 (1974) 443-451.

[9] R. Maus, F. Löffler, Size analysis of airborne bacteria by an optical particle counter and an impactor, Journal of Aerosol Science, 25 (1994) 141-142.

[10] B. Sachweh, H. Büttner, F. Ebert, Improvement of the resolution and the counting accuracy of an optical particle counter by fast digital signal recording, Journal of Aerosol Science, 20 (1989) 1541-1544.

[11] E.W. Max Born, Principles of Optics, Cambridge University Press, Cambridge, 1999.

[12] S. Hochrainer, Further developments of the particle counter sizer PCS-2000, Journal of Aerosol Science, 31 (2000) 771-772.

[13] B.S. Li, Z.L. Ji, L. Mei, An approach for correcting particle size distributions measured by the optical particle counter welas 2000, Aip Conference, 2010, pp. 686-689.

[14] J. Stier, M. Quinten, Simple refractive index correction for the optical particle counter PCS 2000 by Palas, Journal of Aerosol Science, 29 (1998) 223-225.

[15] H. Umhauer, Particle Size Distribution Analysis by Scattered Light Measurement Using an Optically Defined Measuring Volume, Journal of Aerosol Science, 14 (1983) 765-770.

[16] B. Sachweh, H. Umhauer, F. Ebert, H. Buttner, R. Friehmelt, In situ optical particle counter with improved coincidence error correction for number concentrations up to 107 particles cm-3 quantification and elimination of the effect of particle shape and structure, Journal of Aerosol Science, (1998) 1075-1086.

[17] D.I.J. Raasch, D.I.H. Umhauer, Errors in the Determination of Particle Size Distributions Caused by coincidences in optical particle counters, Particle & Particle Systems Characterization, 1 (1984) 53-58.

[18] H. Buettner, Measurement of the Size of Fine Nonspherical Particles with a Light-Scattering Particle Counter, Aerosol Science & Technology, 12 (1990) 413-421.

[19] D. Holve, S.A. Self, Optical particle sizing for in situ measurements Part 2, Applied Optics, 18 (1979) 1632-1645.

508245 An Experimental Investigation of Single Droplet Drying Above Boiling Point

Abstract:

An experimental method has been developed to enable dried particle structure, properties and drying rate to be predicted based on droplet drying history, particularly focusing on effects driven by boiling including bubble nucleation, growth and collapse. The method involves a single droplet drying rig which will be used to investigate drying kinetics above boiling point and across a broad range of drying conditions, drop size and composition space. The drying rig was improved as compared to previous apparatuses reported in literature, in that it is designed to measure droplet drying rates and morphological changes at high temperatures and under conditions that more closely resemble those found in industrial applications. The aim is to conduct drying experiment for droplets as small as 300μm diameter, in an air stream which temperature may be as high as 350°C.

Drying kinetics and morphological evolution will be determined, particularly focusing on the second drying stage as well as inflation/deflation periods. Initially, the constructed drying rig is checked and evaluated by conducting experiments on droplets of pure water. Afterwards, experiments will be conducted on solutions of different materials such as Hydroxypropyl Methylcellulose (HPMC), sodium sulphate and sucrose. An assessment of morphologies, transformations and drying behaviour under isothermal drying would be completed at a wide space of process condition, initial water content, initial drop size and composition space. The investigation will look to link material properties, particle size, initial moisture content and drying history to morphology. Results will be compared with single droplet drying models reported in literature.

An initial investigation has focused on mapping the morphological drying behaviour of HPMC, sodium sulphate and sucrose. It has shown that the three materials show different behaviour; sucrose showed deflation and inflation cycles, sodium sulphate droplets supersaturated and crystallized while HPMC showed inflation and rupture. Differences in drying behaviour is due to difference in material and transport properties. Therefore, the next stage of this work is focused on conducting some physical properties measurement at different water concentrations and temperature. In addition, different materials will be dried on the rig and the drying behaviour will be determined. The experimental data from this work will be used to develop, a universal morphology regime map which will link material properties, particle size, initial moisture content and drying history to morphology.

508482 Characterising Powder Flow in Dynamic Processes

Abstract:

Flow inconsistency of cohesive powders is a major issue for process industries. It is often difficult to predict and control flow of cohesive powders to enable efficient and reliable process operations particularly for dynamic processes such as mixing and filling. Many quantitative and qualitative flow measurement techniques exist that are used for measuring powder flowability under quasi-static conditions. However, there are few available techniques capable of assessing powder flowability under dynamic conditions. The FT4 Powder Rheometer (Freeman Technology), which drives a rotating blade along the vertical axis through a column of powder contained within a cylindrical vessel, is one of the few available devices that is capable of measuring powder flow in a dynamic state [1]. The flow resistance of the rotating blade as it penetrates the powder bed is characterised by flow energy, however, there is lack of understanding in how to relate this measurement to process design.

This work focuses on determining the relationship between the torque and shear stress in the FT4. The effect of particle size on the external flow profile around the blade has been investigated experimentally for glass beads using a high speed camera. The material investigated shows a decrease in downward specific flow energy with increasing particle size; indicating that the increased shear resistance in the bulk for smaller particle sizes exists under dynamic conditions. Particle Image Velocimetry (PIV) analysis indicates an increase in the volume of particles displaced by the blade with increasing particle size for both upward and downward blade movement. An increase in impeller tip speed leads to a reduction in flow energy for this material, and a reduction in the volume displaced by the blade. The prevailing stresses and strains as well as the volume of particles displaced by the blade are estimated by Discrete Element Method (DEM). DEM allows for the determination of a relationship between the flow energy and/or shear stress with particle properties and blade velocity and direction. Furthermore, Positron Emission Particle Tracking (PEPT) analysis of the FT4 has been carried out to determine the radial and axial displacement of particles with respect to the bed height, and to validate the DEM results.

508671 Impact of Non-Spherical Projectiles on Granular Media

Abstract:

Recently, several studies involving numerical and experimental methods have focused on the study of impact dynamics in both dry and wet granular media. An example of such instances can be an animalâ€™s foot impacting sand during its locomotion. Understanding the physics behind the interaction between granular media and animal feet can be helpful in optimizing locomotion parameters for bio-inspired robots. However, most of the studies involving impact dynamics considered spherical projectiles under different conditions, while characteristic models should involve more complex shapes. Impacting different geometries with conserved density, volume, and velocity on a granular bed may experience contrasting drag forces upon penetration. This is due to the difference in the surface areas coming in contact with the granular media. In this work, non-spherical geometries of following shapes were considered: cylinder, triangular prism, hexagonal prism and cuboid. These geometries were impacted on a loosely packed non-cohesive dry granular bed with the same impact velocities. The effect of surface area in contact with the granular media was analyzed for impact, with the help of penetration depth measurements. We also tested whether the power law for spherical projectiles is applicable for non-spherical projectile impacts.

508664 Computational and Experimental Shear Cell Study with Rigid Cylindrical Particles

Abstract:

Flow properties of particles are commonly assessed through shear cells. Both the Schulze ring shear tester (SRST) and Jenike shear tester (JST) measure the interparticle stresses which are important to the design of many granular flow processes. This work uses precision-cut rigid cylinders of various aspect ratios and sizes to compare the measured stresses between both testers. It is found that the shearing results for both testers are similar, which is important for future work involving flexible biomass materials.

Previous work has sought to use shear cell results to validate various aspects of discrete element method (DEM) simulations. The SRST is recreated in a DEM simulation with rigid glued-sphere cylindrical particles to assess the contact detection method by comparing the experimental and computation results. Also, a parameterization study is performed to examine the effects of coefficient of restitution, coefficient of friction, and other particle properties on the measured steady-state stress. Overall, this serves as an efficient validation of the contact detection method which will be used to simulate complex particle shapes.

508865 Defluidization Behaviour of Industrial Reactive Powders at High Temperature

Abstract:

Fluidized bed reactors are used in a wide range of industrial chemical processes, and they are particularly useful in high temperature systems thanks to their ability to provide high heat transfer rates and rapid solids mixing which lead to isothermal and more controllable conditions. Their applications spread the environmental, chemical, energy and process industries. For example, they are a key technology in the petroleum industry, catalytic chemicals synthesis processes, combustion and gasification of solid fuels (coal, wastes and biomass), metals productions and many more.

Use of fluidized beds in combustion-type operations may lead to particles sintering and bed agglomeration, which again may result in defluidization, unscheduled process downtime and additional costs. However, the details of the physical mechanisms behind such phenomena are still not entirely understood. Despite the countless researches that have been done so far, such processes still disclose unrevealed features and challenge the researchers worldwide.

The objective of this paper is to assess the effect of high temperature on the flow properties and fluidization behaviour of mixtures of carbon coke and two different titanium ores, namely synthetic and natural rutile. To this end, a simple approach for the early detection of agglomeration in the regimes of incipient fluidization -where the carbon combustion causes defluidization- will be followed. Such a monitoring approach is based on the simultaneous measurements of local temperatures and the pressure drop across the particles bed.

507739 Manufacturing and Characterization of Spherical Blend PBT-PC Particles for Additive Manufacturing

Abstract:

Polymer powder-based additive manufacturing technologies often yield excellent mechanical and functional part qualities. One of these technologies is selective laser sintering (SLS), a process employing micron-sized polymer particles, which are selectively fused by a laser [1]. While hardly any boundaries regarding part design exist, there are quite some restrictions concerning the variety of commercially available SLS materials. For the time being, the most widely used polymeric material for SLS is polyamide 12 (PA12) with a market share of roughly 95% [2]. Hardly any blend materials are available; mostly , the processing of multi-material systems via SLS is achieved by simply mixing different polymer powder materials [3].

In this contribution, a new approach to blend particles is introduced. Via co-grinding and subsequent rounding, manufacturing of spherical particles consisting of semi-crystalline polybutylene terephthalate (PBT) and amorphous polycarbonate (PC) thermoplastic is possible. The polymers are jointly comminuted and mixed in a planetary ball mill and the product particles are rounded in a heated downer reactor [4,5]. The thereby achieved intermixed, but not transesterified, PBT-PC composite particles are characterized regarding size and morphology via electron microscopy. Intermixing of the polymers is investigated via staining techniques and Raman spectroscopy. Thermal properties of the blend particles are analyzed via DSC, also in regard to transesterification in SLS process-like conditions. Furthermore, flowability and size distribution of the blend particles are examined. This way, we show the high potential of our process route [4] for the production of novel polymer blend particle systems for SLS.

References:

[1] I. Gibson, D. Rosen, B. Stucker, Additive manufacturing technologies, 2. ed., Springer, New York, NY, 2015.

[2] T.T. Wohlers, Wohlers Report 2016, Wohlers Associates, Fort Collins, Col., 2016.

[3] D. Drummer, K. Wudy, F. Kühnlein, M. Drexler, Polymer Blends for Selective Laser Sintering: Material and Process Requirements, Phys. Procedia. 39 (2012) 509–517. doi:10.1016/j.phpro.2012.10.067.

[4] J. Schmidt, M. Sachs, C. Blümel, B. Winzer, F. Toni, K.-E. Wirth, W. Peukert, A novel process route for the production of spherical LBM polymer powders with small size and good flowability, Powder Technol. 261 (2014) 78–86. doi:10.1016/j.powtec.2014.04.003.

[5] J. Schmidt, M. Sachs, S. Fanselow, M. Zhao, S. Romeis, D. Drummer, K.E. Wirth, W. Peukert, Optimized polybutylene terephthalate powders for selective laser beam melting, Chem. Eng. Sci. 156 (2016) 1–10. doi:10.1016/j.ces.2016.09.009.

506658 Creating Tuneable Agglomerates via 3D Printing

Abstract:

It is essential to offer a universally accepted way to systematically evaluate agglomerate breakage propensity and mechanisms. Discrete Element Method (DEM) simulation is a commonly used tool but is limited by the lack of identical, controlled agglomerates to test and validate simple models, let alone replicate the complex structure of real industrial agglomerates.

Our research presents a systematic approach to produce 3D printed agglomerates with tuneable properties. Agglomerates with different mechanical properties and structures were designed and printed by an Objet500 Connex 3D printer. The tensile and shear strength of inter-particle bonds was characterized using Instron tester. Breakage results of agglomerates with simple cubic structures and the design of agglomerates with irregular structures were presented. This 3D printing method will allow more rigorous testing of agglomerate breakage models. Discrete Element Method (DEM) has been used to simulate the agglomerate crushing using the Timoshenko Beam Bond Model (TBBM) with bond properties matching the 3D printed agglomerates. For both two agglomerate structures used in this research, the simulation and experiment showed similar breakage patterns. However, the simulation results significantly underestimated the compressive loads of spherical random structured agglomerates. In order to improve the simulations of compressive granule breakage, the influences of bond geometry definitions and non-linear behaviour of the bond contacts need to be explored.

508699 Effect of Particle Size Distribution and Wettability on Penetration Behavior, Granule Formation and Granule Properties in Single Drop Granulation

Abstract:

In wet granulation, single drop granulation (where one drop forms one granule) can simplify the complexity and reveal the fundamentals of the interaction between the liquid and the powder bed. Combining drop penetration time, granule formation mechanisms and the characterization of granule properties is a novel study to achieve a comprehensive map of the single drop granulation process, which is relevant in particulate applications like pharmaceuticals, detergents, agricultural chemicals and catalysts.

In this work, single drop impact of liquid on a static powder bed was studied to investigate the granule formation mechanism, droplet penetration time, and the characterization of granule properties (morphology, internal structure and content uniformity). Water was used as the liquid binder and two pharmaceutical powders, microcrystalline cellulose (MCC) (mean particle size ~54.8 Î¼m, contact angle with water ~36Â°) and acetaminophen (APAP) (mean particle size ~9.74 Î¼m, contact angle with water ~89°), were mixed to make heterogeneous powder beds with varying compositions. The complete drop impact and penetration was recorded with a high-speed camera. The granule morphology, its internal structure and content uniformity were characterized by the prism method with image analysis, micro-CT and UV-vis spectrometry, respectively. Besides the current MCC and APAP system, MCC and APAP with other particle sizes and relative contact angles will also be studied. It is believed that the mean particle size of the powder bed is predominant in influencing the granule formation mechanism, drop penetration time, and granule properties.

508017 Atomically Deposited Sintering Aids: Assessing the Effects of Alumina Particle ALD on the Sintering and Performance of SOFC Electrolytes and Dental Ceramics

Abstract:

The crystal structure of zirconium oxide (ZrO_2) can be tailored by the incorporation of aliovalent dopants, most commonly, yttrium oxide (Y_2O_3). The concentration of dopant dictates the thermodynamically stable crystal phase where 3 mol%
Y
conducting cubic phase. To achieve suitable mechanical properties and ionic conductivity, Y-TZP and YSZ ceramics must be near theoretical density, requiring high sintering temperatures. Previous researchers have successfully lowered the required Y-TZP and YSZ sintering temperatures through incorporation of aluminum oxide (Al_2O_3) particles. While this technique has proven to reduce the sintering temperature in both systems, the presence of particulate Al_2O_3 inclusions in densified parts reduces the homogeneity of the material which may have deleterious effects on densification and microstructure. In this work, we report on the use of atomic layer deposition (ALD) to conformally coat individual Y-TZP and YSZ particles with the desired amount of Al_2O_3. Constant-rate-of-heating (CRH) experiments were conducted using a horizontal push-rod dilatometer to extract the activation energy of sintering and also gain mechanistic insights into the active diffusion mechanisms as a function of ALD film thickness. Transmission electron microscopy (TEM) and scanning electron microscopy (SEM) were used to characterize the as-deposited thin Al_2O_3 films and the effects of the Al_2O_3 films on grain growth and microstructure during densification. Ionic conductivity and flexural strength tests were conducted to determine the effect of Al_2O_3 thin films on relevant material properties. Al_2O_3 incorporation by ALD is shown to have unique phenomenological effects on the densification of Y-TZP and YSZ as compared with conventional Al_2O_3 incorporation using particle-based approaches.

508644 Effect of Particle Size Distribution and Operating Parameters on Conduction and Convection Heat Transfer Mechanisms in Rotary Drums

Abstract:

Granular materials undergo process steps that include transportation, drying, heating and chemical or physical conversion. In several instances, these materials need to be heated or cooled during processing, and rotary drums (kilns) are the most commonly used process equipment for this purpose. It is not always possible to conduct experiments using different materials and consider all process conditions to study the thermal mechanism due to the inherent complexity of these processes. Therefore, computational modeling and simulations have been extensively used to understand the heat transfer mechanism in rotary drums. To maintain simplicity and to overcome the challenges of simulating polydispersed particles, a common practice is to assume that the polydispersed particles with a narrow size distribution behave like monodispersed particles thermally. However, no effort has been made to validate this assumption. This work focuses on understating the effect of polydispersed particles on the conduction and convection heat transfer mechanisms in indirectly heated rotary drums. It also aims to investigate the effect of rotation speed, fill level, particle size and their interaction effects on the conduction and convection heat transfer mechanisms in rotary drums. For this, the discrete element method (DEM) (using MFIX-DEM, an open-source computational fluid dynamics solver suite) simulation technique is used to analyze the thermal behavior of spherical silica beads, which are the most common catalyst support material.

507974 Selective Hydrogenation of Citral over Supported Pt Catalysts on Various Substrates

Abstract:

Highly dispersed platinum (Pt) nanoparticles (NPs) were deposited on various substrates by atomic layer deposition (ALD) for the reaction of citral hydrogenation. The substrates included multi-walled carbon nanotubes (MWCNTs), silica gel (SiO_2), commercial Î³-Al_2O_3, and ALD-prepared porous Al_2O_3 particles (ALD-Al_2O_3). Based on TEM and CO-chemisorption analysis, the Pt NPs on all substrates were around 1.3 nm, and Pt/MWCNTs, Pt/ALD-Al_2O_3 and Pt/Î³-Al_2O_3 showed similar Pt dispersion (75%-85%). But the Pt dispersion on SiO_2 was low (63%). In the reaction of citral hydrogenation, Pt/SiO_2 showed the highest activity due to the strong acidity of SiO_2 and the conversion of citral reached 82% after 12 hr with a selectivity of 58% of unsaturated alcohols (UA), including geraniol and nerol. Among all catalysts, Pt/MWCNTs showed the highest selectivity of UA (65%) due to its unique structure and electronic effect. In the cycling test, Pt/MWCNTs and Pt/ALD-Al_2O_3 catalysts were more stable than Pt/SiO_2 which could be attributed to the different interactions between the catalysts and the solvent.

508345 Numerical Simulation of Particle Classification in a Classifier Based on Coanda Effect

Abstract:

Efficient and precise classification of particles has been an important issue in many industrial applications as well as in academia. A relatively new particle classifier based on 'Coanda effect' was developed and reported to have many favorable features (Liang 2010). They include simultaneous multiple classification, simple design with no moving parts, and easy scale up to higher production rates. However, this classifier has attracted little attention so far, and our understanding of underlying mechanism associated with classification performance needs to be improved.

The purpose of this study is to investigate the detailed fluid flow and particle transport characteristics in a Coanda classifier using CFD (Computational Fluid Dynamics). The main focus is placed on the effects of working conditions (inlet flow velocity, particle diameter) and numerical method for turbulence dispersion of particles on the classification performance. As a numerical method, three different approaches are considered: k-ε model, k-ε model combined with DRW (Discrete Random Walk) model, LES (Large Eddy Simulation). In the first approach, standard k-ε model is adopted to obtain the mean flow and then particle trajectory is calculated based on the mean flow without consideration of turbulence dispersion. However, in the second approach using DRW model, the instantaneous flow velocity is constructed under the assumptions of isotropic turbulence in order to include turbulence effect. Finally, in LES, the instantaneous flow field is dynamically and accurately obtained at each time step. Numerical simulations are carried out for two-dimensional mean flow in case of k-ε model, but for three-dimensional unsteady flow in LES. Particles are tracked using a Lagrangian approach based on one-way coupling. We consider talc particles with density of 2,860 kg/m^3 and diameter of 1 to 20 μm.

The Coanda classifier is mainly composed of three inlet channels, three outlet channels, and Coanda block. Inlet channels are divided into one main inlet and two control inlets. The main air jet containing particles is supplied with the velocity of 35 m/s through the main inlet. Two control air jets with velocities of 32 m/s (lower) and 40 m/s (upper) are also supplied through the control inlets for enhanced control.

The trajectories of moving particles obtained from the mean flow (k-ε model) are analyzed for different diameters. The main jet containing particles deflects toward the curved Coanda block due to Coanda effect as expected. Fine particles (d≤2μm) follow the flow streamlines and enter the lower outlet channel. However, coarse particles (d≥8μm) are detached from the streamlines and go into the upper outlet channel. This observation agrees well with the previous result (Liang, 2010).

Even though the mean flow and corresponding particle trajectories provide useful information, realistic particle passage do not take a form of single curve but they change in time due to flow turbulence. Therefore, particle trajectories are simulated again, turbulence dispersion being considered with DRW model. Due to turbulence dispersion, particles of the same diameter are scattered around the particle pathline obtained from the mean flow. As a result, particles of the same diameter can not be collected into a single outlet. For example, some particles (about 5%) of 1 μm enter the middle outlet even though most of them (about 95%) enter the lower outlet designed for collecting fine particles. This means that flow turbulence deteriorates the accuracy of classification.

Finally, LES simulations are performed to understand dynamic behaviors of fluid flow and particle transport. Instantaneous flow fields such as vorticity contours clearly show unsteady fluid motions inside the classifier. It is also shown that the generation of unsteady vortices leads to the scattering of particle trajectory.

A more detailed analysis including the classification efficiency will be given in the final presentation.

Acknowledgements

This research was supported by the National Research Foundation of Korea (NRF) under grant number 2017M2A8A4018482 and 2015R1D1A1A0105 9675.

References

Liang, S. (2010). *Chem. Eng. Comm.*, **197**, 1016-1032.

503491 Numerical Simulation of the Secondary Air Distribution Layer in the Gas Solid Fluidized Bed Based on Flunt Software

Abstract:

Due to the continuity beneficiation in the practical production, the gangue would accumulate on the distributor forming the gangue layer, which was called Secondary Air Distribution Layer and has vital effect on the gas distribution. In this paper, the ventilation characteristics of the secondary air distribution layer was analyzed. In addition, FLUENT software was used to study the effect of the modeled layer for the fluidity characteristics (e.g., pressure drop, bed height, volatility, etc.) in the gas-solid fluidized bed with the complex layer. The results showed that the complex layer could promote the secondary distribution of the airflow in the bed, which effectively inhibited the production of large bubbles and reduced the bed pressure drop fluctuation. Thus, itâ€™s of significance to provide synergy effect on the fluidization stability in the fluidized bed. Based on the above analysis, the introduced complex structure would be convenient to achieve the coal separation effectively.

507949 CFD-DEM Simulation of the Fluidization of Non-Spherical Particles in Fluidized Bed

Abstract:

In chemical and energy industries, fluidization concerning non-spherical particles (such as rod-like particle, disk-like particles, ellipsoidal particles, etc.) is often encountered, but research in this area is minimal. Yet understanding fluidization of non-spherical particles is crucial in many relevant industrial process. To investigate the fluidization of non-spherical particles, a CFD-DEM coupling algorithm is employed in this research to simulate a bubbling fluidized bed. The non-spherical particles are modeled by super-ellipsoids, multi-sphere models, and other particle models. The simulations results show that, the fluidization of non-spherical particles has significant differences from that of the spherical particles. The local void fraction and particle's projected face area perpendicular to fluid flow direction are directly related to the orientations of the non-spherical particles, which has a significant influence on the particle drag force and fluidization. Due to the significance of particle orientation, we further investigate the factors that influence the distribution of particle orientations. The results show that particle orientation depends on the fluidization time, fluidization velocity, and structure of the fluidized bed.

508320 CFD and DEM Simulation of the Cold Spray Process for Surface Coating with Fine Particles

Abstract:

The coating and functionalization of technical surfaces has been become increasingly important in the last two decades for various applications, since of its manifold possibilities for process optimization, as friction reduction, corrosion resistance or improved wetting behavior. Cold spray is a thermal spraying method for surface coating. It was developed in the late 1980s in Novosibirsk [1] and has gone through continuous optimization on certain applications since then. For example showed our previous study a reduction of the friction properties on steel pins, microstructured with fine Titania particles by an adapted spraying method [2]. In the cold spray process particles are dispersed in a gas stream, which is heated up to several hundred degrees and accelerated to supersonic velocities in a Laval nozzle. The particles impact on a substrate with about 800 m/s and form a strong bonding due to the high kinetic energy which is transferred to deformation and heat in the contact zone. In contrast to other thermal spraying methods, the temperatures occurring during the process are far below the melting temperatures of the used materials, which makes the process suitable also for thermal sensitive powders. The quality of the produced coating layers depends highly on the properties of the powder and substrate as well as on the nozzle design and the process parameters as initial temperature and particle velocities. Due to the high gradients, the high occurring velocities and the small size of the particles, typically in the range of 1 to 50 μm, the contact zone of impacting particles can only be measured indirectly and with destructive methods. Therefore, still not all phenomena which lead to a strong bonding in the contact area and a high quality coating are completely understood. For the study of these mechanisms two different simulation approaches have been used:

a) Simulation of particle and gas flow with Computational Fluid Dynamics (CFD)

For a better understanding of the spray process, the nozzle, the adjacent flow areas and the substrate were computed with a CFD method. The trajectories of single particles were calculated by an Euler-Lagrange approach. For the flow the k-Ï‰ SST model was used. Local temperatures in the substrate were calculated by the law of heat conduction. The impact conditions dependent on the nozzle geometry and the particle properties. With the local temperature distribution in the substrate and in the particle, precise impact conditions could be estimated. A study has been conducted to vary the parameters of the spray process, temperature and pressure as well as the particle properties as shape factor, density and size. Moreover the nozzle geometry has been varied in the simulation to optimize the spray process. The simulation results are compared to images of coated surfaces obtained by Scanning Electron Microscopy (SEM) for validation.

b) Simulation of particle-surface impact behavior with Discrete Element Method (DEM)

The DEM is widely used for investigations of particular processes. The advantage of this method is the detailed calculation of particle-particle or particle-wall contacts to obtain the important parameters as forces and deformations. The contact force in the particle wall contact is also dependent on the surface morphology. Therefore the surface topography of two real technical surfaces, a polished and an unpolished Titanium substrate were measured by Scanning Probe Microscopy (SPM) imaging with a nanoindenter (TI Premier, Hysitron). The surface topography was discretized and implemented in a DEM model. A parameter study has been conducted to determine the distribution of forces and contact angles occurring at the particle impact. Different particle sizes and impact velocities were studied.

The cold pray process dependents on various parameters which affect each other. Since of the high gradients and the small scales occurring during the process experimental measurements are often difficult. With this numerical study different parameters could be considered uncoupled of the other parameters. The results can be used for a better process understanding and an optimization of commercial cold spray systems.

[1] Alkhimov, A. P., Papyrin, A. N., Kosarev, V. F., Nesterovich, N. I., Shuspanov, M. M.: Gas dynamic spraying method for applying a coating; US Patent, 5302414; 1994

[2] Buhl, S., Schmidt K., Sappok D., Merz, R., Godard, C., Sauer, B., Kerscher, E., Kopnarski, M., Antonyuk, S., Ripperger, S.: Surface structuring of case hardened chain pins by cold-sprayed micro particles to modify friction and wear properties; Particuology 21, 2015

508568 Two-Fluid Validation of Constituive Models for the Simulation of Cylindrical Particles

Abstract:

Two-fluid simulations are a popular tool in the modeling of large processes. The constitutive models that close the governing equations are based in granular kinetic theory and typically assume that the particles are spherical. In reality, very few processes utilize spherical particles, hence there is a need for constitutive models that describe more complex shapes. Previous work used discrete element method (DEM) simulations to create a solid stress[1,2] and collisional dissipation rate[3,4] model for cylindrical particles that can replace the current spherical models. These new models have been added to MFiX and this work focuses on their validation. Cylindrical particles of various aspect ratios and solid densities were used in cratering and hopper discharge experiments. Simulations are designed to recreate the experiments in order to compare the results for crater depth and hopper discharge rate. The model effectiveness is assessed with regards to assumptions made concerning particle rotation and preferential alignment.

1. Guo Y, Wassgren C, Ketterhagen W, Hancock B, James B, Curtis J. A numerical study of granular shear flows of rod-like particles using the discrete element method. *J Fluid Mech*. 2012:1-26.

2. Berzi D, Thai-Quang N, Guo Y, Curtis J. Stresses and orientational order in shearing flows of granular liquid crystals. *Phys Rev E*. 2016;93(4):40901.

3. Buettner KE, Guo Y, Curtis JS. Using the Discrete Element Method to develop Collisional Dissipation Rate Models that Incorporate Particle Shape. *AIChE J*.

4. Berzi D, Thai-Quang N, Guo Y, Curtis J. Collisional dissipation rate in shearing flows of granular liquid crystals. *Phys Rev E*. 2017;95(5):50901.

509002 The Comparsion of Coarse Grained CFD-DEM for Simulating the Dense Bubbling Fluidized Bed" Requires Immediate Action

Abstract:

CFD-DEM is an effective tool to study the hydrodynamics in dense bubbling fluidized bed. However, the expansive cost of calculating the particles motion is not easy for its application in industrial application or dense particles system. Thus several coarse grained models (CGM) were put forward to simplifying the calculation of particles and decreasing the particles motion, such as PSM, SPA, MSPA,CGEE, EMMS-DPM where a group of particles are set as a parcel or coarse particle to reduce the computation. In this study the experiment and simulation test conducted to test five coarse grained methods. The result indicated that the models increasing the gravity force, pressure gradient force and drag force linearly with the coarse grained ratio (CGR) can be effective to ensure similar macro behavior between original system and coarse graining system. The simulation data of the solid fraction distribution and bed expansion ratio reflected good agreement as experiment. But the bubbles structure became less clear as increasing the CGR. Furthermore, the power spectrum density (PSD) and Wavelet analysis of the pressure signal were used and showed that SPA and EMMS-DPM gave better agreement result in term of dominant frequency and energy ratio at mesoscale feature. Finally the computational efficiency for different models was compared and the models of EMMS-DPM, PSM and CGEE have higher computation efficiency than SPA and MSPA relatively if the collision model of spring-damping were adopted.

509153 Stability Analysis of Uniform Gas Solids Flow

Abstract:

Uniform distribution of gas solids flow should be ensured to gain sufficient performance of fluidized beds. The highly chaotic entropy dissipation across solids bed will trigger non-uniform fluidization, although distributor plays an important role in obtaining homogenous gas stream. Hence, total solids pressure loading (Î¦$_T$), the pressure drop ratio between solids bed and distributor, is introduced here to present their interaction. The boundary between uniform and non-uniform fluidization can be detected by stability analysis based on Prigogineâ€™s minimum of entropy production principle, and this criterion for *N* parallel paths can be simplified into calculating the second derivative of pressure drop through one path with respect of Î¦$_T$. Furthermore, a phase diagram, illustrating the effects from operational parameter (Î¦$_T$), geometrical characteristics of distributor (C_d) and property of solids (*Ga*) on instability of uniformity, is attained to draw some useful conclusions of robust uniform fluidization.

509309 Numerical Simulation of Particle Sedimentation Related to Nuclear Safety By CFD-DEM Algorithm

Abstract:

Numerical Simulation of Particle Sedimentation related to Nuclear Safety by CFD-DEM Algorithm

Byoungcheol Hwang[1], Kiyofumi Moriyama[1], and Hyun Sun Park[1*]

[1] *Division of Advanced Nuclear Engineering (DANE), POSTECH, Pohang, Korea*

**corresponding author: hejsunny@postech.ac.kr*

Abstract

The Discrete Element Method (DEM) is a technique to treat particle-particle interactions, and is used in many industries such as civil engineering, powder processing, pharmacy, and so on. Coupling of this method with Computational Fluid Dynamics (CFD) is needed in problems where the particles interacts with fluid significantly, such as bubble column reactor, mineral processing, and so on.

The simulation of important phenomena in nuclear power plant accidents that involves significant core damage and melting, i.e. so called severe accidents, is also one of the topic where the CFD-DEM is useful. After the Fukushima accident in 2011, the severe accident analysis and preparation of counter measures to mitigate the accident consequences became regulatory requirements in many countries. One of the measures for the cooling and stabilization of the molten core is the flooding of the cavity under the reactor vessel where the dropping core melt is caught in a water pool. It is expected that the molten core breaks up into particles of several millimeter sizes and accumulates on the floor, then makes a particle bed, which is hopefully coolable. In this situation, the assessment for the coolability of the particle bed is very important, and a number of experimental and analytical approached suggest that the internal/external structure of the particle bed affects strongly the coolability. The CFD-DEM is a promising candidate for an analysis method that can clarify such details of the particle bed formation.

We tested the applicability of the multiphase CFD-DEM code, an open-source developed by CFDEM®coupling in 2011, for the simulation of particle sedimentation and particle bed formation under water. Our focus in the present work is on the prediction of the particle falling characteristics and particle bed shape. Although the real condition during a severe accident includes very high temperature particles (>1000K) and boiling water (two-phase flow) in a large scale, we started with a simulation of much simple experiment cases without heat transfer or gas phase, with spherical particles only as the first step for the validation of the method. The experimental data with solid particles of various diameters (4 mm, 6 mm) and densities (3600, 6000, 7800 kg/m^3), particle release nozzle diameters (20, 30, 40 mm) were referred to. The results of numerical simulation were compared with the experimental data in terms of particle falling time, maximum particle dispersion angle, particle average falling velocity, particle bed distribution, and so on. The particle falling time was well predicted by the simulation with the maximum deviation 5%. The maximum particle dispersion angle by the simulation significantly deviated from the experimental data probably due to the influences of initial disturbances. The particle average falling velocity was overestimated. The final particle bed profile by the simulation showed agreed with the experimental observation accurately in the peripheral region, but the height of the center part was underestimated. The fluid dynamic drag model and the particle-particle contact model would need improvement for better accuracy. The present work confirmed the validity of CFD-DEM algorithm on the application for the particle sedimentation in a simple liquid pool. The particle-pool interaction including the gas phase will be the next step.

Keywords: CFD-DEM; particle-fluid interaction; particle sedimentation; nuclear safety; severe accident

509440 Application of Extedned Discrete Element Method to the Melting Process of a Packed of Particles

Abstract:

Application of eXtedned Discrete Element Method to the melting process of a packed of particles

Mehdi Baniasadi, Maryam Baniasadi, Bernhard Peters, Xavier Besseron

Université du Luxembourg, Faculté des Sciences, de la Technologie et de la Communication, Luxembourg

e-mail: mehdi.baniasadi@uni.lu, maryam.baniasadi@uni.lu, bernhard.peters@uni.lu, xavier.besseron@uni.lu, web page: http:/www.xdem.de

ABSTRACT

The phenomena of melting takes place in many industrial processes, such as metal processing, environmental engineering and thermal energy storage systems. The problem of modeling of solid-liquid-gas flow including physical and chemical conversions of particles such as melting is the subject of intense research in industrial processes. These systems are difficult to model with numerical analysis due to the complex interaction between the different fluids and granular phases condering heat, mass and momentum excahnge. Basically models used to predict multiphase flows with particles are Eulerian-Eulerian and Lagrangian-Eulerian. The difference between the two models is based on different treatment of the movement of the particles. The particles are tracked individually in Lagrangian-Eulerian approach while they are considered as a continuum medium in Eulerian-Eulerian. Lagrangian-Eulerian is closer to the real physical processes rather than the Eulerian-Eulerian model for granular flows with particleâ€“particle interaction[1].

In this paper, our main focus is on the modelling of the melting process for a packed bed of particles subject to a fluid flow using a numerical technique based on Lagrangian-Eulerian so-called eXtended Discrete Elemnet Method (XDEM)[2]. In this approach, the particles are resolved as discrete phase coupled via heat, mass and momentum transfer to the surrounding fluid phase. The Discrete Element Method (DEM) coupled by heat transfer is used in order to evaluate temperature distribution, melting rate as well as velocity, oriantion and position for each particle individually. In addition the approach allows to be linked with continues numerical approaches such as Computational Fluid Dynamics (CFD) in order to solve fluid fields. The model has been validated by comparing predicted results with existing experimental data for melting of a single ice particle as well as a packed bed of ice particles in the horizontally flowing water. Fluid-particle and particle-particle intractions has also been considered. In addition, the model has the capability to be extended to the packed bed of particles with different size and properties to produce different liquid phases.

REFERENCES

[1] Dierich, F., Nikrityuk, P. A. & Ananiev, S. 2D modeling of moving particles with phase-change effect. *Chem. Eng. Sci.* **66,** 5459â€“5473 (2011).

[2] B. Peters, Thermal Conversion of Solid Fuels (WIT Press, Southampton, 2003).

514248 Real-Time Interpretation of Chaotic Characteristics and Instability of Dense Gas Solid Fluidization

Abstract:

The establishment of instability monitoring system targets at analyzing the chaotic characteristics and evaluating the instability of the fluidization process for dense gas solid fluidization. Experimental pressure fluctuation and pressure drop measurements were interpreted from dense gas solid fluidization system based on chaotic time-series analysis. The pressure fluctuation signals were filtered in terms of attenuating the noise signals. The filtered pressure signals were analyzed with respect to attractor reconstruction. Based on the reconstructed attractor, the maximum Lyapunov exponent and entropy were investigated and estimated under different bed heights and air velocities. Results show that the maximum Lyapunov exponents udner different bed heights and air velocities are positive, which indicates that dense gas solid fluidization exhibits chaotic characteristics. With the increase of bed height, estimated entropy of dense gas solid fluidization increases with the increase of bed height under identical air velocity, while it varies slightly with the increase of air velocity when the air velocity alters from minimum fluidization velocity to 1.36 times of the minimum fluidization velocity.

508510 Glucan Particles As Potential Carriers of Natural Flavonoids for Treatment of Idiopatic Inflammation Diseases

Abstract:

One of the novel methods of pharmaceutical formulation is the loading of the drug into biocompatible microparticles, preferably extracted from natural sources. This approach offers several potential advantages, *e.g.* increased bioavailability (the ratio of the administered drug, which actually enters the bloodstream) compared to conventional drug delivery methods, enabling the option of oral administration even for parenteral drugs. Multiple principles of the loading process are possible – drug encapsulation, adsorption onto the particle surface or precipitation inside the pores a hollow particle. The presented work focuses on precipitation of a drug from an organic solvent inside glucan particles. Glucan microparticles are ellipsoid porous shells, which are obtained from cell walls of *Saccharomyces cerevisiae* (baker´s yeast). The particle size is 2-5 micrometers. Cell walls of glucan particles are mainly composed of β-1,3-glucans. In order to prevent inflammatory reaction in organism, the yeast organelles are removed by a series of alkaline and acidic extraction steps. Subsequently, present lipopolysacharides are extracted using different organic solvents.
Lymphatic cells are mainly present in inflammation area in organism where they are responsible for progress of inflammation. Glucan particles (GPs) interact with lymphatic cell´s receptors on their surface and after that β-1,3-glucans are phagocytosed by the lymphatic cells. Therefore, GPs represent an effective oral delivery system for transporting of biological active substances. Curcumin is an anti-inflammatory flavonoid, which occurs in plants of family *Zingiberacaea*. In our project, curcumin was delivered to an inflammation in organism using GPs. Loading of curcumin into GPs was enabled via precipitation of curcumin from organic solvents. Curcumin dissolved in organic solvents was added to the GPs and through pores penetrated into cavity of glucan shell. Next, the solvent was removed and the curcumin was precipitated in glucan microparticles. GPs with the flavonoid inside were observed using confocal microscopy and SEM (scanning electron microscopy). The achieved content of curcumin was successfully analyzed by other methods, which can not be mentioned directly due to article in preparation. Furthermore, the curcumin-loaded GPs were tested *in vitro* on macrophage like cells and *in vivo* on laboratory mice in order to evaluate their effectivity. The obtained results show that these composites are promising materials for biomedical applications such as efficient treatment of inflammatory diseases (e.g. idiopatic inflammatory bowel disease, rheumatoic arthritis).

508601 Use of Advanced Imaging Techniques in Tablet Disintegration Study

Abstract:

Certified testing of tablet disintegration and dissolution provides us useful information. However, there are many more that are hidden due to techniques' restrictions. Such information is e.g. mechanism of disintegration, change of surface morphology or uniformity of the substances' distribution in the tablet. Therefore, there are many techniques used to obtain the remaining information such as Magnetic Resonance Imaging (MRI), Fourier Transform Infra-Red (FTIR) or Raman spectroscopy and microscopy, Terahertz spectroscopy, Scanning Electron Microscopy (SEM) and other.

Presented work uses tablet made by direct compression of binary mixture of ibuprofen and lactose (based on previous results there was 0.2 - 0.5 w/w of ibuprofen content used) pressed at different compaction pressures (low, medium and high). The tablets were subjected to several solutions and were studied by techniques that have insight into the tablet (MRI) and ones that can see the surface (FTIR and SEM).

MRI experiments provided us with several interesting and several expectable results. As the content of ibuprofen was increasing so was increasing the dissolution time and the wetted layer was decreasing. Similar results were observed when the compaction pressure was increased. When the tablets were submerged to dissolution limiting solutions (saturated lactose solution and solution saturated with both lactose and ibuprofen) the disintegration was severely limited and some tablets began to increase their volume. However, in case of 0.2 w/w ibuprofen tablets it didn't matter whether there was dissolution discriminating solution or not, it disintegrated anyway.

FTIR microscopy showed, that not only the amount, but also the shape and size of the particles from which the tablets were compacted can be determined. These findings were further supplemented by the dry state information obtained by Scanning Electron Microscopy after different dissolution times. Further the FTIR-ATR imaging technique helped to understand the basic mechanisms of the tablet disintegration and to study the dissolution and water penetration into the model ibuprofen-lactose tablets.

508632 Correlation Study between Liquid Penetration and Mechanical Properties of Pharmaceutical Tablets

Abstract:

Main phenomenon influencing treatment of patient in pharmacy is amount of active pharmaceutical ingredient dissolved in medium such as water or blood. Even though this isa key step to achieve desirable medical effect it is not the first one occurring.The first step is governed by surface properties such as wettability and roughness and the first contact point is followed by liquid penetration which is highly dependent not onlyon type of used liquid, ionstrength, liquid density but also on mechanical and structural properties of tablets such as porosity, structure strength or particle bonds.Through many analytical methods we can describe each of thesesteps individually. Aim of this study is to find proper correlation amongseveralmethods such as sessile drop method, texture analysis (TA) and magnetic resonance imaging (MRI) of tablet during dissolution.

For experiments binary mixture of tabletwas used. Tablets consist of lactose (hydrophilic) and ibuprofen (hydrophobic)with variable content (from 0.2 to 0.5 w/w ratio) and were also variable in compaction pressure used on them (from low â€“ 5 kN through 10kN as middle to 18 kN as high).Three different solutions were used according to their saturation with primary substances to prevent the tablet from dissolving. Surface properties were evaluated on contact angle and soaking rate of liquid, both measured by sessile drop method. Texture analysis was performed by recording the material resistance during dry and wet state of tablet. Then the data was processed and trends were obtained.Tablets during the MRI experiments were submerged into dissolution medium and the proceeding wetting front was measured. Later the images were processed via image analysis and trends were obtained.

Results obtained from surface experiments show that higher the ibuprofen content is the more hydrophobic surface appears to be and therefore the contact angle is increasing. Contact angle is independent on manufacturing pressure but soaking rate on the other hand depends strongly.MRI images show that this technique is not suitable for rapidly disintegration tablets, where the calculated error is within experiment conditions. On the other hand, tablet with prolonged dissolution are well measured. Results from MRI experiments also state that local saturation with substances supress dissolution of tablet. Texture analysis of tablets show that they undergo immediately structural weakening after being poured with solution independently on type of solution. Rate of this structural change mainly depends on solution's acidity. Thenthe tablet decreases its resistance in time which can be interpreted as relaxation rate. With increment of ibuprofen or saturation with primary substances this rate decrease, which means that dissolution is suppressed. Correlation between MRI and relaxation rate was done. Results shows that the correlation is mainly dependent on texture analysis measurement. On the other hand, correlation between MRI and soaking rate appears to be more tentative.

Experiment characterising surface, structural and liquid penetration properties were investigated by three different techniques. Obtained results were evaluated and put to mutual correlation. From liquid penetration point of interest there is correlation between MRI and contact angle method. Also, a correlation between MRI and texture analyser was found.

508642 In Situ Drug Amorphisation By Microwave Irradiation Stabilized By Mesoporous Silica

Abstract:

One of the common issues in the pharmaceutical industry is the poor solubility of the newly developed active pharmaceutical ingredients (APIs). The solubility might be significantly enhanced by changing the form of the API from crystalline to amorphous form. The amorphous state is characterized by lower thermodynamic stability and thus better dissolution properties leading to better bioavailability. In addition, the amorphous API in a dosage form (e.g. tablets) must also remain stable and not recrystallize for at least the shelf life of the drug product (usually 2 - 3 years). Various methods are being utilized to prepare and stabilize amorphous formulations in the industry, however they usually involve difficult formulation steps or suffer from poor efficiency, low stability or can be patent obstructed.

In this work a stable amorphous formulation is prepared simply using the standard wet and dry granulation process or direct tableting. Crystalline form of API is used during the initial process. The amorphous form is then prepared *in situ* and â€œon demandâ€ either during manufacturing, or by the end user by applying microwave radiation (MW).

To stabilize the API in an amorphous form after the MW, multi-structured mesoporous silica particles were added to the formulation. The particles were prepared using an emulsion synthesis of octylamine in tetraethyl orthosilicate. Using ethanol as an inhibitor instead of the commonly used nitric acid as a catalyst reduces the silica primary nano-particles nucleation rate. The formed nuclei grow larger and then self-assemble around the octylamine droplets and form hollow micro-particles. The prepared microparticles were characterized using the Scanning electron microscopy (SEM) and the Transition electron microscopy (TEM). To characterize the porous structure, nitrogen sorption (BET) analysis was conducted. The results clearly show the particle assemblies contain macropores (cavities), mesopores and micropores and possess very high specific surface area and also very large pore volume. Together with the thermal and chemical stability of silica these properties make the particles well suitable for preserving the melted API in an amorphous form after the MW.

As a proof of concept, initially powder mixtures and later simple tablets composed of Ibuprofen (model API), silica micro-particles and a filler excipient were experimentally investigated. The *in situ* amorphization of ibuprofen was conducted using a commercial microwave oven. The heating properties of the oven, i.e. the spatial distribution of microwave heating was characterized. Various microwave power input and varying composition of the mixtures (excipient/Ibuprofen ratio) were tested with respect to the melting point of the Ibuprofen. The change from the crystalline to amorphous structure due to MW heating was analyzed by x-ray diffraction (XRD), differential scanning calorimetry (DSC) and FTIR-ATR techniques. The dissolution profiles of ibuprofen from both MW untreated and treated tablets were measured using the High Performance Liquid Chromatography (HPLC).

Based on these initial results, fully formulated tablets were prepared with various API dosage strength using a tablet press. Same compression force and same dwell time was used for all of the tablets. The final tablet composition includes excipient capable of absorption of the MW, which ensures a uniform heating in the whole volume of the tablet. Next it contains the silica microparticles, which by capillary forces absorb the API once melted and stabilize it after cooling down.

Lastly, the tablets contain standard filler, lubricant and disintegrant excipients. After the MW application the tablets change their structure due to the melted API intake into the silica micro-particles, forming visible pores in the tablet structure. This has a notable impact on the dissolution and hardness of the resulting tablets. For the final tablet characterization SEM, Texture Analysis, FTIR and Magnetic Resonance Imaging were used.

508646 Manufacture of Personalised Medicines By API Printing on Porous Tablets

Abstract:

Personalised medicine offers an opportunity to produce tailor-made medicaments for patients. In order to realize this vision, it is necessary to find an alternative approach to the conventional tablet design and production. One that will replace the existing mass production of drugs offering only a few dose options of the active pharmaceutical ingredient (API), while at the same time it must meet the strict reproducibility and a defined relationship between composition and release rate of active substances requirements.

One of such approaches is to print the API on demand on a solid substrate directly when it becomes needed. Appropriate dosing of API and the possibility to even combine several APIs into a single medicament improves the variability of the dosages available for prescription and reduces the overall number of pills the patient has to take. The goal of this work is to find a simple, automated and cost-effective method of preparing personalised medicaments, which could take place directly in the pharmacy shops or at home.

The method in this work consists of a precise dosing of a dissolved API into an already prepared substrate â€" porous placebo tablets with high sorption capacity. The tablets are prepared using direct compression in a tablet press from a simple physical mixture of a binder, filler, disintegrant and lubricant. As a filler, mesoporous silica particles were used. The high pore volume provides the free space for API sorption into the tablets and the small size of the mesopores ensure the API is deposited in an amorphous form.

The prepared particles used in the formulation were characterized using Transmission electron microscopy (TEM), Dynamic light scattering (DLS) and nitrogen sorption analysis (BET). The prepared tablets hardness and uniformity was characterized both prior and after loading. The tablets after loading were characterized using the X-ray powder diffraction and the Fourier transform infrared spectroscopy to determine the API form and deposition uniformity, followed by dissolution tests to confirm the applied dosage strength and to measure the release profile. So far, several BCS Class II APIs were successfully infused into the substrates using precise on-demand dosing. Impregnation of various doses of API were accurate. Later characterization revealed no crystalline API were present and the dissolution profiles were very satisfactory.

509441 Understanding Phase Transition of Acetaminophen in the Bulk and Surface of Acetaminophen

Abstract:

Pharmaceutical solids processing often lead to the generation of new phases in the bulk and on the surface of the processed drugs. Phase transformation of the API can potentially change the physiochemical properties and influence the drug performances including solubility, dissolution and stability. To detect these changes, analytical tools such as XRD are usually employed. However, these techniques remain suitable for bulk analysis. Whereas, phase transition on the surface remains hard to detect due to the lack of literature, researches in this area and the absence of surface analytical techniques. The aim of this work is to investigate different phases (pure crystalline, partially crystalline partially amorphous and amorphous) of Acetaminophen drug by combining bulk and surface analytical techniques. For bulk analysis, FTIR and XRD have been employed. The former shows no significant differences between Acetaminophen various phases. Whereas the latter confirmed the presence of different phases indicated by a remarkable variation in the absorbance intensity the peak width. Where, high and sharp peaks obtained for the pure crystalline Acetaminophen, fewer peaks observed for semi crystalline Acetaminophen and a single broad peak observed for the amorphous Acetaminophen. For the surface analysis, Nitrogen K near-edge X-ray absorption fine structure (NEXAFS) spectra of a series of NEXAFS indicates that the precise determination of IPs and Ïƒ* shape resonance energies is sensitive enough to distinguish between the different phases of Acetaminophen. DFT calculations provides information on the molecular orbitals involved during electronic excitation occurred as a result of NEXAFS experiment.

509599 Treatment of Cystinosis through Vitamin E Modified Silicone Hydrogel

Abstract:

Cystinosis is an orphan disease caused by a genetic mutation that leads to deposition of cystine crystals in many organs including cornea. Ophthalmic manifestation of the disease can be treated with hourly instillation of cysteamine eye drops. The hourly eye drop instillation is tedious to the patients leading to poor compliance and additionally, significant degradation of the drug occurs within one week of opening the bottle, which further complicates this delivery approach. Our work focuses on designing a contact lens to treat the disease with improved efficacy compared to eye drops. Our goal is to design a lens that is safe and that can deliver a daily therapeutic dose of cysteamine to the cornea while retaining drug stability. We show that cysteamine diffuses out rapidly from all lenses due to its small size. The rapid release of cysteamine and other ophthalmic drugs can be prolonged significantly by integration of vitamin E aggregates inside of the polymer matrix of silicone hydrogels. The increase in release duration increases as square of the volume fraction of the loaded vitamin E fraction which leads to high efficacy and suggests that the mechanism of attenuation in drug release is formation of nanosized barriers. The increase in duration depends on the type of the contact lens. The barriers are not effective in single-phase hydrogel lenses while the effect in the bi-phasic silicone hydrogels depends on the type of the contact lens. Vitamin E incorporation increases the release duration of both ACUVUE®OASYS® and ACUVUE® TruEye™ but the effect is more pronounced in TruEye™ likely due to the low solubility of vitamin E in the lens matrix and higher aspect ratio of the barriers. This result along with the high aspect ratio seen through SEM imaging in silicone hydrogels suggests that barriers could be forming at the interface of the silicone and hydrogel phases. The presence of vitamin E has an additional beneficial effect of reduction in the oxidation rates, likely due to a transport barrier between the oxygen diffusing through the silicone channels and drug located in the hydrogel phase. Based on this study, both Acuvue®OASYS® and ACUVUE® TruEye™ can be loaded with vitamin E to design a cysteamine eluting contact lenses for effective therapy of cystinosis. The lenses must be worn for about 4-5 hr. each day, which is less than the typical duration of daily-wear. Vitamin E loaded contact lenses could be very useful for treating many other ophthalmic diseases as well.

517683 Systems Integration for Dry Granulation Based Continuous Pharmaceutical Tableting

Abstract:

Over the past decade, the pharmaceutical industry has seen significant progress in the mechanistic understanding of relevant unit operations, availability of multiple sensor technologies for inline monitoring of critical quality attributes and advances in data analytics to support manufacturing decision-making. These have enabled a systematic realization of continuous manufacturing for pharmaceutical tablets using the direct compaction route. However, the limitations and difficulty in handling of some particulate materials provide the impetus for investigating granulation routes to the continuous manufacturing of tablets.

In this work, the integration of dry granulation into a continuous tableting line is investigated. Specifically, the Process Analytical Technology (PAT) tools required for inline and at-line monitoring of the tableting line, and the semi-mechanistic models for the roller compactor and the tablet press are discussed. While the Kawakita model is used for the tablet press, the roller compaction process model is consisting of an integration of material balance and actuator dynamics from Hsu et. al (2010) with the modified Johansonâ€™s model presented by Liu & Wassgren (2016). The PAT tools, mechanistic equations, and automation system are integrated for measurement validation and data accuracy using data reconciliation in a structured ISA-95 network using DeltaV. The experimental data is presented from the operation of the continuous dry granulation line-based pilot plant at Purdue University.

509248 Modeling Granular Material Mixing and Segregation Using a Multi-Scale Model

Abstract:

A multi-scale model is presented for predicting the magnitude and rate of powder mixing and segregation in blenders. The model combines diffusion and percolation correlations from small-scale discrete element method (DEM) simulations with advective flow field information from large-scale finite element method simulations. The multi-scale model predictions closely match results from both large-scale DEM-only simulations and experiments, but they are much more computationally efficient than DEM-only simulations. Therefore, in contrast to DEM, the multi-scale model is well suited for predicting blending in industrial-scale systems.

Modelling Granular Media using Dynamical Density Functional Theory

T. Hurst, B. Goddard & R. Ocone

Motivation

Granular media is ubiquitous in industrial and physical problems. One of the characteristic qualities of granular media at the microscopic level is inelastic hard body collisions, which drive macroscopic behaviour. Modelling granular media is a complex and interesting problem. Solid-like phenomena make it difficult to approximate a granular system with a continuum model, while chaotic behaviour and inelastic collapse in hard particle systems introduce modelling challenges on the microscopic scale. We investigate uses of dynamical density functional theory (DDFT), augmented with data gained from microscopic models, for granular media simulation.

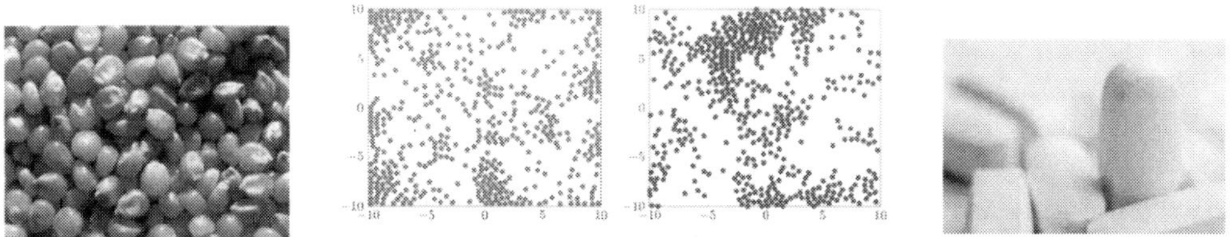

Figure 1: Inelastic systems with reflective (left) and periodic (right) boundaries

Dynamical Density Functional Theory

To derive a DDFT, we start with a set of Langevin equations for N hard particles of diameter σ in d dimensions with positions $\boldsymbol{r}^N$ and momenta $\boldsymbol{p}^N$:

$$\frac{d\boldsymbol{r}^N}{dt} = \frac{\boldsymbol{p}^N}{m}, \quad \frac{d\boldsymbol{p}^N}{dt} = -\nabla_{\boldsymbol{r}^N} V(\boldsymbol{r}^N, t) - \gamma \boldsymbol{p}^N + (m k_b T \gamma)^{1/2} \boldsymbol{G}(t), \tag{1}$$

on the hard sphere domain, $\mathcal{D}_N = \{(\boldsymbol{r}^N, \boldsymbol{p}^N) \in \mathbb{R}^{2dN} : \forall i, j, \|\boldsymbol{r}_i - \boldsymbol{r}_j\| \geq \sigma\}$, where $V(\boldsymbol{r}^N, t)$ includes external and interparticle potentials, and $\boldsymbol{G}(t)$ is an optional stochastic force term. The Kramer's equation is then a partial differential equation for its distribution function $f^{(N)}(\boldsymbol{r}^N, \boldsymbol{p}^N, t)$:

$$\frac{\partial f^{(N)}}{\partial t} + \frac{1}{m}\boldsymbol{p}^N \cdot \nabla_{\boldsymbol{r}^N} f^{(N)} + \nabla_{\boldsymbol{r}^N} V \cdot \nabla_{\boldsymbol{p}^N} f^{(N)} = \gamma \nabla_{\boldsymbol{p}^N} \cdot \boldsymbol{p}^N f^{(N)} + \gamma k_b T \nabla^2_{\boldsymbol{p}^N} f^{(N)}. \tag{2}$$

We integrate out all but one position and momenta in the Kramers equation, to find:

$$\frac{\partial}{\partial t} f^{(1)} = \mathcal{L}_1 f^{(1)} + \mathcal{L}^{2,\dots,N}_{\text{part}}(f^{(k)}) + \mathcal{L}^{(1)}_{\text{diss}} f^{(1)} + \mathcal{L}_{\text{coll}}(f^{(2)}), \tag{3}$$

where $\mathcal{L}_1$ introduces effects from external potentials, $\mathcal{L}^{(1)}_{\text{diss}}$ effects due to dissipation and friction, $\mathcal{L}^{2,\dots,N}_{\text{part}}$ effects from interparticle potentials, and

$$\mathcal{L}_{\text{coll}}(f^{(2)}) = -\sigma^{d-1} \int_{\mathbb{R}^d} d\boldsymbol{p}_1 \int_{\mathbb{R}^d} d\boldsymbol{p}_2 \int_{S_1^{d-1}} (\boldsymbol{p}_1 - \boldsymbol{p}_2) \cdot \omega f^{(2)}(\boldsymbol{r}_1, \boldsymbol{r}_1 + \sigma\omega, \boldsymbol{p}_1, \boldsymbol{p}_2, t) d\omega, \tag{4}$$

is a general collision term that arises due to the hard-sphere domain, when assuming only binary collisions. By taking the first two $\boldsymbol{p}$-moments and upon additional assumptions we arrive at a closed set of equations to model the dynamics of the system, a DDFT [1, 2]:

$$\frac{\partial}{\partial t}\rho = -\nabla \cdot (\rho \boldsymbol{v}), \tag{5}$$

$$\left(\frac{\partial \boldsymbol{v}}{\partial t} + \boldsymbol{v} \cdot \nabla \boldsymbol{v}\right) + \gamma \boldsymbol{v} = -\frac{1}{m}\nabla \frac{\delta \mathcal{F}[\rho]}{\delta \rho} - \sigma^{d-1} g_\sigma \int_{S_1^{d-1}} \rho(\boldsymbol{r} + \sigma\boldsymbol{\omega}/2)\rho(\boldsymbol{r} - \sigma\boldsymbol{\omega}/2)$$
$$\times \boldsymbol{\omega} \cdot \left\{\frac{k_B T}{m}\mathbf{1} + \boldsymbol{v}(\boldsymbol{r} + \sigma\boldsymbol{\omega}/2) \otimes \boldsymbol{v}(\boldsymbol{r} + \sigma\boldsymbol{\omega}) - \boldsymbol{v}(\boldsymbol{r} + \sigma\boldsymbol{\omega}/2) \otimes \boldsymbol{v}(\boldsymbol{r} - \sigma\boldsymbol{\omega}/2)\right\} \mathrm{d}\boldsymbol{\omega}, \tag{6}$$

where $\mathcal{F}$ is the well-studied Helmholtz free energy operator, which can include particle-particle interactions and external potentials, ρ is the density, $\boldsymbol{v}$ is the local velocity and g_σ is the value of the correlation function at σ, which must be determined. We choose to model g using statistical data from microscopic simulation.

DDFT Simulations

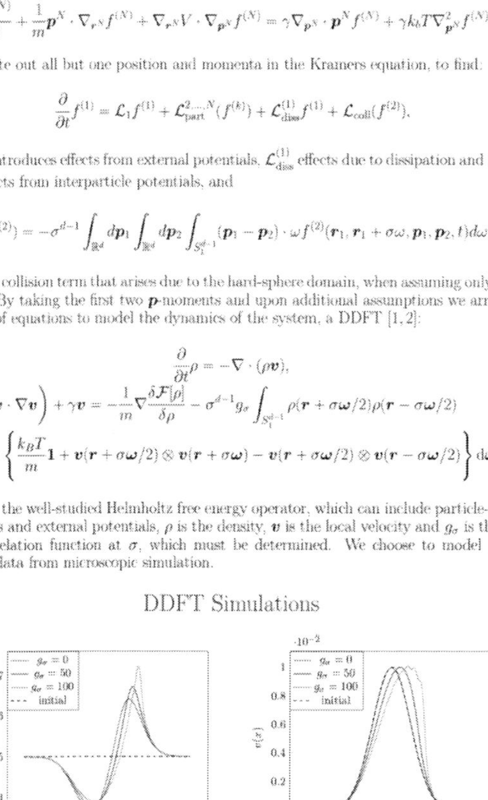

Figure 2: A simple 1D system neglecting all but the collision operator in (5-6), for different values of g. Left: density $\rho(x)$, right: local average velocity $v(x)$.

Data-Driven Modelling of g

A system of N hard spheres of diameter σ in d dimensions with positions $\boldsymbol{x}^N$ and velocities $\boldsymbol{v}^N$ obey

$$\frac{d\boldsymbol{r}^N}{dt} = \boldsymbol{p}^N, \quad \frac{d\boldsymbol{p}^N}{dt} = V(\boldsymbol{r}^N, t), \tag{7}$$

with a pairwise collision condition: if particles i and j are in contact, i.e. $\|\boldsymbol{r}_i - \boldsymbol{r}_j\| = \sigma$ and $(\boldsymbol{r}_i - \boldsymbol{r}_j) \cdot (\boldsymbol{p}_i - \boldsymbol{p}_j) < 0$, we instantaneously change the velocities via the rule:

$$\boldsymbol{p}_i^{out} = \boldsymbol{p}_i^{in} - \frac{1+\alpha}{2}\boldsymbol{\nu}_{ij} \cdot (\boldsymbol{p}_i^{in} - \boldsymbol{p}_j^{in})\boldsymbol{\nu}_{ij}, \quad \boldsymbol{p}_j^{out} = \boldsymbol{p}_j^{in} + \frac{1+\alpha}{2}\boldsymbol{\eta}_{ij} \cdot (\boldsymbol{p}_i^{in} - \boldsymbol{p}_j^{in})\boldsymbol{\eta}_{ij}, \tag{8}$$

where $\boldsymbol{\eta}_{ij} = \frac{\boldsymbol{r}_i - \boldsymbol{r}_j}{\|\boldsymbol{r}_i - \boldsymbol{r}_j\|}$ and $\alpha \in [0, 1]$ is the coefficient of restitution. For simple V, the Newton equations can be modelled using Event Driven Particle Dynamics (EDPD). The cell method and bounded increasing priority queues [3] reduce the computational effort to $\mathcal{O}(1)$ per collision. Using statistical methods for synthetic data (e.g. emulation [4]) we gain an understanding of the correlation function g near equilibrium, which we can include in (6-7).

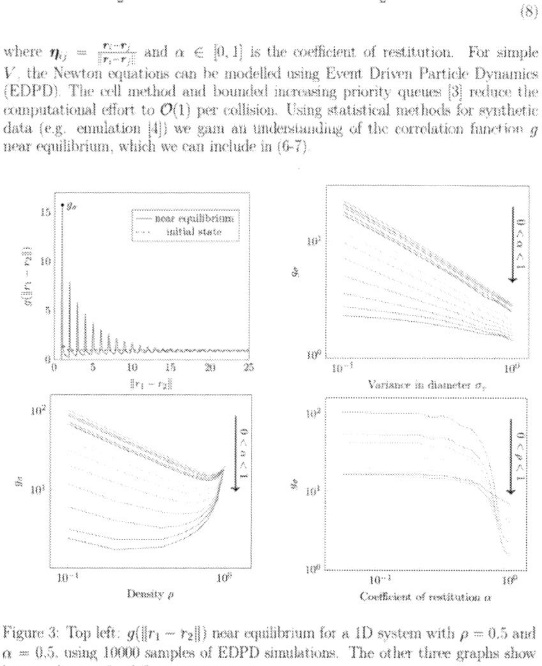

Figure 3: Top left: $g(\|r_1 - r_2\|)$ near equilibrium for a 1D system with $\rho = 0.5$ and $\alpha = 0.5$, using 10000 samples of EDPD simulations. The other three graphs show how g_σ changes for different parameters.

Future Work

We wish to include additional moments in the DDFT derivation, simulate DDFT in higher dimensions including interparticle potentials, and use EDPD in higher dimensions to calculate g.

[1] Archer, A., *J.Chem.Phys.*, **130**, 014509, (2009).
[2] Goddard, B. et. al., *Phys.Rev.Lett*, **109**, 1-5, (2009).
[3] Paul, G., *J.Comp.Phys.*, **221**, 615-625, (2011).
[4] O'Hagan, A., *Reliab.Eng.Syst.Saf.*, **91**, 1290-1300, (2006)

The support from EPSRC (grants EP/N034066/1, EP/L025159/1 and EP/L016508/01) is kindly acknowledged.

509555 A Study on Partially Wetted Particle Collisions with a Wet Wall

Abstract:

For the production of poly-ethylene, a gas-phase polymerization reaction is executed in a fluidized bed reactor. The polymerization reaction produces an excess of heat and the reactor is unable to efficiently remove the excess heat. The introduction of an Inert Cooling Agent (ICA) solves the temperature problem but also has a downside. Introducing the ICA in the reactor influences the particle collisions, leading to substantially different flow behavior. Besides the reduction of the restitution coefficient, the introduction of the liquid can result in adhesion instead of collision. This adhesion results in the formation agglomerates in the reactor and reduces the efficiency of the process. Due to the complex physics of this problem, Computational Fluid Dynamics (CFD) is a suitable method to gain insight in this phenomenon.

In order to study this complex system of partially wetted particle collisions a Direct Numerical Simulation (DNS) is used for incompressible flow. To model the coupling between the solid phases and the fluid phase, a second order accurate Immersed Boundary Method (IBM) is used (Deen et al.,2012). The interface between the fluids is tracked by a Volume Of Fluid (VOF) method (Baltussen et al., 2017). The contact-line dynamics are solved by imposing a contact angle as a boundary condition (Patel et al., 2017).

The simulated results will be compared with dry particle collisions on a wet wall (Tang et al., 2017) and with wet particle collisions on a wet wall.

References:

1. Deen, N.G., Kriebitzsch, S.H., van der Hoef, M.A. and Kuipers, J.A.M., Direct numerical simulation of flow and heat transfer in dense fluidâ€"particle systems. Chemical Engineering Science, 81 (2012): 329-344.

2. Baltussen, M. W., Segers, Q. I. E., Kuipers, J. A. M., and Deen, N. G., Cutting bubbles with a single wire. Chemical Engineering Science 157 (2017): 138-146.

3. Patel, H.V., Das, S., Kuipers, J.A.M., Padding, J.T. and Peters, E.A.J.F., A coupled Volume of Fluid and Immersed Boundary Method for simulating 3D multiphase flows with contact line dynamics in complex geometries. Chemical Engineering Science, 166 (2017): 28-41.

4. Tang, Y., Buck, B., Heinrich, S., Deen, N.G., Kuipers, J.A.M., Interface-resolved simulations of normal collisions of spheres on a wet surface. American Institute of Chemical Engineers, (2017).

508853 Strategies to Establish Mechanistic Liquid Transport Models for Wet Granular Matter

Abstract:

Strategies to Establish Mechanistic Liquid Transport Models for Wet Granular Matter

Johannes G. Khinast and Stefan Radl

Wet granular matter plays a significant role in both nature and industry. Unfortunately, these materials are intrinsically hard to model due to the fact that three phases interact. Examples documenting these challenges are (i) the counter-intuitive fluid depletion in sheared granular materials,[1] or (ii) the spontaneous formation and growth of agglomerates in wet fluidized beds.[2] Relevant phenomena arise due the interplay of (i) liquid-particle interactions (e.g., the redistribution of liquid on particles), (ii) particle-particle collisions (i.e., aggregation and separation dynamics[3]), and (iii) heat and mass transfer (i.e., evaporation of liquid).

We summarize recent advances related to the above phenomena, and that ultimately aim on predicting the amount of liquid adhering to individual particles. Specifically, in Part I we discuss experimental routes, focusing on the observation of individual liquid-particle interactions. In Part II, modeling strategies are discussed, including (i) sub-particle-scale models (i.e., models on the droplet and pore scale), (ii) inter-particle models necessary to describe clustering and aggregation dynamics (a focus will be on bridge filling models[4,5]), as well as (iii) equipment scale models that account for evaporation and heat exchange[6] in industrially-relevant devices. Part III is devoted to simulation strategies that are needed to solve the model equations. Specifically, topics related to (i) improving numerical schemes (e.g., the voidage calculation algorithm), as well as (ii) improving the speed of calculations (e.g., by using multi-core CPU clusters or graphics processing units) are discussed.

Finally, we provide an outlook, focusing on advanced sub-particle-scale models that consider the liquid morphology on individual particles. Also, we summarize the most recent activities in the field by providing a tentative regime map for liquid transport rates in wet granular matter. This map is thought to be useful for predicting the behavior of wet fluidized beds without a rigorous model or simulation. Extensions to large-scale systems and simulations by our groups are shown as well.

References

1. Mani, R. A., Kadau, D., Or, D. & Herrmann, H. J. Fluid depletion in shear bands. *Phys. Rev. Lett.* **109,** 248001 (2012).

2. Boyce, C. M. *et al.* Growth and Breakup of a Wet Agglomerate in a Dry Gas-Solid Fluidized Bed. *AIChE J.* **63,** 2520â €"2527 (2017).

3. Donahue, C. M., Davis, R. H., Kantak, A. A. & Hrenya, C. M. Mechanisms for agglomeration and deagglomeration following oblique collisions of wet particles. *Phys. Rev. E - Stat. Nonlinear, Soft Matter Phys.* **86,** 21303 (2012).

4. Wu, M., Khinast, J. G. & Radl, S. The Effect of Liquid Bridge Model Details on the Dynamics of Wet Fluidized Beds. *AIChE J.* **submitted,** (2017).

5. Mohan, B., Kloss, C., Khinast, J. G. & Radl, S. Regimes of Liquid Transport through Sheared Beds of Inertial Smooth Particles. *Powder Technol.* **264,** 377â€"395 (2014).

6. Askarishahi, M., Salehi, M. & Radl, S. Full-Physics Simulations of Spray-Particle Interaction in a Bubbling Fluidized Bed. *AIChE J.* **63,** 2569â€"2587 (2017).

509664 Multi-Scale Characterization of Powder Compaction Spanning Single Particle, Single Tablet Fabrication, and Industrial Rotary Tablet Press

Abstract:

In a continuous tableting line, hundreds of tablets are produced every minute with the overall plant throughput being multiple kilograms per hour. Tablets need to have certain Critical Quality Attributes (CQAs) at the end of the production line; hence a Distributed Control System (DCS) plays the crucial role of controlling various unit operations by using a network of different sensors. The DCS system needs predictive, robust and computable models to inform changes in upstream and downstream process parameters. These models must be able to capture the dynamics of the system induced by variations in the blend composition, material flow properties, production rate, etc. Therefore, it is important to characterize the powder and the compaction process thoroughly.

We perform this process characterization at three distinct length scales, namely the particle scale, the single tablet scale, and the bulk production scale. Specifically, we characterize individual pharmaceutical powder particles under diametrical compression using a Shimadzu MCTâ€W 510 micro-compression tester, at small deformation rates. We characterize the fabrication of single tablets using a Gamlen D Series compaction simulator. Even though the compactor simulator is much more highly instrumented than most industrial rotary tablet presses, the compaction rate is smaller than that of tablet presses. We also use a Sotax AT4 tester to assess in-line the CQAs of a few selected tablets made in a continuous tableting line using a Natoli BLP-16 press.

It is worth noting that we collect data at different length and time scales and thus special care needs to be taken in characterizing the compaction process and in developing predictive models for the DCS system. In this talk, we will discuss the similarities and differences between the three components of this multi-scale characterization approach. We will also show that models at various length scales account for different physical mechanisms and all are needed to robustly capture the dynamics of the system induced by changes and perturbations in the continuous tableting line, such as variations in the blend composition, material flow properties, and production rate.

508888 Experiments and Simulations on Particle Impregnation By Metal Solutions for Industrial Catalysts: From Fundamentals to Scale up

Abstract:

Impregnation of active metals onto a porous support is a crucial step in the preparation of industrial heterogeneous catalysts. The process is typically performed in rotating vessels with a spray-nozzle to distribute the liquid until the pore volume is reached. The inter-particle variability of the impregnated liquids inside the particles can significantly affects the activity and selectivity of the resulting catalyst. Current scale-up practices lead to poor fluid distribution and inhomogeneity in metal content. The aim of this work is to understand the dynamic behavior of the particles under the spray nozzle, which is essential to achieve a desired content uniformity and to develop a scale-up model for the catalyst impregnation process. In this work, impregnation was explored by applying Discrete Element Method simulation techniques in conjunction with experimentation. The simulations were validated by experiments utilizing both a geometrically identical double cone and a cylindrical blender fixed with a single nozzle impregnator. In addition, we developed a novel fluid transfer model of metal solutions from the nozzle to the particles and between particles in the bed. Results obtained from our model show good agreement with experiments in the particle metal content along the axis of rotation. We also studied spheres and cylinders of different sizes and aspect ratios. Axial mixing analysis and liquid distributions are used to investigate the propagation of the fluid throughout the particle bed with the goal of understanding the effect of operational and material parameters and ultimately to improve fluid content uniformity in systems with particles of different morphologies. The fluid content uniformity is characterized by the relative standard deviation (RSD) of the liquid content from all the particles in the system. Our results show that cylinders always take less time to physically mix than spheres of the same diameter and mixing times are also shorter for cylinders of higher aspect ratios when compared to cylinders of smaller aspect ratio. Likewise, the times to reach good fluid content uniformity are shorter for cylinders with higher aspect ratios as compared with cylinders with lower aspect ratios. We found a strong linear correlation between the times to achieve good axial mixing and the times to achieve good fluid content uniformity in the entire particle bed, which suggests that mixing in the axial direction controls fluid uniformity in the entire particle bed. Lastly, we performed scale up studies on the impregnation process using two dimensionless numbers to characterize the system: Froude number (Fr) and flow rate number (C_Q). In general, smaller flow rate numbers tend to give more homogeneous liquid distribution during and after the total liquid is sprayed; however, the there is a limit on how small a Froude number can be. We found optimal values for Froude and flow rate numbers. An optimal Fr number value is found when the RSD is minimal, which gives the best mixing performance. Using larger C_Q numbers the spraying can be completed in shorter times, but the particles show greater inhomogeneity of the fluid content in the particle bed. An optimal value for the flow rate number C_Q allows the impregnation process to finish in an acceptable time frame, and the final product to have a uniform distribution of metal. The ability to control catalyst impregnation and establish effective models for large batches can significantly reduce the amount of time required per batch while simultaneously achieving a more homogeneous liquid distribution.

Multi-Scale Computational Modeling of Selective Laser Melting

Daniel Moser
Sandia National Laboratories
dmoser@sandia.gov

Jayathi Murthy
University of California Los Angeles

Abstract

Selective Laser Melting (SLM) is an additive manufacturing technique used to rapidly create parts by selectively fusing successive layers of powder with a laser. Defects can arise in SLM parts due to poor selection of processing parameters. Accurate models of the SLM process have the potential to reduce the testing and experimentation required when using SLM. However, simulating the complete build of an SLM part is challenging due to the difference in scale between the size of the part being built and the size of the particles and laser being used. Bed-scale continuum models, in which the powder bed is represented as a continuous medium, are typically used to reduce computational cost. Continuum models require as inputs volume-averaged effective properties of the powder which are often unknown and difficult to obtain by experimental means. In this work, we describe a multi-scale modeling approach in which particle-scale models, where individual particles are resolved in a small, representative, domain, are used to predict effective properties of a powder with uncertainty. These properties are then used in a part-scale continuum model to predict temperature history, melt percentage, and thermal stresses with uncertainty in an SLM build.

Introduction

Free-form fabrication techniques reduce the cost of creating prototypes or small batch parts by producing parts directly from CAD models without part-specific tooling. Selective Laser Melting (SLM) is a promising free-form fabrication process as it works with a wide variety of materials, particularly metals. SLM produces a solid object by selectively fusing successive layers of powder. A thin layer of powder is deposited on top of a piston. The surface of the powder is then scanned by a laser with a modulated power, fusing the powder to itself and the layer below where the cross-section is intended to be solid and leaving it loose where it is not. When the scan of the layer is complete, the piston holding the part is lowered, a new layer of powder is deposited on top and the process repeats. After the build is complete, the loose powder is removed, leaving the final part.

Despite its promise, SLM is a process still in its infancy and suffers from a number of limitations. Control of SLM processing parameters, such as laser power, laser speed, scanning pattern, and powder pre-heating temperature is critical for the creation of a high-quality part. Improper control of these parameters can lead to geometric errors, incomplete melting, where layers are not melted all the way through and do not adequately fuse to the layer below, and over melting, where material is vaporized by the laser, creating pores. Furthermore, large thermal stresses can cause deformation, reduce part performance, and even result in build failure [5]. However, it is difficult to determine what the optimum processing parameters should be for a given part geometry and material. Thus, experimentation is often required when using new materials or geometries in order to determine the parameters needed to produce an acceptable part (eg. [11]). Also, no technique currently exists to predict whether or not produced parts will meet imposed quality standards. Comprehensive and accurate computational modeling of the SLM process has the potential to reduce the amount of manual experimentation and testing required when producing and certifying a part. Models could theoretically be used to determine optimum processing parameters for a new material or geometry, predict the quality of a part produced using a given set of parameters, and to inform control schemes [4].

There are broadly two approaches to modeling the SLM process, bed-scale models and particle-scale models. Bed-scale models approximate a powder bed as a continuous medium, neglecting individual particles. Particle-scale models resolve individual particles in the domain and often encompass detailed physics. While using a particle-scale model to simulate a full-scale part build would be too computationally expensive, particle-scale models of small portions of SLM powder beds have had success in qualitatively explaining some phenomena occurring at the particle level [2]. However, it is often unclear how particle-scale model results can be effectively used to make predictions about the bed-scale. This work presents a framework in which particle-scale simulations are used to determine effective parameters for use in a bed-scale model. Uncertainties calculated for effective parameters are propagated through the bed-scale model using polynomial chaos expansions, allowing the effects of uncertainties at the particle-scale to be assessed at the bed-scale.

Modeling Approach

Mathematical Model

For the bed-scale model, heat transfer within the SLM powder bed is governed by the enthalpy equation [13].

$$\frac{\partial H}{\partial t} = \nabla \cdot (k_{eff} \nabla T) + f(x, y, z, t) \tag{1}$$

k_{eff} is the thermal conductivity, $T(x, y, z, t)$ is the temperature, $H(x, y, z, t)$ is the total enthalpy of the powder and $f(x, y, z, t)$ is the heat source. k_{eff} is generally unknown for a given powdered material and will be informed from particle-scale model results. Mixed convective-radiative boundary conditions are applied to the top boundary of the domain. The remaining boundaries are insulated.

Enthalpy and temperature are coupled through the relationship [13]

$$H = (1 - g(T)) \int_{T_{ref}}^{T} \rho c_s dT + g(T) \int_{T_{ref}}^{T} \rho c_l dT + g(T)\rho L \tag{2}$$

$g(T)$ is the melt fraction, ρ is the density, c_s and c_l are the specific heats of the solid and liquid phases, and L is the latent heat of fusion. The lever rule is used as the relation between melt fraction and temperature, $g(T) = \frac{T-T_s}{T_l-T_s}$, for T between T_s, the solidus temperature, and T_l, the liquidus temperature.

The laser heat source, $f(x, y, z, t)$, is modeled as a Gaussian [10].

$$f = \alpha I_0 \beta exp\left(-2\frac{(x-x_l)^2 + (y-y_l)^2}{\omega^2} - \beta z\right) \tag{3}$$

α is the optical absorptivity of the powder bed, I_0 is the laser intensity, x_l and y_l are the x and y positions of the laser, ω is the laser characteristic radius and β is the extinction coefficient. Like k_{eff}, α and β are often not known for powder materials and will be informed from particle-scale model results.

All domain boundaries are taken as zero flux except for the top boundary where mixed convective-radiative boundary condition is used.

In order to predict residual stresses in built parts, a bed-scale mechanical model is coupled to the thermal model. In the mechanical model, the incremental strain, the change in strain over a time increment, $\Delta\boldsymbol{\epsilon}$, is related to the incremental stress through Hooke's law for elastic, isotropic materials as given in Equation 4.

$$\Delta\boldsymbol{\sigma} = 2\mu(\Delta\boldsymbol{\epsilon} - \Delta\boldsymbol{\epsilon}^p - \Delta\boldsymbol{\epsilon}^t) + \lambda \boldsymbol{I} tr(\Delta\boldsymbol{\epsilon} - \Delta\boldsymbol{\epsilon}^p - \Delta\boldsymbol{\epsilon}^t) \quad (4)$$

λ and μ are Lamé's parameters for the material, $\Delta\boldsymbol{\epsilon}^e$ is the elastic strain increment, $\Delta\boldsymbol{\epsilon}^p$ is the plastic strain increment, and $\Delta\boldsymbol{\epsilon}^t$ is the thermal strain increment. The thermal strain increment is calculated from the temperature solution using Equation 5 [9].

$$\Delta\boldsymbol{\epsilon}^t = \int_{T_1}^{T_2} a(T)dT \quad (5)$$

T_1 and T_2 are the temperatures at the beginning and end of a time increment and a is the coefficient of thermal expansion.

For the plastic strain increment, $\Delta\boldsymbol{\epsilon}^p$, a J2 plasticity model is employed, in which $\Delta\boldsymbol{\epsilon}^p$ is calculated by enforcing the inequality that the von Mises equivalent stress not exceed the yield strength.

$$\sqrt{\frac{3}{2}\boldsymbol{s} : \boldsymbol{s}} \leq \sigma_y \quad (6)$$

$\boldsymbol{s}$ is the deviatoric of the total stress, $\boldsymbol{\sigma} + \Delta\boldsymbol{\sigma}$, and σ_y is the yield strength of the material. Finally, the stress increment is then related to the displacement increment through Newton's law for statics.

$$0 = \nabla \cdot \Delta\boldsymbol{\sigma} \quad (7)$$

All domain boundaries are taken as zero stress except for the bottom boundary where a zero displacement condition is used to simulate a large baseplate. For the stress model, $\boldsymbol{\sigma}$ and $\boldsymbol{\epsilon}$ is the output quantity of interest and the temperature, T, provides coupling to the thermal model.

Implementation Details

The governing equations are discretized using a finite volume discretization scheme. The computational domain is divided into regular, hexahedral cells and the governing equations are integrated over the cell volumes to form a system of linear equations which are solved iteratively. Since the mechanical governing equations depend on the solution to the thermal governing equations, but not visa-versa, the equations are solved segregated, with the thermal solution being calculated first and the computed temperature passed to the mechanical solver.

The bed-scale model is a multi-phase system consisting of powder, sold metal, and a background gas (generally air). Domains are initialized with a layer of powder on top of a solid metal baseplate, as shown in Figure 1. Material properties are then assigned in the domain based on location, with cells that lie in the powder layer assigned the properties of powder and cell that lie in the plate assigned properties of solid metal. As powder melting occurs, powder volume fractions are converted to a combination of solid volume fractions and air volume fractions, depending on the prescribed porosity of the powder. This is shown in Figure 2. Material properties of cells that

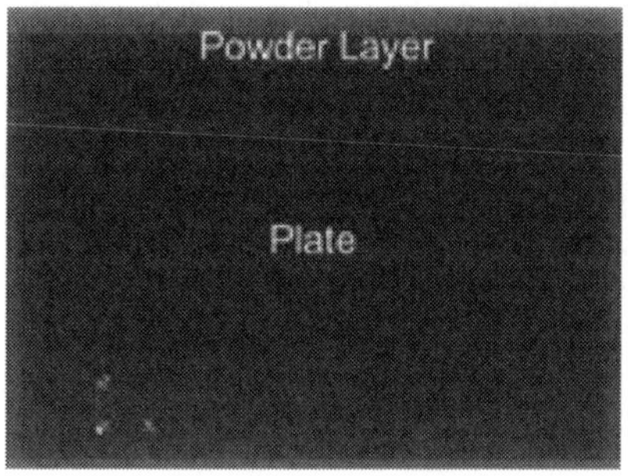

Figure 1: Initial domain configuration of bed-scale model showing single layer of powder on a baseplate

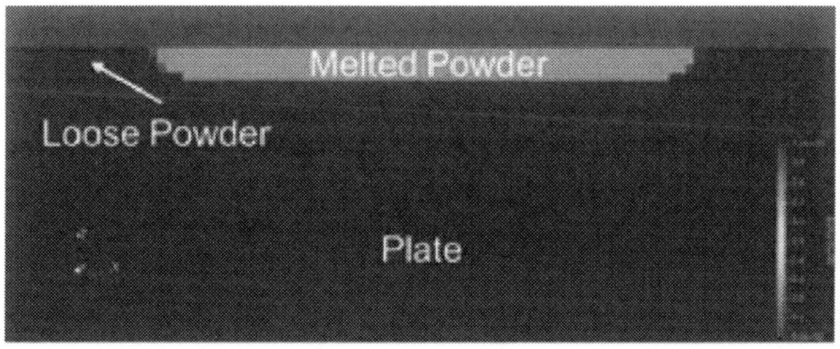

Figure 2: Consolidated powder on top of baseplate. Colors indicate volume fraction of metal. Red shows the plate with a metal volume fraction of 1. Blue shows powder with a metal volume fraction of 0. Orange shows melted powder converted to a combination of metal and air volume fractions

contain a mixture of materials are computed using volume averaging as given in Equation 8.

$$\zeta_{total} = \gamma_{powder}\zeta_{powder} + \gamma_{solid}\zeta_{solid} + (1 - \gamma_{powder} - \gamma_{solid})\zeta_{air} \tag{8}$$

γ is the volume fraction of each component and ζ is a given material property.

Both the thermal and mechanical model equations are nonlinear. The thermal nonlinearity due to the enthalpy term is solved using Newton's method with a line search. The plastic nonlinearity in the mechanical model is solved using the elastic predictor-plastic corrector method in which a trial stress is computed assuming a fully elastic response and the von Mises equivalent stress computed for each cell. For cells in which the inequality of Equation 6 is violated, a plastic strain increment is calculated to return the equivalent stress to the yield surface. Equation 4 is then used to update the stress increment for the next iteration and the solution iterated until the relative change in the displacement field with each iteration is below a prescribed tolerance.

The mechanical model also requires that Lamé's parameters, μ and λ, be specified for the powder and the air when these quantities are only well defined for a solid. To get around this, Lamé's parameters for air and powder are set to be 3 orders of magnitude smaller than the values for the solid material, as it is determined that reducing them further has little impact on the solution and slows down convergence. Thus, air and powder cells provide very little resistance to deformation and allow the solid cells they border to deform freely. Powder and air are also set to have thermal expansion coefficients of zero and arbitrarily large yield strengths so they do not undergo thermal expansion or plastic deformation.

Results

The bed-scale thermo-mechanical model is compared against the experimental results of Yadroitsava et al. [15]. Yadroitsava measured experimentally the residual stress present in a single layer of 316 stainless steel powder melted onto a stainless steel substrate using SLM. The layer thickness used is $40\mu m$ and the size of the area being melted is 1cm x 1cm. Residual stresses were measured in directions parallel and perpendicular to the laser scan direction in the plane of the powder bed. Measurements were made at the center of the part, both at the bottom of the layer, near the baseplate, and at the top, near the free surface [15].

In order to compare to Yadroitsava's results, simulations are conducted using a domain consisting of a single layer of stainless steel powder. Thermal properties for 316 stainless steel are given by Khairallah et al. [3, 2] and summarized in Table 1. Mechanical properties are given by Hodge et al. [1]. As properties such as Young's modulus and yield strength are temperature dependent, polynomial curve fits are applied to Hodge's mechanical property values to create functions that can be used by the simulation. Mechanical properties used in the simulation are summarized in Table 2. Laser parameters used in the simulation are chosen to match those used by Yadroitsava and are summarized in Table 3.

Table 1: 316 Stainless Steel Thermal Properties

Property	Value
Density	7.43 g/cm^3
Thermal Conductivity	20.0 W/m K
Specific Heat	320.3 + 0.379T J/kg K (T in K)
Emissivity	0.3-0.6 (uniform random)
Latent Heat	270 KJ/kg
Solidus Temperature	1648 K
Liquidus Temperature	1673 K

Table 2: 316 Stainless Steel Mechanical Properties

Property	Value
Young's Modulus	2.304e11 - 9.3795e7T Pa (T in K)
Coefficient of Thermal Expansion	1.2989e-5 + 8.4443e-9T - 2.8339e-12T^2 1/K (T in K)
Yield Strength	4.2094e8 - 6.004e5T + 3.9859e2T^2 - 0.1143T^3 Pa (T in K)
Poisson's Ratio	0.3

The computational domain is meshed using a uniform mesh size of $10\mu m$ and a constant time

Table 3: Laser Properties for Stainless Steel Stress Calculations [15]

Property	Case 1
Power	50 W
Speed	10 cm/s
Beam Diameter (FWHM)	70 μm

step size of 0.0001s is used. This is selected as further mesh and temporal refinement results in variations of less than 5% for the maximum value of both the temperature and stress solutions.

While some effective powder properties are calculated by voluming averaging the properties of steel and air, three properties, extinction coefficient, thermal conductivity, and absorption coefficient, are not accurately approximated using this method [6]. These parameters are instead calculated using particle-scale models. Moser et al. [7] developed a method for modeling interactions between a laser and powder particles. The method uses the Discrete Element Method (DEM) to represent powder particles and ray-tracing to simulate a laser. Powder beds are generated and laser rays propagated through them by calculating ray-sphere intersections. On every ray-sphere interaction, a portion of the ray energy is absorbed and a portion reflected to propagate further in the domain. By tracking the deposition of energy in the domain, effective absorptivites and extinction coefficients for powders can be obtained. Based on this analysis, a correlation was derived between effective powder absorptivity and solid metal absorptivity and is given in Equation 9.

$$\alpha_{eff} = 0.053 + 1.37\alpha - 1.04\alpha^2 + 0.399\alpha^3 \pm 0.02 \tag{9}$$

Similarly, a correlation between the effective powder extinction coefficient, solid metal absorptivity, and the average radius of the powder particles was derived and is given in Equation 10.

$$\beta R = 0.325 + 1.03\alpha - 1.22\alpha^2 + 0.587\alpha^3 \pm 0.02 \tag{10}$$

However, the absorptivity of 316 stainless steel is itself an uncertain quantity which varies based on the surface properties of the metal. As given in Table 1, this quantity is approximated as a uniform random variable between 0.3 and 0.6. In oder to calculate the probability distributions of the effective powder absorptivity and extinction coefficient, samples are randomly drawn from the stainless steel absorptivity distribution and each sample is used to calculate a powder absorptivity and extinction coefficient using Equations 9 and 10. The uncertainties of 0.02 in each of the equations are approximated as Gaussian random variables with means of zero and standard deviations of 0.02. After an effective absorptivity or extinction coefficient is calculated for a sample, a sample is drawn from the Gaussian distribution and added to the result. This process is repeated for subsequent samples to create a histogram of effective absorptivity and extinction coefficient values. These histograms are then normalized by the total number of samples drawn to create an approximation of the probability density functions of these quantities, as shown in Figure 3.

Moser et al. [8] also developed a method for calculating effective thermal conductivities of powders using the DEM to represent powder particles. Powder beds are generated and thermal gradients imposed across them. Relations for heat transfer between particles due to particle-particle contact, conduction through the background gas, and radiation are used to calculate the resulting heat flux

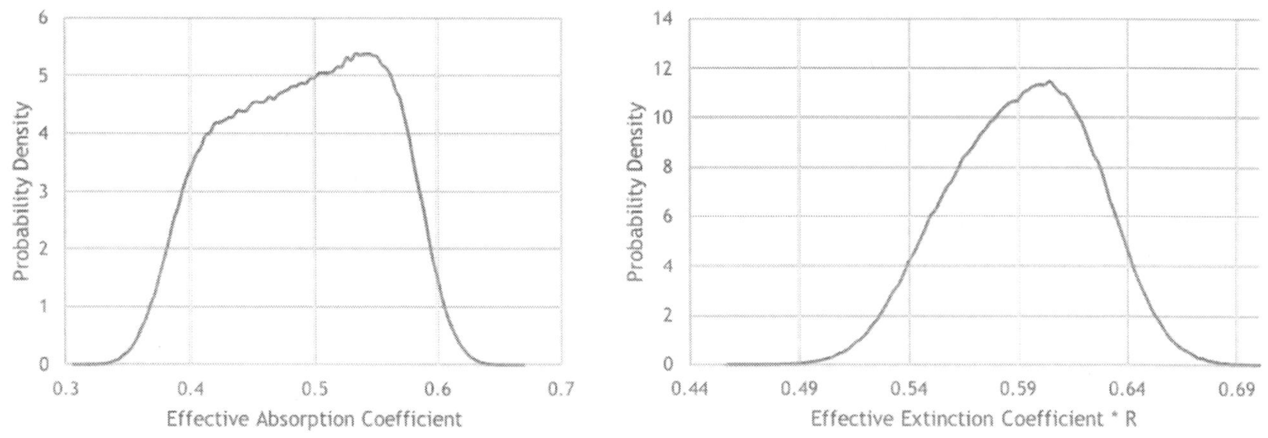

Figure 3: Probability density functions for powder absorptivity and extinction coefficient

and thus the effective thermal conductivity. Based on this analysis, a correlation was developed between the metal conductivity (k_s), gas conductivity (k_g), particle diameter (D_p), temperature (T), and powder conductivity (k_{eff}) and is given in Equation 11.

$$\frac{k_{eff}}{k_s} = -2.44\theta^2 + 15.2\theta\kappa + 3.57\theta - 25.2\kappa^2 + 11.7\kappa + 0.001 \pm 11\% \tag{11}$$

Where θ and κ are non-dimensional quantities given by: $\theta = \frac{\sigma T^3 D_p}{k_s}$ and $\kappa = \frac{k_g}{k_s}$. As the bed-scale thermo-mechanical model does not currently consider temperature-dependent thermal conductivities, the powder effective conductivity is computed at a temperature of 974K, halfway between room temperature (300K) and the solidus temperature of 316 stainless steel. Powder conductivity is then taken to be a Gaussian random variable with mean 0.32 and standard deviation 0.04

Once the bed-scale model input probability distributions are known, a stochastic expansion technique called polynomial chaos expansion (PCE) [14] is employed to propagate theses probability distributions through the model and estimate the probability distribution of the output quantities of interest.

The model output is expressed as a polynomial expansion in the input random variables with unknown coefficients as given in Equation 12 [14].

$$X(\theta) = \Sigma_{j=0}^{\infty} a_j \Phi_j(\zeta(\theta)) \tag{12}$$

X is a quantity of interest, θ is a vector of model inputs, a_j are the coefficients of the expansion, Φ_j are the basis polynomials, and ζ is a vector of random variables representing the input probability distributions. Φ_j are constructed using a tensor product basis for multiple uncertain inputs.

A stochastic collocation technique [14] is used in which the model is solved deterministically at selected collocation points, sparsely distributed in the space of possible input parameters [12]. These collocation points are then used in interpolation schemes to reconstruct the coefficients of the polynomial expansion. In this work, Smolyak sparse grids are employed [12].

Once the coefficients of the expansion are known, samples are drawn from the input distributions and used to evaluate the response surface to predict a probability density function of the output. The calculated standard deviation of the output is then a measure of the overall uncertainty in the model prediction resulting from uncertainties in the inputs.

In order to compare bed-scale model results to the data of Yadroitsava et al. [15], the quantities of interest for this work are the residual stresses parallel and perpendicular to the path of the laser both near the powder-baseplate interface and near the top of the domain. The action of the laser in the simulation creates overlapping melt tracks of melted powder. Figure 4 shows the temperature field as the laser performs a line scan along the powder layer. Repeated, overlapping laser line

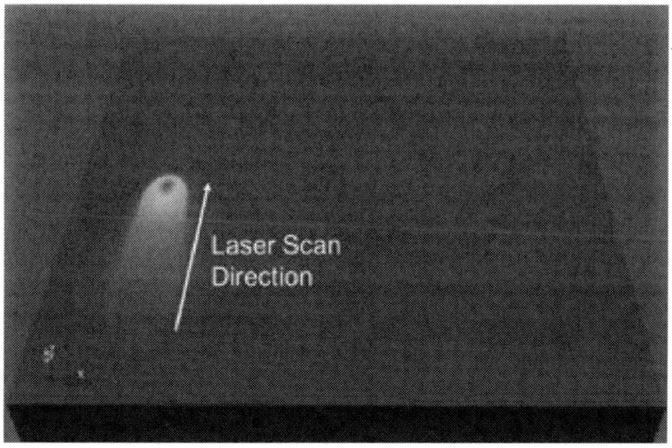

Figure 4: Temperature response due to laser line scan

scans result in a recurring pattern of residual stress distributions as shown in Figure 5. Since it is

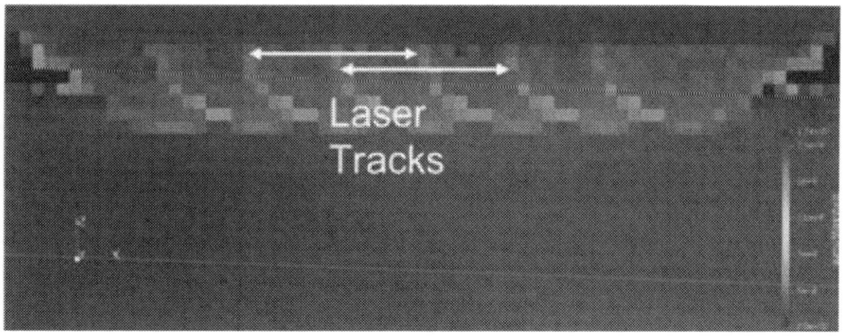

Figure 5: Residual stress cross section showing results of overlapping laser scans

not known exactly where the measurements of Yadroitsava et al. were taken relative to the laser scans, calculated residual stresses are averaged across a laser track to calculate mean residual stress values as well as standard deviations of those means. These standard deviations due to position are added in quadrature with the standard deviations due to the uncertain model input parameters calculated using the PCE method discussed previously to determine an overall standard deviation on the residual stress predictions. The predicted stress profiles in the consolidated powder layer from baseplate to free surface are shown in Figure 6. Error bars indicate one standard deviation. The range of Yadroitsava's measurements near the plate and near the free surface are shown as horizontal lines.

As can be seen, the model predictions fall within one standard deviation of the experimental data near the baseplate, but predict significantly lower residual stresses near the free surface. This is likely due to the fact that the bed-scale model does not account for the motion of the metal and

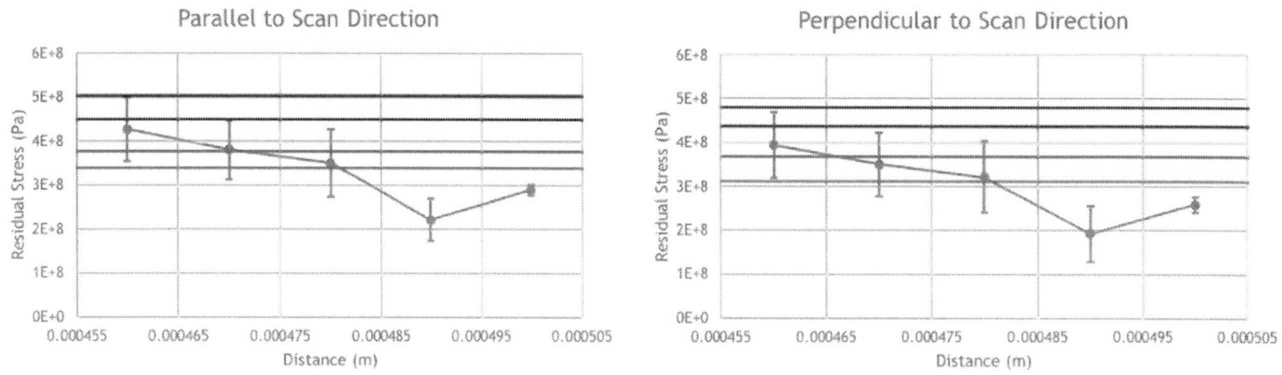

Figure 6: Predicted residual stress profiles parallel to and perpendicular to laser scan direction. Baseplate is at X = 0.00046m. Horizontal lines show the range of Yadroitsava's measurements near the plate and near the free surface

air that occurs when the powder melts. While in reality the two materials will separate, with the metal settling on to the baseplate and the air escaping out, the model keeps the melted powder as a mixture of powder and air. This could lead to a modeled melt track that is not as stiff as a real melt track, leading to lower residual stresses, an effect that would become more pronounced further from the baseplate. Additionally, more particle-scale physics, such as micro structures, may need to be included to allow for an accurate prediction. Finally, residual stress data for additively manufactured materials is difficult to obtain and scarce in the literature. More data to compare against would allow a better assessment of the model's predictive capability.

Conclusion

A bed-scale thermo-mechanical model of the SLM process is presented. Particle-scale SLM models are used to calculate parameters of the developed bed-scale model. This demonstrates a framework by which continuum, bed-scale models can be informed with particle-scale physics, providing a mechanism for computationally expensive particle-scale models to be used to improve predictions at the bed-scale. Additionally, by using the PCE technique to propagate uncertainties through the bed-scale model, the impact of particle-scale uncertainties can be calculated for the bed-scale prediction. This allows a quantitative comparison to experimental data. The developed bed-scale model is able to predict residual stresses near the baseplate that agree with data in the literature, but predictions away from the baseplate under-predict residual stress. Additional particle-scale physics may need to be included using this framework to improve the accuracy of the prediction. However, more residual stress data for additively produced materials is needed to fully explore the model's predictive capability.

Acknowledgments

Sandia National Laboratories is a multimission laboratory managed and operated by National Technology & Engineering Solutions of Sandia, LLC, a wholly owned subsidiary of Honeywell International Inc., for the U.S. Department of Energys National Nuclear Security Administration under contract DE-NA0003525. The views expressed in the article do not necessarily represent the views of the U.S. Department of Energy or the United States Government.

References

[1] N E Hodge, R M Ferencz, and J M Solberg. Implementation of a thermomechanical model for the simulation of selective laser melting. *Computational Mechanics*, 54(1):33–51, 2014.

[2] Saad A. Khairallah, Andrew T. Anderson, Alexander Rubenchik, and Wayne E. King. Laser powder-bed fusion additive manufacturing: Physics of complex melt flow and formation mechanisms of pores, spatter, and denudation zones. *Acta Materialia*, 108:36–45, 2016.

[3] Saad A. Khairallah and Andy Anderson. Mesoscopic simulation model of selective laser melting of stainless steel powder. *Journal of Materials Processing Technology*, 214(11):2627–2636, 2014.

[4] W E King, A T Anderson, R M Ferencz, N E Hodge, C Kamath, S A Khairallah, and A M Rubenchik. Laser powder bed fusion additive manufacturing of metals; physics, computational, and materials challenges. *Applied Physics Reviews*, 2(4):41304, dec 2015.

[5] Peter Mercelis and JeanPierre Kruth. Residual stresses in selective laser sintering and selective laser melting. *Rapid Prototyping Journal*, 12(5):254–265, oct 2006.

[6] Daniel Moser, Scott Fish, Joseph Beaman, and J Y Murthy. Multi-Layer Computational Modeling of Selective Laser Sintering Processes. In *Proceedings of the ASME 2014 International Mechanical Engineering Congress & Exposition.*

[7] Daniel Moser, Sreekanth Pannala, and Jayathi Murthy. Computation of Effective Radiative Properties of Powders for Selective Laser Sintering Simulations. *JOM*, 67(5):1194–1202, 2015.

[8] Daniel Moser, Sreekanth Pannala, and Jayathi Murthy. Computation of Effective Thermal Conductivity of Powders for Selective Laser Sintering Simulations. *Journal of Heat Transfer*, 138(8):82002, may 2016.

[9] Mehrdad Negahban. *The Mechanical and Thermodynamical Theory of Plasticity.* Boca Raton, FL, 1 edition, 2012.

[10] J Christian Nelson, Samuel Xue, Joel W Barlow, Joseph J Beaman, Harris L Marcus, and David L Bourell. Model of the selective laser sintering of bisphenol-A polycarbonate. *Industrial & Engineering Chemistry Research*, 32(10):2305–2317, oct 1993.

[11] Brock Partee, Scott J Hollister, and Suman Das. Selective Laser Sintering Process Optimization for Layered Manufacturing of CAPA® 6501 Polycaprolactone Bone Tissue Engineering Scaffolds. *Journal of Manufacturing Science and Engineering*, 128(2):531–540, sep 2005.

[12] S. A. Smolyak. Quadrature and interpolation formulas for tensor products of certain classes of functions. *Soviet Mathematics, Doklady*, 4:240–243, 1963.

[13] C R Swaminathan and V R Voller. A general enthalpy method for modeling solidification processes. *Metallurgical Transactions B*, 23(5):651–664, 1992.

[14] D Xiu and J Hesthaven. High-Order Collocation Methods for Differential Equations with Random Inputs. *SIAM Journal on Scientific Computing*, 27(3):1118–1139, jan 2005.

[15] I. Yadroitsava and I. Yadroitsev. Residual Stress in Metal Specimens Produced by Direct Metal Laser Sintering. In *Proceedings of Solid Freeform Fabrication Symposium*, pages 614–626, 2015.

ACCELERATED HEAT TRANSFER SIMULATIONS USING COUPLED DEM AND CFD

Marina Sousani, DEM Solutions Ltd, Edinburgh, UK

Andrew M. Hobbs, Astec Inc, Tennessee, USA

Adam Anderson, ANSYS UL Ltd, Sheffield, UK

Richard Wood, DEM Solutions Ltd, Edinburgh, UK

Abstract

This work presents an accelerated simulation of heat and mass transfer by coupling Discrete Element Methodologies (DEMs) and Computational Fluid Dynamics (CFD), utilising Graphics Processing Unit (GPU) technology. The presented model is a continuation of previous work[1] and focuses on demonstrating the capabilities and effectiveness of implementing the GPU combined with the Central Processing Unit (CPUs) technologies to run a complex industrial simulation. Furthermore, different flighting configurations have been used to investigate the influence of the design to the drying process. A model of an aggregate drum dryer was used to produce hot mix asphalt and different configurations have been implemented to investigate the effect of GPU-CPU technology in such a complex simulation. Commercial codes from ANSYS and DEM-Solutions were coupled to simulate heat transfer from the hot gases to the aggregate particles. Fluid flow and particle-fluid interactions are solved by the CFD solver which exchanges information at regular intervals. The results showed that the coupled model captures accurately the convective heat transfer from the fluid to the solid phase and demonstrated significant improvement in terms of simulation time. The proposed model will have a significant impact in industrial applications as it describes a methodology to simulate large-scale applications rapidly and accurately.

Keywords: coupled DEM and CFD; GPU; aggregate dryer; heat and mass transfer

Introduction

Particle simulations have been well established as a means to model complex physical systems in a number of fields and are becoming increasingly more important in others. Much research has been done in a variety of areas such as molecular dynamics, plasma physics, fluid dynamics and rock mechanics [2-4], where particle simulations were coupled with Computational Fluid Dynamics (CFD)[5]. The Oil and Gas industries have been using CFD software for many years to model fluid behaviour in reservoirs, saturation of residual oil, well performance, fluidisation, sedimentation and others [6-11]. Lately the ability to simulate larger number of particles, especially for applications where particle-particle interaction and particles of different shape must be considered, has given a boost to the demand for more coupled DEM-CFD simulations.

Recently this type of modelling shows an increasing interest in the process manufacturing industry too. Hot mix asphalt (HMA) is the most common pavement surface in the U.S., which comprises around 94% of all roads. It is also increasingly popular in airport runways in the U.S. due to its efficiency on takeoffs and landings, its low maintenance and fast construction time. It has been approved by the Environmental Protection Agency (EPA) and is thus considered a practical solution for water storage, flood control, erosion and hazardous-waste landfills worldwide [12-14]. HMA is a composite material that consists of mineral aggregates (sand and various sizes of crushed rock) mixed together with liquid asphalt cement binder (a product of crude oil), laid in layers and compacted. For an efficient mixing process, the asphalt binder must be heated at temperatures above 180°C, to decrease its viscosity, and the aggregate must be dried prior to mixing [13]. Most aggregate is stored in stockpiles where the minimum moisture content will be that of the ambient air. A typical value for aggregate moisture content is 5% by mass meaning that for a nominal production rate of 320,000 kg/h of mix, 15,875 kg/h of H_2O must be removed. Typically, half of the total heat input is required to dry the aggregate making drying efficiency of critical importance [1].

The aggregate drying process is accomplished in a counter flow drum that is positioned on a 4.7° inclination and contains internal metal slats (flights). The drum is heated by a direct fire burner and the flights lift and veil the aggregate material through the hot gases to heat and dry the aggregate particles. Wet aggregate is fed into the uphill end and is moved along the drum towards the burner (Figure 1).

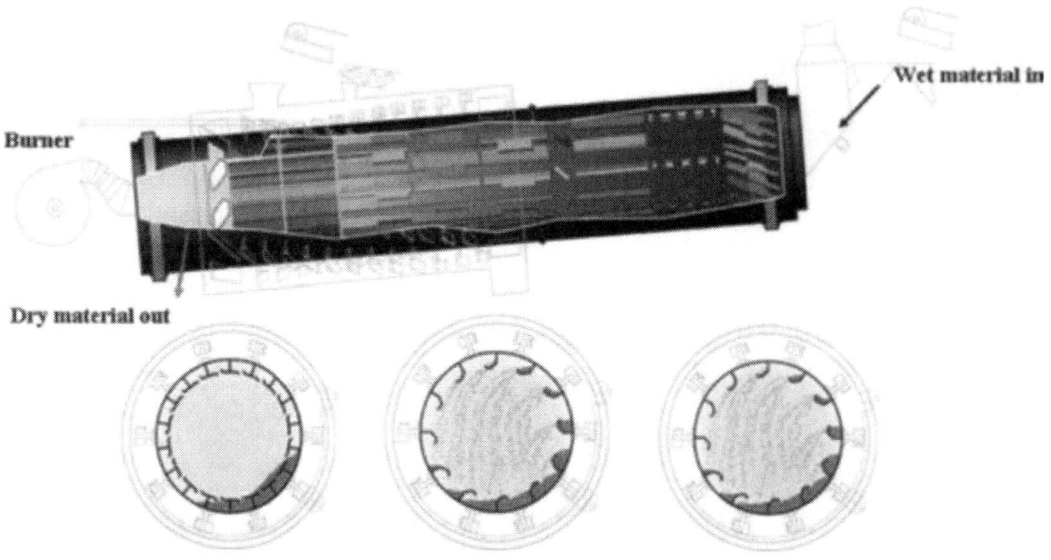

Figure 1 *Schematic of the Counter-flow aggregate dryer drum. Hobbs [1].*

A key challenge associated with this process is the fact that veiling performance is tied to many different variables including flight shape, number of flights, drum rotational speed, and fill level. At the same time, direct observation of this phenomenon is impossible due to the harsh internal environment of the drum. Part of the aim of this work is to propose a methodology that enables not only the monitoring of the material's behaviour but the optimization of the flight design for increased drying efficiency. Furthermore, this paper presents an innovative way of accelerating such demanding simulations by performing DEM calculations in GPU mode while solving for the fluid in CPU mode.

Numerical methods

CFD is a very popular method amongst researchers that simulates fluid phenomena and provides qualitative prediction of fluid flows. A first appearance of CFD methods dates back to 1947 from the work of Kopal [15] of supersonic flows on sharp cones and the first generation of CFD solutions the period between the 1950s and early 1960s [16-19]. Since then, the role of CFD in engineering predictions has been so strong that nowadays is considered a standard aspect in fluid dynamics, beyond the experimental and theoretical elements. A wide range of industrial applications employ CFD methods to study critical fluid phenomena and enhance their operations [1, 20-23]. There are commercial software that have been used for modelling multiphase, combustion and heat transfer for particle flows including densely packed systems such as fluidized bed reactors. However, in some applications it is critical to account for the shape and the particle-particle interactions of the granular material, to provide a meaningful insight.

DEM is a method that describes the overall (macroscopic) mechanical behaviour of assemblies of particle systems as the result of the interaction of their constitutive individual elements. It allows displacements, rotations and complete separation of the discrete elements, and recognizes any new contacts that are developed during the simulation [24, 25]. Initially models, which were based on the particle-scale, were developed in order to simulate the micromechanical behaviour of soils and sands (non-cohesive materials) [26], but since then it has been widely utilised in a number of studies. DEM has been applied in several engineering fields, such as slope stability and mining to powder technology, minerals and process manufacturing [3, 4, 27-32], and is considered a very efficient engineering tool.

In nature particles rarely operate within systems that are fluid free, hence there is a need for a simulation methodology that can capture both the solid and the fluid mechanisms as well as the interaction between the two. Researchers have worked towards combining CFD techniques with the Newtonian particle methods used in DEM. Approaches that incorporate CFD with the DEM have been presented by Tsuji [33], Tsuji [34] and Xu and Yu [35], where the interaction between the solid and gas phases have been modelled by solving Newton's second law of motion, with respect to the motion of the particles, and the Navier-Stokes equation with respect to the motion of the gas. However, the majority of the CFD-DEM approaches involve in-house codes making them inaccessible and expensive to industry. This paper presents an improved version of an existing commercial product that couples the CFD solver (ANSYS Fluent) with the DEM solver (EDEM) from DEM Solutions. The EDEM-CFD coupling is employed to simulate heat transfer from a reacting natural gas flame to the particle phase in an aggregate dryer.

Moreover, complex industrial applications that involve hundreds of thousands of particles are so computationally expensive that not even the most efficient couplings are able to support. This paper presents an innovative way of accelerating computationally intensive workloads by performing DEM calculations in GPU mode while solving for the fluid in CPU mode. The presented solver provides greatly reduced computational times compared to simulations on multi-core CPUs. This process has been performed with the use of the EDEM GPU solver engine, which has been created by DEM Solutions.

DEM – CFD Coupling Methodology

CFD

In previous work [1] particle positions and interactions were solved by the DEM solver, whereas the volume fraction, drag, and heat transfer were calculated by the coupling interface between the solvers. This required significant computation outside of the solvers making this approach less efficient for simulations that required large number of particles. Advantage has now been taken of developments in Fluent since the previous coupling and the latest Fluent solver presents a more sophisticated approach. Specifically, in the new Dense Discrete Phase Model (DDPM) coupling the EDEM particle data are converted into a Discrete Phase Model (DPM) injection and the volume fraction, drag, and heat transfer are calculated directly by Fluent, making for a more efficient calculation.

The volume fraction (the space occupied by each phase), in any given fluid cell, is calculated through an algorithm that utilises an octree method and detects those cells containing particles during the simulation. The flow of the fluid, around the particles, is then determined by the continuous phase flow while taken into account the blockage from the solid phase. Here, the volume fraction represents the space occupied by each phase, thus the laws of conservation of mass and momentum are satisfied by each phase individually. The mass and momentum conservation equations for the particle phase p are given by:

$$\frac{\partial}{\partial t}(\alpha_p \rho_p) + \nabla \cdot (\alpha_p \rho_p \vec{V}_p) = \sum_{q=1}^{nphases} (\dot{m}_{fp} - \dot{m}_{pf}) \tag{1}$$

where subscripts p and f refer to the solid and fluid phase, respecively α_p is the volume fraction, ρ_p is the density of the individual phase, $\vec{V}_p$ is the velocity of the p phase and $\dot{m}_{pf}$ is the mass transfer from the pth to f^{th} phase, respectively. Similarly, the momentum conservation equation for the solid phase is given by:

$$\begin{aligned}\frac{\partial}{\partial t}(\alpha_p \rho_p \vec{V}_p) + \nabla \cdot (\alpha_p \rho_p \vec{V}_p \vec{V}_p) \\ = -\alpha_p \nabla_p + \nabla \cdot \left[\alpha_p \mu_p \left(\nabla_{\vec{V}_p} + \nabla_{\vec{V}_p^T}\right)\right] + \alpha_p \rho_p \vec{g} \\ + F_{vm,lift,user} \\ + \sum_{q=1}^{nphases} (\vec{K}_{pf}(\vec{V}_p - \vec{V}_f) + \dot{m}_{fp} \vec{V}_{fp} - \dot{m}_{pf} \vec{V}_{pf}) \\ + \vec{K}_{DPM}(\vec{V}_{DPM} - \vec{V}_p) + S_{DPM,explicit}\end{aligned} \tag{2}$$

where μ_p is the shear viscosity of the particle phase p, $\vec{K}_{fp}$ is the interphase momentum exchange coefficient between the fluid and particle phases, $F_{vm,lift,user}$ is a lift force (due to velocity gradients in the primary phase flow field), $\vec{V}_p$ and $\vec{V}_f$ are the particle and fluid velocities, respectively. $\vec{V}_{fp}$ is the interphase velocity and is dependent upon the mass transfer $\dot{m}_{fp}$. $S_{DPM,explicit}$ is the explicit component of the particle sink term, and $\vec{V}_{DPM}$ and $\vec{K}_{DPM}$ are the implicit terms of the particle averaged velocity of the considered discrete phase and interphase momentum exchange coefficient, respectively. Note that the equivalent mass/momentum conservation equations for the fluid phase follow the same logic.

The fluid-solid drag function follows the Gidaspow model [36] and is a combination of the Wen and Yu model [37] and the Ergun relation [38]. When the volume fraction, α_p is greater than 0.8 then the Wen-Yu equation is used, whereas when the volume fraction is less than 0.8 the Ergun relation is used [39]. Relevant modifications have been made by ANSYS for the Ergun relation (Eq.4). The fluid-solid exchange coefficient is given by:

α_p>0.8 $$K_{sl} = \frac{3}{4} C_D \frac{\alpha_s \alpha_f \rho_f |\vec{v}_f - \vec{v}_p|}{d_p} \alpha_l^{-2.65} \quad (3)$$

α_p<0.8 $$K_{sl} = 150 \frac{(1-\alpha_f)^2 \mu_f}{\alpha_p d_p^2} + 1.75(1-\alpha_f) \frac{\rho_f}{d_p} |\vec{v}_f - \vec{v}_p| \quad (4)$$

where C_D is the drag coefficient, d_p is the diameter of the particles in the solid phase, α_p and α_f are the volume fractions of the solid and fluid phases, respectively, and Re_s is the Reynold's number at the terminal settling condition for a single particle and is expressed as:

$$Re_s = \frac{\rho_f d_p |\vec{v}_p - \vec{v}_f|}{\mu_f} \quad (6)$$

where $\rho_f$and μ_f are the fluid density and the dynamic viscosity, respectively.

The drag coefficient C_D is directly dependent on the Reynolds number and it uses the following criteria:

$$C_D = \begin{cases} 24/Re_s & R_e < 0.55 \\ 24(1.0 + 0.15{Re_s}^{0.687})/Re_s & 0.55 < R_e \leq 1000 \\ 0.44 & R_e > 1000 \end{cases} \quad (7)$$

For the transfer of the DEM particle phase into a DPM injection, Fluent first checks the coordinates of each DEM particle with the fluid cells positions and applies the effects of the particle to the cell that is found to contain it. However, since the particle may overlap more than one fluid cell, a Node Based Averaging (NBA) algorithm is used to smooth the loading, distributing the particle's effects to neighbouring mesh nodes. This reduces grid dependency and computational instability by spreading the particle's effects smoother across the neighbouring cells.

Furthermore, as particles move through the fluid, thermal energy is also exchanged. The temperature change of each particle over time is calculated based on Equation (8). The summation of the convective and conductive heat fluxes [40] is given by:

$$m_P C_P \frac{dT}{dt} = \sum Q_{heat} \quad (8)$$

where m_P, C_P, and T are the mass, specific heat and temperature of the particle material, respectively.

$$\sum Q_{heat} = Q_{PF} + Q_{P1P2} \tag{9}$$

$$Q_{PF} = h_{PF} A_p \Delta T_{PF} \tag{10}$$

$$Q_{P1P2} = h_c \Delta T_{P1P2} \tag{11}$$

$$h_c = \frac{4k_{P1}k_{P2}}{k_{P1} + k_{P2}} \left(\frac{3F_N r^*}{4E^*}\right)^{1/3} \tag{12}$$

$$h_{PF} = \frac{k_F N_u}{d_p} \tag{13}$$

where, h_{PF} is the convective fluid-particle heat transfer coefficient, A_p is the particle surface area, ΔT_{PF} is the temperature difference between the fluid and the particles and Q_{P1P2} is the inter-particle heat flux, respectively. h_c is the conductive heat transfer coefficient between two particles and ΔT_{P1P2} is their temperature difference, k_P is the thermal conductivity of the particles, F_N is their normal force, r^* is the geometric mean of the particles radii and E^* is the effective Young's modulus. The bracketed term of the equation models the contact area between two particles. Finally, K_F is the gas thermal conductivity of the fluid, N_u is the Nusselt number, and d_p is the particle diameter. The presented model does not consider conductive heat transfer from the particles to the geometry.

The Nusselt number has been computed based on the results from Ranz and Marshall [41] for R_e<200. Detailed information regarding the process can be found in Hobbs [1]. The Nusselt number is given by:

$$N_u = 2 + 0.6{R_e}^{1/2} Pr^{1/3} \qquad R_e < 200 \tag{14}$$

$$Pr = \frac{C\mu}{k} \tag{15}$$

where R_e is the Reynold's number based on the diameter of the individual phase and the relative velocity $|\vec{u}_p - \vec{u}_f|$ and the Pr is the Prandtl number for the subsequent phase.

DEM

In DEM simulations, the main computational challenge is the detection of contacts. To address this EDEM uses an algorithm that utilises a Cartesian grid to search for particle contacts. In particular, the domain is divided into grid cells of specified size, then the algorithm checks every cell but only analyses those that contain two or more elements (active cells). The idealized length of a grid cell is 2-6Rmin where R_{min} is the minimum particle radius in the simulation. This reduces the simulation time significantly, while the results remain unaffected by the number of grid cells. Furthermore, the EDEM solver utilises the Hertz–Mindlin contact model [42-44] to solve the contact forces (normal and tangential) between particles, including damping coefficients. For each identified contact the resultant force is calculated taking into consideration the body forces. Particle velocities and new positions, as a result of the forces acting on them, are updated using an explicit time marching scheme. The DEM detection algorithm and calculation cycle are shown in Figure 2(a) and (b), respectively.

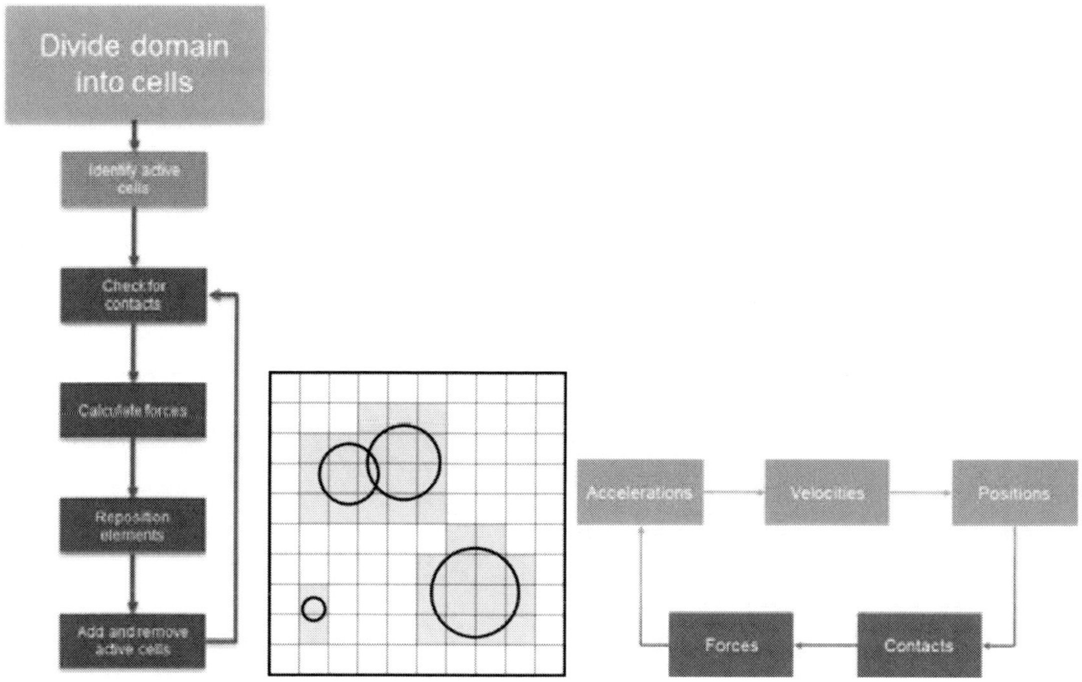

(a) (b)

Figure 2 *Schematic of the detection algorithm steps applied on a representative 2D cell grid and the calculation cycle used in the DEM.*

Choosing the right time-step is critical when reducing the simulation time. In EDEM a time-step is chosen as a percentage of the Rayleigh time-step value, which refers to the time taken for a shear wave to propagate through a solid particle, and it is calculated based on the smallest particle of the simulation. The normal range is between 10-40% of the Raleigh time-step and is given by the following equation:

$$T_R = \frac{\pi R \left(\frac{\rho}{G}\right)^{1/2}}{(0.1631v + 0.8766)} \quad (16)$$

where R is the particle radius, ρ the particle density, G the shear modulus and v is the Poisson's ratio.

Figure 3(a) and (b) describe the full coupling methodology and sequence of actions between the two software. EDEM provides a Coupling Interface that includes all the relevant functions that exchange information with Fluent, while Fluent EdemUDF is a user defined function that is loaded into Fluent and handles all drag and heat transfer calculations. Specifically, after setting up the DEM model, EDEM calculates the particles positions and velocities and the DEM-CFD coupling initiates in which the Fluent UDF updates the particle properties taken from EDEM and converts them such that they are passed into Fluent. The latter calculates the volume fraction, drag forces and heat transfer, which are then passed back into EDEM through the UDF and the coupling interface. In general, the DEM time-step tends to be smaller than in CFD and can have a difference between 10 to 1000 times. Therefore, EDEM performs the necessary iterations in order to catch up with the next Fluent time-step and UDF repeats its calculations while passing the data back to Fluent.

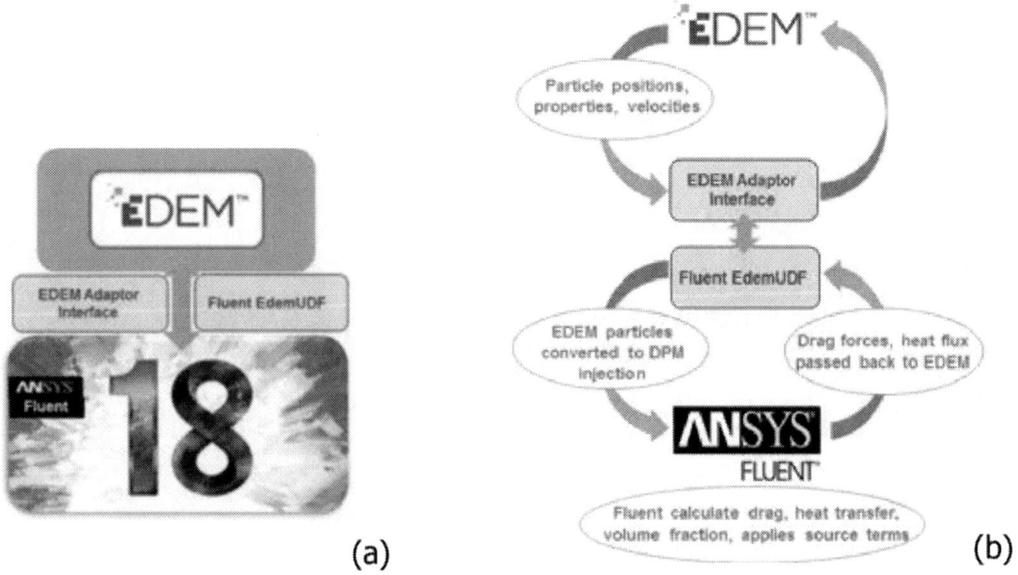

Figure 3 *Schematic of the EDEM-Fluent coupling logic.*

GPU Technology

This project presents an innovative way of accelerating large scale industrial simulations by performing DEM calculations in GPU mode while calculating fluid flow in CPU mode. The EDEM GPU solver provides higher clock speeds compared to simulations on multi-core CPUs. The presented work has been performed with the use of the EDEM GPU solver engine, created and distributed by DEM Solutions LTD. The EDEM GPU simulation engine uses the Open Computing Language (OpenCL) [45] in combination with the AMD Radeon R9 Fury X graphics card for increased accuracy. OpenCL is an open industry language for general purpose parallel programming across CPUs, GPUs and other processors, that gives complete freedom to users. For example, using OpenCL allows a programmer to write general purpose programs that execute on GPUs without the need to map their algorithms onto a specific vendor's 3D graphics API. Finally, OpenCL creates an efficient programming interface, consisting of an API for coordinating parallel computation across heterogeneous processors, and a cross-platform intermediate language with a well specified computation environment.

Different parallel programming tools exist but the major advantage of using OpenCL is the fact that it fully supports all graphic cards, without tying the DEM user to a single GPU vendor. For example NVIDIA [46] is restricted in using CUDA [47], while OpenCL encapsulates NVIDIA, AMD, Intel Xeon Phi as well as CPUs under a common programming environment. The execution speed and accuracy of arithmetic operations in a computing unit is directly related to the precisión used. Most computers use the floating-point format, which is an internal representation of numeric values stored in computer memory. This format can be categorized into the single and double precision, where the former requires 4 bytes (32 bits) and the latter 8 bytes (64bits). Even though increasing the number of bits results in increased accuracy, double precision models have much higher memory requirements. Thus, it is critical to select a fitting precision type for a specific application, so that a practical balance between accuracy and simulation time/memory is maintained. The presented results for both graphic cards used single precision.

The presented work uses both the EDEM GPU and CPU solver engines and has been repeated multiple times under different configurations but only selected combinations will be presented herein. The set of simulations, shown in Table 1, used 4CPU cores for the Fluent and 30, 40 CPU cores for EDEM, respectively. Moreover, the same simulation was repeated twice by using the EDEM GPU solver engine with double precision and two different graphics cards; the AMD Radeon R9 Fury X and the NVIDIA Quadro GP100.

Table 1 *Configuration combinations for the coupled simulation with the use of the AMD Radeon R9 Fury X and the NVIDIA Quadro GP100 graphic cards, respectively.*

CPU mode	**GPU mode**	
	AMD Radeon R9 Fury X	**NVIDIA Quadro GP100**
4 Fluent/30 EDEM	8 Fluent/8 EDEM	12 Fluent/12 EDEM
4 Fluent/40 EDEM		

Simulation setup

For this simulation similar geometry and setup has been used as presented in Hobbs [1]. Moreover, the same drum but different flight designs were used for this simulation in order to evaluate the heat transfer performance of each design. Figure 4(a),(b) demonstrates the 3D model of the aggregate dryer with its internal features (conveying screw flights, veiling flights, combustion flights). A CFD tetrahedral mesh was created by removing all the internal features and preserving mesh density in the burner and combustions zones, whereas a less dense mesh was used in the rest of the geometry.

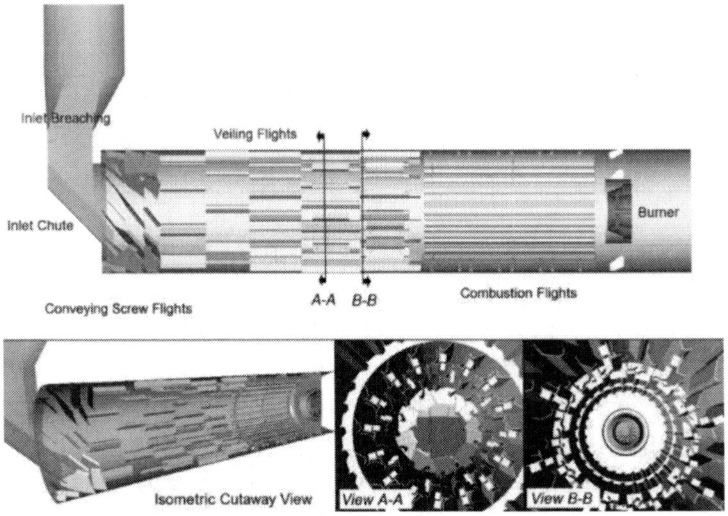

(a) – Geometry option A

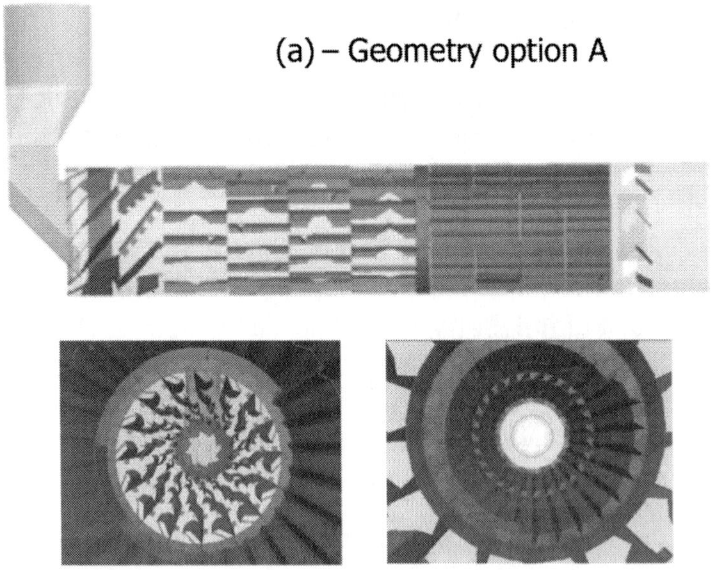

(b) – Geometry option B

Figure 4 *CAD geometries of the aggregate dryer used for the coupled DEM-CFD simulation [1].*

The full set of the CFD input simulation parameters are shown in Table 2.

Table 2 *CFD Input parameters*

CFD Input parameters	
Type of fluid	Air
Fluid density (kg/m^3)	1.225
Fluid viscosity (Pa·s)	1.789E-05
Turbulence model	Standard k-ε
CFD cell type	Hybrid tetrahedral hexahedral
Cell number	7.2E+05
CFD time-step (sec)	0.02

The complete 3D drum dryer model was also imported into EDEM. A factory was defined that generated 539.942 discrete elements, comprised of bi-sphered particles of the same size, with a normal distribution and an initial temperature of 300K. The simulated results refer to 1 second of real physical time, after it had reached a quasi-static condition (100sec of run time). A number of trials were performed to determine the optimum time-step and simulating parameters, such as the cell size of the simulator grid. The full set of EDEM inputs are shown in Table 3.

Table 3 *EDEM Input parameters*

EDEM Input parameters	
Physics	Hertz-Mindlin contact model
Particle density (kg/m^3)	2900
Particle diameter (mm)	18
Particle Shear modulus (Pa)	1.38E+07
Particle Poisson's ratio	0.25
Geometry density (kg/m^3)	7800
Geometry Shear modulus (Pa)	7.50e+07
Geometry Poisson's ratio	0.30
Particle-Particle coefficient of restitution	0.45
Particle-Particle static friction coefficient	0.55
Particle-Particle rolling friction coefficient	0.05
Particle-Geometry coefficient of restitution	0.50
Particle-Geometry static friction coefficient	0.45
Particle-Geometry rolling coefficient	0.05
Particle generation rate (kg/sec)	80.64
EDEM time-step (sec)	1E-04
Cell size (mm)	31.49

Results and discussion

To post-process the CFD-DEM results, Ensight from CEI and EDEM Analyst were used. Figure 5 shows a side view of the aggregate dryer that demonstrates the swirling flow and the fluid-particle temperature distributions ranking between 300-3000K and 300-1000K, respectively. The figure also includes cross sections of the drum with the different flight configurations at specified points (A,B,C) showing the convective heat transfer. The purpose of the dryer drum is to get heat into the particles as efficiently as possible which is translated into maximizing the convective heat flux. If the heat fails to go into the particles it is lost in the exhaust gases resulting in wasted energy. It can be observed that the neighbouring to the burner particles have higher temperatures and the fluid temperature is almost a 3-step decrease of flow. This indicates that as the particles are showered through the hot gases, thermal energy is exchanged, and the heat is transferred from the fluid to the particles thus the fluid temperature drops. Comparing the different designs it can be observed that the Geometry option B gets less heat into the particles, which means that the drum is less effective.

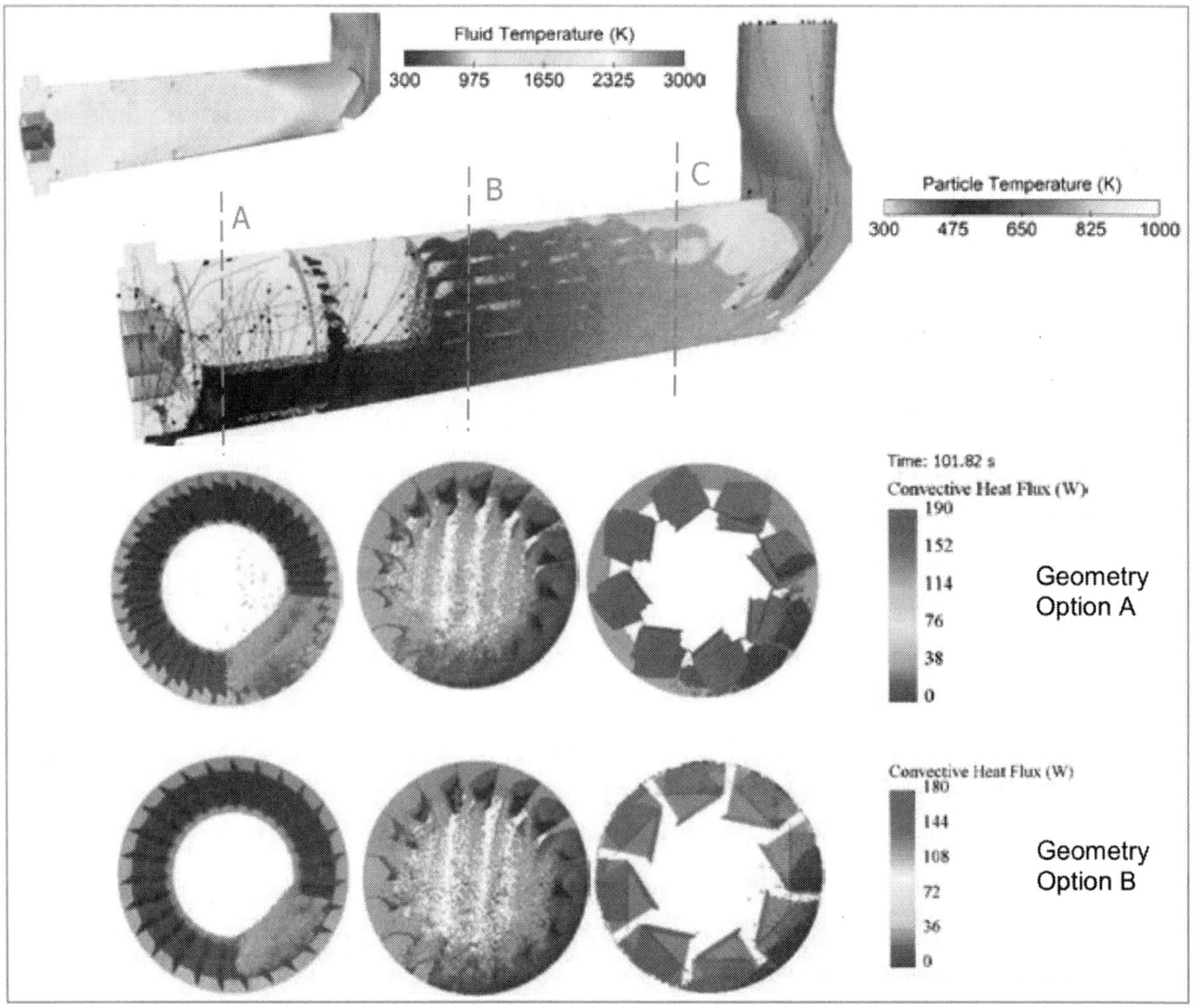

Figure 5 *Coupled CFD-DEM simulation of an aggregate dryer demonstrating the fluid - particle temperature distribution for both geometry options.*

Furthermore, figure 6 demonstrates an inside view of the heat transfer from the fluid phase to the particle phase for the Geometry option A. It can be observed that the convective heat flux reaches a max value of around 190W and 180W, respectively, and the simulated model accurately captures the dynamic state of the heat transfer along the length of the drum.

Figure 6 *Particle temperature distribution. Cross section at the regions close to the burner.*

The following figures illustrate the net computational time, under different configurations, for both the EDEM/Fluent coupling and the EDEM part of the simulation. Some representative configurations have been selected for discussion. Figure 7 demonstrates the total computational time of the coupled simulation with and without the GPU solver. Specifically, the selected simulation results in CPU mode (grey columns) used 8, 12 and 32CPU cores for the Fluent part of the simulation and 30 or 40CPU cores for the corresponding EDEM part, respectively. It can be observed that in the first case the simulation took around 1.5 hours, while the simulation times dropped to 1.43 and 1.33 hours, respectively. This shows that even though a large number of cores have been selected, the overall simulation time is only slightly reduced. In contrast, the same simulation run in GPU mode with the use of the AMD Radeon R9 Fury X or the NIVIDIA Quadro GP100 graphic cards leads to reduced simulation times of 1.07 hours and 57minutes, respectively. This means that the NVIDIA Quadro GP100 simulation was 1.5 times faster on average, resulting in an average reduction time of 32%.

Figure 8 follows the same logic as Figure 7 demonstrating the net computational time of the aforementioned configurations including only the EDEM part of the simulation. The grey columns show that it took between 39 and 39.25min to simulate the discrete phase for the 30 CPU (for EDEM) configuration and 38.88min for the 40 CPU (for EDEM) configuration. The selected configurations show a minor difference in terms of simulation time, whereas the same simulation took just over 6mins when run in GPU mode (orange column). This corresponds to an average 6.0 speed up demonstrating an average time reduction of 83% for the discrete calculations.

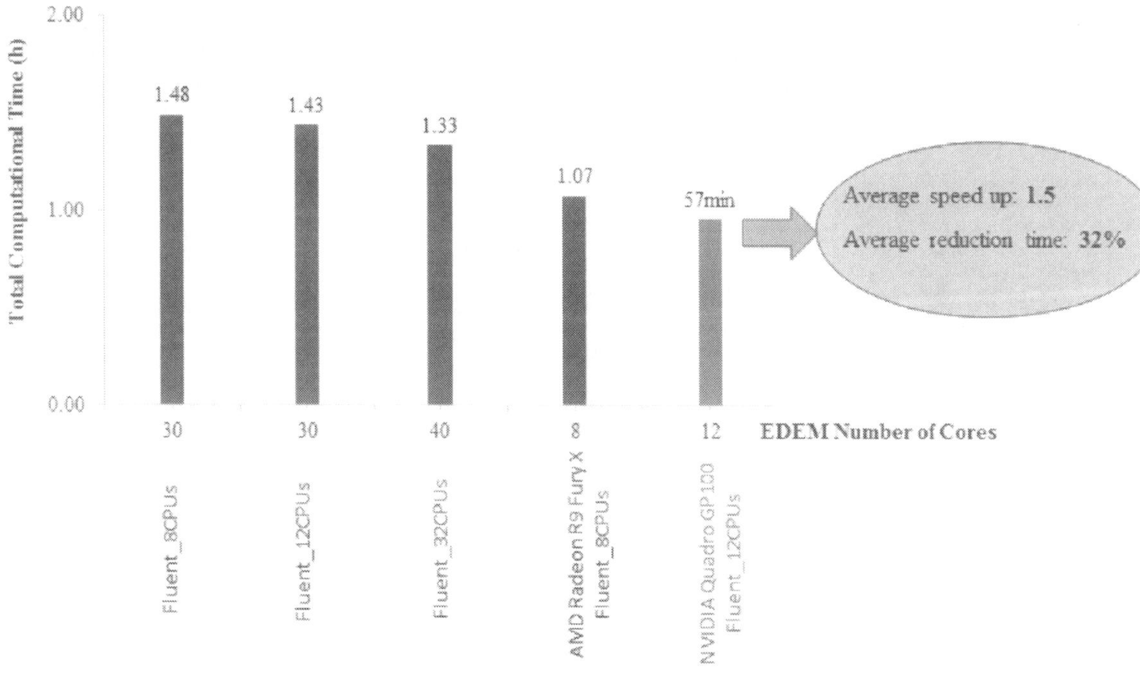

Figure 7 *Net computational time of the coupled simulation under different configurations in CPU mode (grey columns) using the AMD Radeon R9 Fury X graphic card (green), and using the NVIDIA Quadro GP100 (orange).*

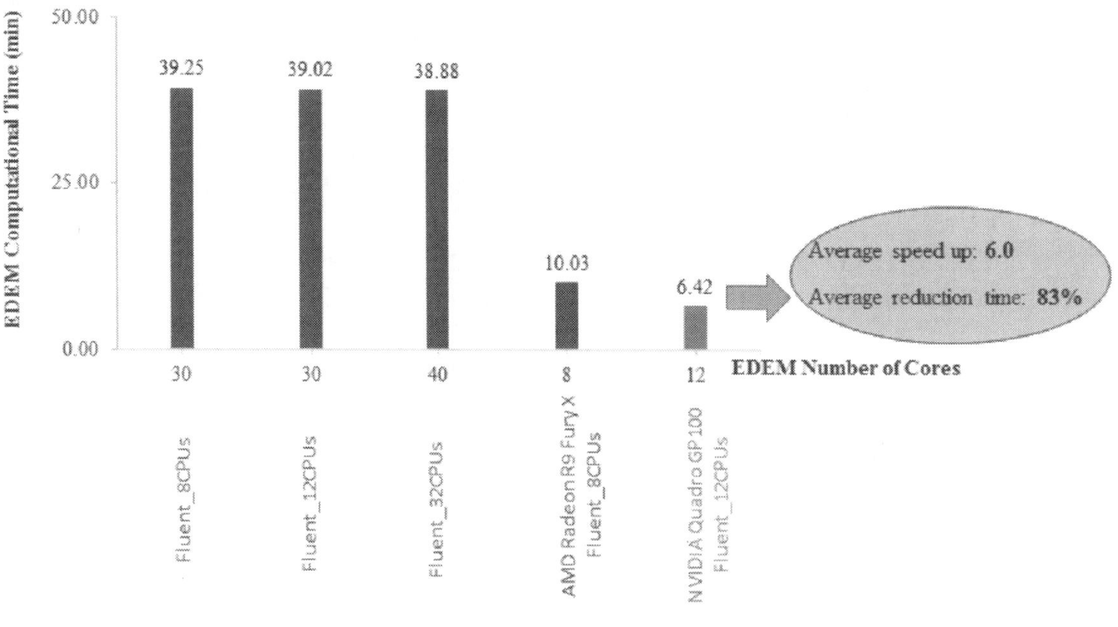

Figure 8 *EDEM computational time of the coupled simulation under different configurations in CPU mode (grey columns) using the AMD Radeon R9 Fury X graphic card (green), and using the NVIDIA Quadro GP100 (orange).*

Conclusions

Many industrial applications involve discrete particle flows. Due to their transient nature, they often require extensive computational times and recalculation of the relevant particle physics, which makes such analyses impractical. Until recently the only option to run such simulations was the use of CPU technologies which put a barrier between efficiency and fast results. The current work demonstrates the capabilities and effectiveness of implementing an innovative GPU solver combined with multiple CPU cores to run a complex industrial simulation that models heat transfer within an aggregate drum dryer. The ANSYS Fluent CFD solver has being coupled with EDEM and run with and without the EDEM GPU solver engine.

The results show that the coupled model not only provides valuable results since the observation in real-time of the particle-fluid behaviour, and thus the heat transfer to particles, is impossible, but also help towards effective design decisions. Furthermore, the simulation features very fast simulation run times. With the use of one of the latest GPU graphic cards (NVIDIA Quadro GP100), the net computational time of the coupled DEM-CFD simulation was reduced by 32% corresponding to a 1.5 speed up. Furthermore, the DEM part of the simulation experienced the greatest reduction in computational time with a net computational time reduction of 83%, making the particle calculations 6.0 times faster. More simulations were run with less CPU cores, for both Fluent and EDEM, and the results showed even larger speed ups and time reductions. This indicates that the GPU technology could be very beneficial in cases where less powerful workstations are used for such simulations.

This work highlights the significance of using GPU technology for challenging coupled CFD-DEM calculations. This kind of analysis was previously difficult or even impossible to perform and thus has a major impact in large-scale industrial applications that require simulations of millions of particles.

References

1. Hobbs, A., *Simulation of an aggregate dryer using coupled CFD and DEM methods.* International Journal of Computational Fluid Dynamics, 2009. **23**(2): p. 199-207.

2. Hockney, R.W., *Computer Simulations Uning Particles.* 2nd ed. 1981, New-York: Routledge.

3. Donzé, F.V., V. Richefeu, and S.-A. Magnier, *Advances in discrete element method applied to soil, rock and concrete mechanics.* Electron J Geotech Eng, 2009. **8**: p. 1-44.

4. Stefanou, I. and I. Vardoulakis. *Stability assessment of SE/E rock corner slope of the Acropolis Hill in Athens.* in *5th GRACM International Congress on Computational Mechanics.* 2005.

5. Date, A.W., *Introduction to computational fluid dynamics.* 2005: Cambridge University Press.

6. Byrne, M.T., et al., *Computational Fluid Dynamics for Reservoir and Well Fluid Flow Performance Modelling*, in *SPE European Formation Damage Conference*. 2011, Society of Petroleum Engineers: Noordwijk, The Netherlands.

7. Wang, S., et al., *Simulation of Flow Behavior of Particles in a Liquid–Solid Fluidized Bed.* Industrial & Engineering Chemistry Research, 2010. **49**(20): p. 10116-10124.

8. Qiu, L.-c. and C.-y. Wu, *A hybrid DEM/CFD approach for solid-liquid flows.* Journal of Hydrodynamics, 2014. **26**(1): p. 19-25.

9. Fayal, Z.F., B. Lakhdar, and N. Zoubir, *Horizontal Well Performance Flow Simulation CFD-Application*, in *SPE Production and Operations Conference and Exhibition*. 2010, Society of Petroleum Engineers: Tunis, Tunisia.

10. Afsharpoor, A., et al., *CFD modeling of the effect of polymer elasticity on residual oil saturation at the pore-scale.* Journal of Petroleum Science and Engineering, 2012. **94–95**: p. 79-88.

11. Weinlader, A., et al., *Prediction of the oil flow distribution in oil-immersed transformer windings by network modelling and computational fluid dynamics.* IET Electric Power Applications, 2012. **6**(2): p. 82-90.

12. National Asphalt Pavement Association, N.A.P.A. *History of Asphalt.* 2017.

13. Asphalt Pavement Alliance, A.P.A. *Pavement Technologies.* Pavement Materials, 2017.

14. National Asphalt Pavement Association *Asphalt for Environmental Liners (PS17).* 1984.

15. Kopal, Z., *Tables of supersonic flow around cones.* 1947, Cambridge: Massachusetts Institute of Technology. 555.

16. Fay, J.A., *Theory of Stagnation Point Heat Transfer in Dissociated Air.* Journal of the Aerospace Sciences, 1958. **25**(2): p. 73-85.

17. Blottner, F.G., *Chemical nonequilibrium boundary layer.* AIAA Journal, 1964. **2**(2): p. 232-240.

18. Blottner, F.G., *Nonequilibrium laminar boundary-layer flow of ionized air.* AIAA Journal, 1964. **2**(11): p. 1921-1927.

19. Hall, H.G., Eschenroeder, A.Q., Marrone, P.V., *Blunt-Nose Inviscid Airflows With Coupled Nonequilibrium Processes.* Journal of the Aerospace Sciences, 1962. **29**(9): p. 1038-1051.

20. Biagioli, F., F. Guethe, and B. Schuermans, *Combustion dynamics linked to flame behaviour in a partially premixed swirled industrial burner.* Experimental Thermal and Fluid Science, 2008: p. Medium: X; Size: page(s) 1344-1353.

21. Dupont, V., et al., *The reduction of NOx formation in natural gas burner flames.* Fuel, 1993. **72**(4): p. 497-503.

22. Griffiths, W.D. and F. Boysan, *Computational fluid dynamics (CFD) and empirical modelling of the performance of a number of cyclone samplers.* Journal of Aerosol Science, 1996. **27**(2): p. 281-304.

23. Ma, L., D.B. Ingham, and X. Wen, *NUMERICAL MODELLING OF THE FLUID AND PARTICLE PENETRATION THROUGH SMALL SAMPLING CYCLONES.* Journal of Aerosol Science, 2000. **31**(9): p. 1097-1119.

24. Sousani, M., Ingham, D., Pourkashanian, M., Sheng, Y., Kenneth, I-I. E., *Simulation of the hydraulic fracturing process of fractured rocks by the discrete element method.* Environmental Earth Sciences, 2015. **73**(12): p. 8451-8469.

25. Sousani, M., Kenneth, I-I. E., Ingham, D., Pourkashanian, M., Sheng, Y., *Modelling of hydraulic fracturing process by coupled discrete element and fluid dynamic methods.* Environmental Earth Sciences, 2014. **72**(9): p. 3383-3399.

26. Cundall, A.P. and O.D.L. Strack, *A discrete numerical model for granular assemblies.* Géotechnique, 1979. **29**(1): p. 47-65.

27. Kim, B.S., S.W. Park, and S. Kato, *DEM simulation of collapse behaviours of unsaturated granular materials under general stress states.* Computers and Geotechnics, 2012. **42**: p. 52-61.

28. McKinnon, S.D. and I. Garrido de la Barra, *Stress field analysis at the El Teniente Mine: evidence for N–S compression in the modern Andes.* Journal of Structural Geology, 2003. **25**(12): p. 2125-2139.

29. Baird, A.F. and S.D. McKinnon, *Linking stress field deflection to basement structures in southern Ontario: Results from numerical modelling.* Tectonophysics, 2007. **432**(1–4): p. 89-100.

30. Pantaleev, S., et al., *An experimentally validated DEM study of powder mixing in a paddle blade mixer.* Powder Technology, 2017. **311**: p. 287-302.

31. Romuli, S., S. Karaj, and J. Müller, *Discrete element method simulation of the hulling process of Jatropha curcas L. fruits.* Biosystems Engineering, 2017. **155**: p. 55-67.

32. Johansson, M., et al., *Cone crusher performance evaluation using DEM simulations and laboratory experiments for model validation.* Minerals Engineering, 2017. **103–104**: p. 93-101.

33. Tsuji, Y., Kawaguchi, T.,Tanaka, T., *Discrete particle simulation of two-dimensional fluidized bed.* Powder Technology, 1993. **77**(1): p. 79-87.

34. Tsuji, T., Yabumoto, K., Tanaka, T., *Spontaneous structures in three-dimensional bubbling gas-fluidized bed by parallel DEM–CFD coupling simulation.* Powder Technology, 2008. **184**(2): p. 132-140.

35. Xu, B.H. and A.B. Yu, *Numerical simulation of the gas-solid flow in a fluidized bed by combining discrete particle method with computational fluid dynamics.* Chemical Engineering Science, 1997. **52**(16): p. 2785-2809.

36. Gidaspow, D., Bezburuah, R., Ding, J., *Hydrodynamics of circulating fluidized beds: Kinetic theory approach.* Conference: 7th international conference on fluidization. 1991, Chicago, US: Illinois Instute of Techonology. 8.

37. Wen, C.Y., Yu, Y. H., *Mechanics of Fluidization.* The Chemical Engineering Progress Symposium Series, 1966. **62**(62): p. 100-111.

38. Ergun, S., *Fluid flow through packed columns.* Chemical Engineering Progress, 1952. **48**(2): p. 89-94.

39. Benzarti, S., H. Mhiri, and H. Bournot, *Drag models for simulation gas-solid flow in the bubbling fluidized bed of FCC particles.* World Academy of Science, Engineering and Technology, 2012. **61**: p. 1138-1143.

40. Chaudhuri, B., F.J. Muzzio, and M.S. Tomassone, *Modeling of heat transfer in granular flow in rotating vessels.* Chemical Engineering Science, 2006. **61**(19): p. 6348-6360.

41. Ranz, W.E. and W.R. Marshall, *Evaporation from drops.* Chemical Engineering Progress, 1952. **48**(3): p. 141-146.

42. Hertz, H., *On the Contact of Rigid Elastic Solids and on Hardness.* Ch 6: Assorted Papers, 1882.

43. Mindlin, R.D., *Compliance of elastic bodies in contact.* J. Appl. Mech., 1949. **16**: p. 259-268.

44. Deresiewicz, H., Mindlin, R. D., *Elastic spheres in contact under varying oblique forces.* 1952, Columbia University, US: Department of Civil Engineering.

45. Khronos OpenCL Working Group, *The OpenCL Specification*, L. Howes, Editor. 2015.

46. nVIDIA. *NVIDIA'S INVENTION OF THE GPU.* [cited 2017 03 April];

47. nVIDIA. *What is CUDA*. [cited 2017 03 April];

508886 Multi-Scale Particulate Process Modelling and Design: Industry-Led Applications (Round-table discussion)

Abstract:

Further R&D advances required to fulfill the future ambitions of customer industries will be highlighted through a joint interactive commentary from the session speakers and the audience.

509573 Optical Single Particle Detection with Wide Dynamic Range for Nano- and Microparticle Sizing and Counting

Abstract:

Dispersions, suspensions, and emulsions are used in different industrial branches and consist of various materials. Two important parameters to characterize such products are particle size distribution and particle concentration. Typical methods used in industry to characterize suspensions are photon correlation spectroscopy also named dynamic light scattering and laser diffraction. But, these methods cannot measure the particle concentration directly and have limited capability of detection of subpopulations in polydisperse systems. Especially small subpopulations and size differences are not detected properly.

To address this issue, we are developing a particle counter for particle sizing and counting with a wide dynamic detection range. The instrument is designed to detect nano- and microparticles in aqueous suspension with simultaneous concentration determination. Particle detection is based on light scattering measured in two directions and hydrodynamic focusing for particle separation. The detection of single particles has the advantage of very good sizing resolution enabling for example determination of particle agglomerates for stability analysis or growth of shell for core shell particle production. Concentration determination is important e.g. for product stability where particle suspensions are stabilized using detergents and for pharmaceutical products where particle concentration must be certified.

We present results for detection of polydisperse particle suspensions in a size range of approximately 100 nm to 3 µm and monodisperse particles made of different materials. The current lower detection limit for polystyrene particles is approximately 65 nm in side scatter and 100 nm for forward scatter. Additionally, the data show a very good resolution in signal intensity enabling accurate particle sizing. The particle concentration can be determined with a measurement uncertainty of less than 1%. However, according to the applied theory of Mie scattering and Rayleigh-Debye-Gans approximation, the lower detection limit for detectable particle size by light scattering strongly depends on the index of refraction of the material which the particles consist of.

ACKNOWLEDGEMENTS

The work is supported by the funding program 'Messen, Normen, Prüfen und Qualitätssicherung' (MNPQ) of the Federal Ministry for Economic Affairs and Energy (BMWi) in Germany

507246 Continuous Synthesis and *inÂ Situ* SAXS Analysis of Silica Nanoparticles in Liquid Phase

Abstract:

Continuous synthesis and *in situ* SAXS analysis of silica nanoparticles in liquid phase

Manuel Meier, Julian Ungerer, Hermann Nirschl

Karlsruhe Institute of Technology, Institute for Mechanical Process Engineering and Mechanics, Karlsruhe, Germany

E-Mail: manuel.meier@kit.edu

Abstract

The synthesis of specific nanomaterials with defined properties (size distribution, morphology etc.) requires detailed knowledge about the particle formation and growth mechanisms. To gain knowledge about the relevant mechanisms, the use of time-resolved particle characterization techniques is helpful. In our previous work, we have developed a laboratory-scaled small angle X-ray scattering (SAXS) device providing time-resolved information on silica particles synthesized in liquid phase [1, 2]. The synthesis was conducted by means of a semi-batch reactor, i.e., samples were analysed *ex situ* by means of SAXS. The obtained results were in a good agreement with results published by others, e.g. Boukari et al. [3], Keefer and Schaefer [4] and Jelinek [5], indicating a gradual densification process dependent on reaction progress. In addition, the porosity of Stöber particles depends on the growth rate that in turn is affected by the reaction conditions: While low growth rates lead to compact structures, high growth rates result in particles with enlarged surfaces.

However, ex situ analyses often lead to unavoidable side effects due to the abrupt change of the ambient conditions of the sample under study, especially for the first minutes. To minimize such errors, the experimental setup was modified in such a way that *in situ* SAXS analyses can be conducted. The modification is based on a laminar flow reactor, which is directly connected to the SAXS device. The reaction time can be varied by a change of the flow rate of the reactants or the reactor length. For validation, compact silica nanoparticles were synthesized via a modified Stöber process [6] and analysed in situ by SAXS. Hereinafter, the obtained results are used to gain new knowledge about the gradual densification process and therefore a deeper understanding how to synthesize microporous silica particles.

Some practical drawbacks had to be overcome in order to ensure proper mixing of the reactants as well as to control properly the reaction temperature. Using the optimized setup, the reaction conditions were varied systematically and the particles were characterized by the *in situ* SAXS technique. In this contribution the optimized experimental setup as well as new insides about the particle formation process of silica nanoparticles will be presented in detail.

References

1. Goertz, V., et al., *Small-Angle X-ray Scattering Study of the Formation of Colloidal SiO2Stöber Multiplets.* The Journal of Physical Chemistry C, 2012. **116**(51): p. 26938-26946.

2. Gutsche, A., et al., *Time-resolved SAXS characterization of the shell growth of silica-coated magnetite nanocomposites.* Journal of Nanoparticle Research, 2014. **16**(7).

3. Boukari, H., J.S. Lin, and M.T. Harris, *Small-Angle X-Ray Scattering Study of the Formation of Colloidal Silica Particles from Alkoxides: Primary Particles or Not?* Journal of Colloid and Interface Science, 1997. **194**(2): p. 311-318.

4. Keefer, K.D. and D.W. Schaefer, *Growth of Fractally Rough Colloids.* Physical Review Letters, 1986. **56**(22): p. 2376-2379.

5. Jelinek, L., et al., *Study of the Stoeber reaction. 1. Properties of colloidal silica spheres prepared via alkoxide hydrolysis.* Langmuir, 1992. **8**(9): p. 2152-2164.

6. Stöber, W., A. Fink, and E. Bohn, *Controlled growth of monodisperse silica spheres in the micron size range.* Journal of Colloid and Interface Science, 1968. **26**(1): p. 62-69.

507364 Advanced Characterization of Non-Spherical Nanoparticles

Abstract:

Excellent product properties of the nanoparticles (NPs) including optoelectronic properties are very often dependent on their disperse properties [1,2]. An accurate evaluation of the disperse properties including shape, size and density therefore becomes a necessity. However, simultaneous characterization of these disperse properties still remains a major challenge. Size evaluation is also quite complicated for non-spherical NPs, where equivalent diameters are used to represent the size.

Nanorods (NRs) and other two dimensional NPs like platelets, nanotubes, ellipsoids etc. require two parameters for complete size characterization. Gold NRs are very popular due to their enhanced optical properties while Zinc oxide NRs are relevant in a number of nanodevices and thin film applications. Ellipsoidal nanoparticles are also well researched for their enhanced surface plasmon resonance. Although electron microscopy based techniques are the obvious choice for their characterization, these methods require analysis of a large number of particles for statistical significance. Dynamic light scattering and similar techniques are more complicated for non-spherical NPs and the hydrodynamic diameters obtained need to be corrected for the effect of rotational diffusion. This cannot be accomplished with only one measurement technique.

There have been commendable advances in the recent past on the characterization of nanoparticles dispersed in colloidal and gas phase systems. Analytical ultracentrifugation equipped with an UV-Vis multiwavelength detector (MWL-AUC) is a powerful and highly accurate tool for the multidimensional analysis of NPs dispersed in colloids [2]. AUC measures the motion of concentration boundaries depending on the applied centrifugal field. The motion of the concentration boundaries can be related to the sedimentation coefficient distribution of the NPs which depends on their mass and hydrodynamic diameter.

Owing to the typical sizes of the NPs, their motion in gas phase systems cannot be represented by continuum equations and kinetic gas theory comes into effect [3]. Scanning mobility particle sizer (SMPS) is the most commonly used measurement technique for characterizing and classifying aerosol NPs. It measures the electrical mobility of the NPs which are often dependent on the average projected area of the NPs in addition to the hydrodynamic diameter. For NRs, platelets and ellipsoidal NPs, AUC and SMPS measurements can be simultaneously analyzed to provide information on the two length parameters, namely the long and the short axes [4]. Colloidal NPs can be transferred to be dispersed in gas phase and this opens up new possibilities in NP characterization. It has been shown that the combination of these methods gives excellent estimates of the average length and diameter for gold and zinc oxide NRs in samples [4].

The method is being extended for nanoellipsoids, nanocubes and nanoplatelets. The combination of the techniques is also used to predict the variation in length and diameter of the nanorods in a sample where the cumulative distributions obtained from AUC and SMPS measurements are combined to provide insights on the length and diameter distributions of the nanorods.

References

1. J. Walter, T. J. Nacken, C. Damm, et al., Small, 2015, 11, 814.
2. J. Walter, G. Gorbet, T. Akdas, et al., Analyst, 2017, 142 (1), 206.
3. C. Zhang, T. Thajudeen, C. Larriba, et al., Aerosol Sci. & Tech.,2012, 46(10),1065.
4. T.Thajudeen, J. Walter, R.Srikantahrajah, et al., Nanocale Horizons, 2017.

507446 Time-Resolved Characterization of Customized Aluminum-Doped Zinc Oxide Nanocrystals By Means of Small-Angle X-Ray Scattering

Abstract:

Abstract

The semiconductor aluminum-doped zinc oxide (AZO) has versatile applications due to its special properties, e.g. low electrical resistance, good mechanical stability and high degree of transparency in the visible range. Due to these properties AZO is considered to be a significant cheaper alternative in order to replace existing and finite indium based oxides in the future for diverse scopes e.g. thin-film solar modules, touch-panels, light-emitting diodes or printing electronics. [1–5] Especially the application of AZO in transparency thin-film solar modules has a high potential with regard to the sustainable energy generation from renewable sources. In the future, glass surfaces, e.g. windows of office buildings, could be used as a renewable energy source.

Therefore, the availability of high quality nanocrystals with fitted sizes and shapes of crystals and a defined doping of aluminum to adjust the optical properties (band gap energy) is required. [2, 4] The major challenge is the link between the particular-based properties (size, shape, grade of agglomeration, morphology etc.) and the versatile applications, because the nanoscale area is difficult to access by measurement systems. Furthermore, the measurement methods to evaluate the structure parameters are frequently limited to discrete process steps and states, e.g. the characterization of powdery nanoparticles, or an elaborate post-processing step is required, e.g. in the electron microscopy.

The small-angle X-ray scattering (SAXS) basically enables a non-invasive determination of structural properties and several states of nanoparticles in individual process steps during the entire process chain (synthesis, stabilization, coating etc.) without the requirement of post-processing steps. Only one measurement technique is favorable to combine the measurement results of any process step referring to the evaluation of the performance of AZO-nanocrystals. Additionally, the SAXS-method offers the possibility to determine several characteristic parameters (particle size distribution, particle morphology and the state of aggregation) within only one measurement.

For synthesizing high-crystalline AZO nanostructures the non-aqueous sol-gel-synthesis is used. Here, particle precursors are solved in aromatic solvents, e.g. benzyl alcohol or benzyl amine, and converted to AZO-nanocrystals with defined sizes, narrow particle size distributions and adjustable morphologies at temperatures of 80°C to 200°C. [6–8] The purpose of the solvent is to control the particle growth and the particle morphology. [8, 9] Even the absence of a hydrolysis step, as occurred in the classical sol-gel synthesis, results in moderate reaction rates. This property enables time-resolved studies of the particle formation and growing process during the synthesis by means of SAXS.

The main objective is to adapt the AZO synthesis on our self-developed SAXS instrument. To obtain detailed knowledge about the particle formation and growth mechanism we run time-resolved in situ measurements. Because of measurement limitations of our SAXS instrument during real-time experiments, adjustments of the AZO synthesis are mandatory. On the one hand, an optimal process control is to be adapted to the measurement system. On the other hand, the chemical reaction needs to be modified with regard to reaction kinetics and particle size. Concerning the process control, a specifically developed ideal batch [10] reactor with computer-assisted online pressure- and temperature monitoring is used to ensure exact reaction settings and high reproducibility of particle properties during the synthesis. The ideal batch reactor, with 358 milliliter in capacity, is fitted specifically to the requirements of the non-aqueous sol-gel-synthesis. The heating and the closed vessel are designed for a maximum temperature of 300°C and a maximum pressure of 16 bar. Regarding to the analysis of the reaction solution during the process, a riser with a needle valve is installed to control the flow rate during the sample extraction. The extraction occurs continuously with low-pulsation due to the vessel pressure through a solvent resistant tube into a quartz glass flow capillary, which is located in the X-ray beam throughout the analysis. Hereby, the instrumental set-up provides ideal conditions for an in situ characterization by means of SAXS without interferences of the synthesis.

In this contribution an appropriate adaptation of the synthesis on our SAXS measurement system, special features of the instrumental set-up and time-resolved studies of particle formation and growth mechanisms, dependent on synthesis parameters, will be reported in detail.

References

[1] X. Jiang, F.L. Wong, M.K. Fung, S.T. Lee, Aluminum-doped zinc oxide films as transparent conductive electrode for organic light-emitting devices, Applied Physics Letters 83 (2003) 1875–1877.

[2] L. Luo, M.D. Rossell, D. Xie, R. Erni, M. Niederberger, Microwave-assisted nonaqueous sol–gel synthesis: from Al: ZnO nanoparticles to transparent conducting films, ACS Sustainable Chemistry & Engineering 1 (2012) 152–160.

[3] T. Minami, Transparent conducting oxide semiconductors for transparent electrodes, Semiconductor Science and Technology 20 (2005) S35.

[4] T. Strachowski, E. Grzanka, W. Lojkowski, A. Presz, M. Godlewski, S. Yatsunenko, H. Matysiak, R.R. Piticescu, C.J. Monty, Morphology and luminescence properties of zinc oxide nanopowders doped with aluminum ions obtained by hydrothermal and vapor condensation methods, Journal of Applied Physics 102 (2007) 73513.

[5] C.R. Ellinger, S.F. Nelson, Selective Area Spatial Atomic Layer Deposition of ZnO, Al2O3, and Aluminum-Doped ZnO Using Poly (vinyl pyrrolidone), Chemistry of Materials 26 (2014) 1514–1522.

[6] G. Garnweitner, M. Niederberger, Organic chemistry in inorganic nanomaterials synthesis, Journal of Materials Chemistry 18 (2008) 1171–1182.

[7] G. Garnweitner, N. Tsedev, H. Dierke, M. Niederberger, Benzylamines as Versatile Agents for the One Pot Synthesis and Highly Ordered Stacking of Anatase Nanoplatelets, European Journal of Inorganic Chemistry 2008 (2008) 890–895.

[8] M. Niederberger, G. Garnweitner, Organic reaction pathways in the nonaqueous synthesis of metal oxide nanoparticles, Chemistry-A European Journal 12 (2006) 7282–7302.

[9] S. Zellmer, A. Kockmann, I. Dosch, B. Temel, G. Garnweitner, Aluminum zinc oxide nanostructures with customized size and shape by non-aqueous synthesis, CrystEngComm 17 (2015) 6878–6883.

[10] Design, construction and manufacturing took place by the Institute for Particle Technology (iPAT) of the Technical University Braunschweig, led by G. Garnweitner.

Multiple particle size distribution characterization techniques applied to certified standards in the nanoparticle size region

Alan F Rawle Malvern Panalytical, Joerg Bolze Malvern Panalytical

Abstract

Polydispersity is the bane of all particle size distribution techniques. For irregular and polydisperse materials the resultant size distribution is dependent on the utilized technique and sizes are generally reported as some form of spherical or circular equivalent. In the nano (~ 1 – 100 nm) region there are relatively few size standards that are available and these are not always applicable to a range of techniques. Standards are important in confirming and verifying any instrument's performance providing confidence both in the technique and the individual instrument selected.

This study compares the following techniques on the NIST gold colloid reference materials RM 8011, 8012, 8013:

- Dynamic light scattering (DLS): see ISO 22412:2017
- Nanoparticle tracking analysis (NTA): see ISO 19430:2016
- Small-Angle X-Ray Scattering (SAXS): see ISO 17867:2015

We will indicate the results obtained on all systems for the different techniques and highlight the reasons why these techniques provide different results on the 'same' particle size 'standards'

Introduction

Particle size distribution techniques present many challenges in measurement and understanding. We note that different techniques measure different properties of the particulate system and interpret the obtained information in different manners [1]. Nowhere is this truer than in the field of nanotechnology where a myriad of techniques each with its own advantages and disadvantages confronts the naïve scientist. This situation is further complicated by unscrupulous vendors pushing their own techniques and technology into applications where it is not suited or where the quality of the obtained data is compromised by poor signal-to-noise and thus liable to misinterpretation. Possibly in no other field is the term 'characterization' misused and abused [2].

In particle size distribution measurements, it is not often that only a mean or median value is sought. A unique or fixed 1-D answer is often required in 2-dimensional space (the size and quantity axes) on a 3-D particle. There is confusion between number and volumes of particles perhaps no more indicative than the EU's recent attempt to define 'nano' based on numbers of particles [3]. Often it is data on numbers relating to the tails of the distribution that is important to elucidate and having tiny amounts of tiny material (how can you measure what you cannot locate and see?) further complicates the issue. The killer word in 'particle size distribution' is the last one.

One route to bring clarity into any field to have measurements conducted in an appropriate manner on materials that have reference or accepted values (together with appropriate tolerances) associated with them. This paper looks at a little of the background literature together with new measurements (including those utilizing newer techniques such as Nanoparticle Tracking Analysis (NTA)) on reference materials with assigned values from prestigious organizations such as NIST in the United States and the Institute for Reference Materials and Measurements (IRMM) in the EU.

A textbook has already been published entitled "Nanotechnology Standards" [4] and thus it is not necessary to rewrite or paraphrase this tome. Much on the background to 'characterization of nanomaterials' can be found in a chapter of that exact title in a publication authored by NIST editors [5].

Selected background

Perhaps one of the best introductions to the complexities of measurement of reference nanoparticle systems can be gleaned by reading one of the certificates ('Report of Investigation') associated with the NIST RM801x (x = 1, 2, 3) gold colloid reference materials nominally 10, 30, and 60 nm in size. For example, if we examine the assigned values and expanded uncertainties for RM8011 we see [6]:

Table 1: Adapted NIST particle size data for RM8011, gold colloid reference material

Table 1. Reference Value Mean Size and Expanded Uncertainty
Average Particle Size (Diameter), in nm

Technique	Analyte Form	Particle Size (nm)
Atomic Force Microscopy	dry, deposited on substrate	8.5 ± 0.3
Scanning Electron Microscopy	dry, deposited on substrate	9.9 ± 0.1
Transmission Electron Microscopy	dry, deposited on substrate	8.9 ± 0.1
Differential Mobility Analysis	dry, aerosol	11.3 ± 0.1
Dynamic Light Scattering	liquid suspension	13.5 ± 0.1
Small-Angle X-ray Scattering	liquid suspension	9.1 ± 1.8

This hints at the most utilized particle size measurement techniques in nanotechnology and perhaps is one of the few documents clearly showing the difference in values obtained by different techniques on a reference material. In the situation where the differences in quoted values is greatest then we learn more about the system by probing it with different tools. It is hard for the amateur or even professional scientist to realize that all stated values are correct but different. Smaller values tend to be associated with number counting techniques such as microscopy and higher values with ensemble techniques based on volume or scattering intensity (Dynamic Light Scattering, DLS, for example). Differences can also result from the nature of the measurement itself. Small-angle X-ray Scattering, SAXS, will probe the electron-dense part of the gold colloid whereas DLS will measure the hydrodynamic diameter of a particle taking into account the surrounding (organic) surfactant or stabilization layer unseen by SAXS. See, for example, a webinar examining this exact comparison [7].

We have examined the RM801x materials using both DLS and SAXS and describe in detail the methods and results briefly indicated in [7].

Hole and coworkers have indicated NTA and DLS results obtained on RM 8013 the nominal 60 nm Au colloid from NIST [8] both 'neat' and in the presence of blood plasma. Results are in line with the NIST specifications and a protein corona enlarges the apparent size in the presence of blood plasma.

A study of citrate-capped Au nanoparticles, ~ 50 nm in diameter, synthesized by HAuCl4.xHO reduction by DLS, SEM, and NTA techniques is described in [9]. This also relates to the importance of the possible aggregation of such materials in the presence of protein materials.

NTA is also described in a thorough study by Jan Mast & Pieter-Jan De Temmerman of CODA-CERVA, [10], in which the RM8102 and RM8013 materials were examined. The publication also indicates work with other standard materials and with other techniques such as electron microscopy.

Table 2 PTA size and uncertainty data for RM8012 and RM8013 measured by NTA

PTA Hydrodynamic diameter (nm)	RM-8012	RM-8013
Mean measured value, C_m	30.3 nm	53.7 nm
Reference value, C_{ref}	27.6 nm [B]	56.0 nm [B]
Repeatability uncertainty, u(r)	6.86%	2.86%
Intermediate precision uncertainty u(ip)	1.60% [C]	1.82%
Measurement uncertainty u(m)	1.91%	1.10%
Combined uncertainty $u_c(x)$	7.30%	3.56%
Expanded uncertainty U(x)	14.59%	7.12%
Δm	2.7 nm	2.3 nm
U(x)- Δm	1.7 nm	1.5 nm

This work is also described in [11].

SAXS data is harder to come by but a study of silica nanoparticles functionalized in different manners and studied via SAXS, TEM, DLS, PTA *(Particle Tracking Analysis – the ISO term for NTA)* and Centrifugal liquid sedimentation (CLS) has been published [12]. In this the statement is made 'The size of the core of the primary particle fraction is best resolved in the SAXS measurements, which show almost no indication of other fractions (agglomerates and smaller particles).

Our work

We have studied the RM801x materials by DLS as part of an ASTM Round Robin exercise [13] used to gather a precision and bias (P & B) statement for the ASTM DLS standard E2490-09 (15), [14]. This International Laboratory Study (ILS) had 26 participants and 7700 results taken to generate the required P & B statement. We used a conventional Malvern Zetasizer ZS with standard polystyrene cuvettes and samples were measured directly without dilution. Measurements of 30 seconds duration were taken. On a TEM image supplied by NIST, we also analyzed the (number) distribution via a proprietary program that forms the heart of the Malvern Morphologi 4 system and earlier Morphologi G3 systems.

The RM801x materials were studied in the Almelo laboratories of Malvern Panalytical. SAXS data were acquired on the Empyrean platform configured with Scatter X[78] in combination with a focusing X-ray mirror and the PIXcel[3D] detector used in 1D mode. Background was a capillary filled with DI water and the acquisition time was 20 minutes.

DLS results

In graphical form:

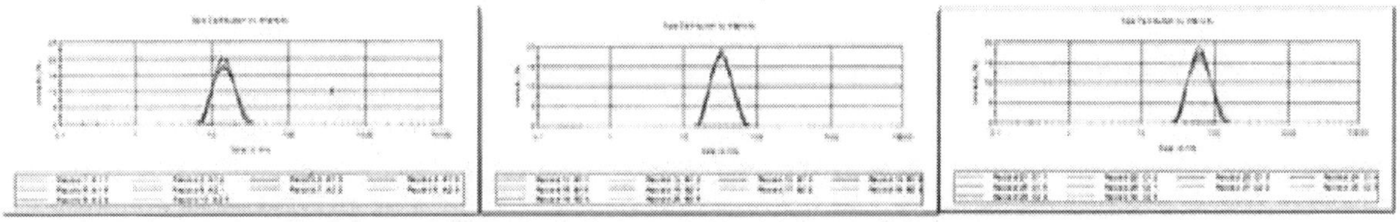

Figure 1: Individual DLS frequency distribution plots for RM8011, RM8012, RM8013

These graphics appear more attractive if the results are placed on a single graph by selecting single typical values:

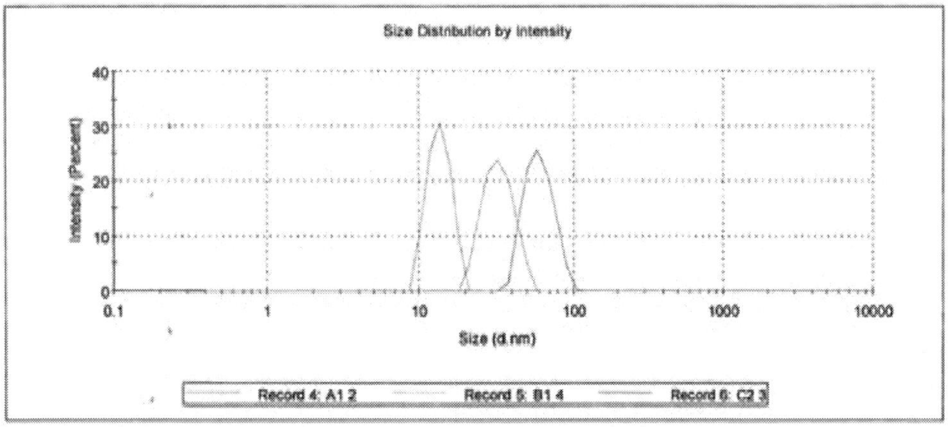

Figure 2: Frequency plots of the 3 RM801x gold colloid standards measured by DLS

SAXS results

In graphical form:

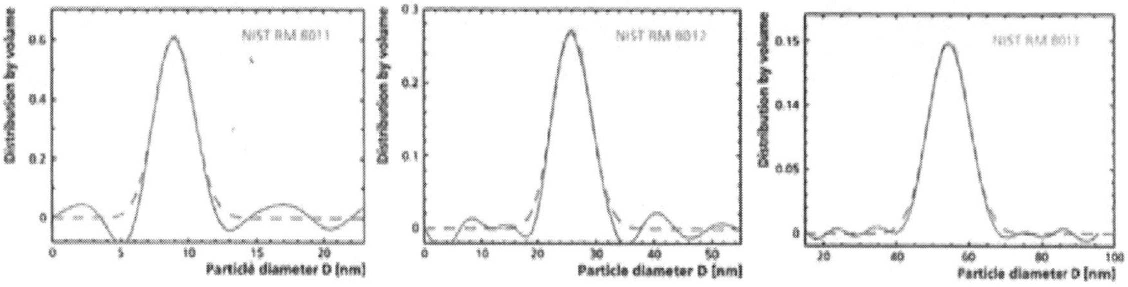

Figure 3: SAXS frequency results for RM8011, RM8012, and RM8013

We summarize all the mean data in tabular form as below:

Table 3 Summary Data for RM801x material measured by different techniques

	RM8011	RM8012	RM8103
DLS - NIST Assigned value (nm)	13.5	28.6	56.6
DLS - NIST Uncertainty value (nm)	0.1	0.9	1.4
SAXS - NIST Assigned value (nm)	9.1	24.9	53.2
SAXS - NIST Uncertainty value (nm)	1.8	1.2	5.3
NIST - AFM	8.5	24.9	55.4
NIST - SEM	9.9	26.9	54.9
NIST - TEM	8.9	27.6	56
Malvern Panalytical DLS	13.7 +/- 0.2	29.5 +/- 0.5	60.0 +/- 0.3
Malvern Panalytical - Image Analysis	Not taken	32.5	Not taken
Malvern Panalytical SAXS	9.0	26.0	54.4

Discussion

The DLS and SAXS results are in excellent agreement, as expected, with the NIST stated values. For the smallest material, RM 8011, then we see the greatest difference (not discrepancy!) between the measurement by SAXS and DLS. This material is the one where the protective surfactant or stabilizing layer makes a significant contribution to the overall diameter of the particle. As the SAXS technique probes the electron-dense part of the particle (the Au part) and the shielding part is basically unseen, we can estimate the thickness of the surrounding layer by comparison with DLS which is measuring the overall hydrodynamic size of the particle.

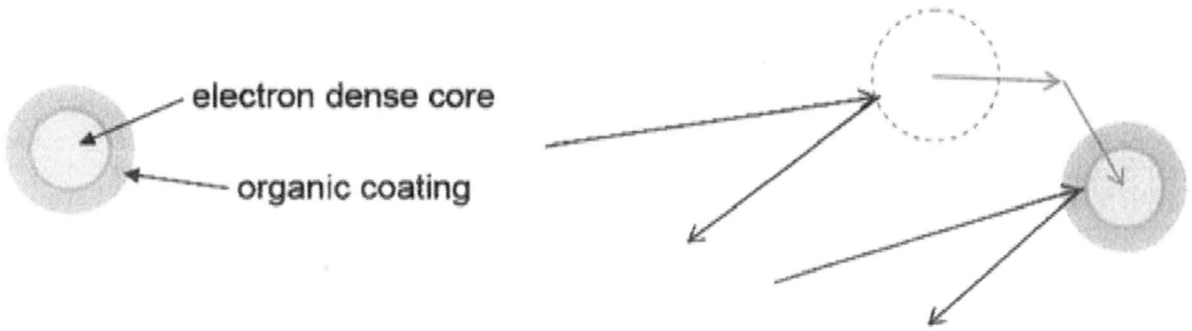

Figure 4: Schematic representation of what DLS and SAXS 'sees' (after V A Hackley)

We have (13.5 – 9.1)/2 or 2.2 nm as this thickness. This is probably a factor of 10 larger than the size of a citrate molecule (if indeed this is the stabilizing agent) indicating that the hydrodynamic diameter must reflect other entities (e.g. molecules from the continuous phase) that travel with the particle itself. Interestingly this calculated thickness is similar for the RM8012 material - (28.6 – 24.9)/2 = 1.85 nm - and the RM8013 material: (56.6 – 53.2)/2 = 1.7 nm.

Conclusions

The measurement and interpretation of reference material standards for the nanomaterial is a vital part of verifying the measurement capabilities of any utilized technique. We gain more information by using more than one tool to probe the system. In the case of DLS and SAXS measurements we can gain an insight into the protective layer present as an essential part of the particulate structure.

References

[1] A F Rawle The Basic Principles of Particle Size Analysis (1993) Available from several sites including: http://www.rci.rutgers.edu/~moghe/PSD%20Basics.pdf

[2] Blog: A F Rawle Characterization is a number (42) - not a goal http://www.materials-talks.com/blog/2014/02/06/characterization-is-a-number-42-not-a-goal/ Accessed 3/27/2018

[3] Commission Recommendation of 18 October 2011 on the definition of a nanomaterial (2011/696/EU) http://eur-lex.europa.eu/legal-content/EN/TXT/?uri=CELEX:32011H0696 Available in many languages and formats Accessed 3/27/2018

[4] V Murashov, John Howard (Editors) Nanotechnology Standards Springer (2011)

[5] A F Rawle Characterization of nanomaterials Chapter 7 in Metrology and Standardization of Nanotechnology: Protocols and Industrial Innovations Editors: E Mansfield, D L. Kaiser, D Fujita, M Van de Voorde Wiley-VCH Verlag GmbH & Co. KGaA DOI:10.1002/9783527800308 (2017)

[6] Taken from: https://www-s.nist.gov/srmors/certificates/8011.pdf

[7] A F Rawle J Bolze The NIST Au colloid nanomaterial, RM 8011: a DLS and SAXS comparative study of the particle size distribution (September 22nd, 2011) https://www.brainshark.com/malvern/vu?pi=251680509&text=M021507&r3f1=

[8] I M-Burgos, D Walczyk, P Hole, J Smith, I Lynch, K Dawson Characterisation of nanoparticle size and state prior to nanotoxicological studies Nanopart Res (2010) 12:47–53 DOI 10.1007/s11051-009-9774-z

[9] S V. Jenkins, H Qu1, T Mudalige, T Ingle, R Wang, F Wang, P C Howard, J Chen, Y Zhang Rapid Determination of Plasmonic Nanoparticle Agglomeration Status in Blood Biomaterials. 2015 May; 51: 226–237. doi:10.1016/j.biomaterials.2015.01.072

[10] J Mast, P-J De Temmerman Protocol(s) for size-distribution analysis of primary MNM objects in powders and liquids for compliance with the EU definition (D2.10) BE mid-term NANoREG meeting February 20, 2015, Brussels, BE

[11] P-J De Temmerman, E Verleysen, J Lammertyn, J Mast Size measurement uncertainties of near-monodisperse, near spherical nanoparticles using transmission electron microscopy and particle-tracking analysis J Nanopart Res (2014) 16:2628 DOI 10.1007/s11051-014-2628-3

[12] C Gollwitzer, D Bartczak, H Goenaga-Infante, V Kestens, M Krumrey, C Minelli, M P´almai, Y Ramaye, G Roebben, A Sikorad, Z Varga A comparison of techniques for size measurement of nanoparticles in cell culture medium Anal. Methods, 2016, 8, 5272

[13] ASTM Research report RR: E56-1001 Interlaboratory Study to Establish Precision Statements for ASTM E2490-09 Standard Guide for measurement of particle size distribution of nanomaterials in suspension by photon correlation spectroscopy (PCS) ASTM International, 100 Barr Harbor Drive, West Conshohocken, PA 19428 1 April 2009

[14] See: https://www.astm.org/Standards/E2490.htm Accessed 3/28/2018

508070 Analytical Ultracentrifugation As a Tool for Multidimensional Characterization of Nanoparticles

Abstract:

The analysis of size, shape and density is essential for nanoparticles (NP), especially as they exhibit a large variety of properties. A combined approach to assess and correlate these parameters is challenging. Conventional particle measurement techniques like scattering, field flow fractionation or microscopy techniques are often limited by statistics, resolution and applicability to broad particle size distributions (PSD).

Analytical Ultracentrifugation equipped with an UV/VIS multiwavelength spectrometer (MWL-AUC) can serve as a powerful tool for the multidimensional characterization of nanoparticles. It combines the detection of optical spectra in the range from 240 to 1000 nm with fractionation by a centrifugal field. Our recent developments regarding instrumentation, data acquisition and evaluation allow us to largely extent the possibilities of MWL-AUC in the context of nanoparticle characterization.

Graphene oxide (GO) is a representative for the emerging class of 2D materials and originates from the exfoliation of chemically modified graphite as a first step in the top-down approach to produce graphene. However, statistically relevant characterization techniques of shape and size distributions with reasonable temporal time effort were not available for these nanoplatelets. Therefore, a new methodology was developed to determine the lateral size distribution directly in solution within a few hours [1]. This unique approach allows to investigate processes like ultrasonication or ball milling of GO and determine the size evolution of nanoplatelets in detail [2]. The thereby determined size reduction rate could be correlated with material properties originating from raman-, solid-state nuclear magnetic resonance and thermal gravimetric analysis measurements, enabling us to identify relevant process parameters for processing of 2D materials. By changing type and amount of functionalization, controlled and scalable production of nanoplatelets becomes possible.

MWL-AUC can be combined with other techniques like the scanning mobility particle sizer (SMPS), where particles are classified according to their mobility diameter in the gas phase [3]. This is especially interesting as the two methods give information on the same particles in two distinct environments allowing for quantitative shape analysis and for validating colloidal stability of e.g. ionomer particles in solution.

Finally, core shell NPs (e.g. ZnO, $CuInS_2$) were addressed as these particle systems cause special challenges to conventional measurement techniques. This is mainly due to the pronounced polydispersity of many nanoparticle systems. Therefore, we developed a new methodology for the simultaneous determination of size and density for polydisperse PSDs based on the parametrically constrained spectrum analysis implemented in the free software tool UltraScan3 [4]. The high resolution both in density and size allows characterizing the core-shell properties of NPs within one single experiment.

MWL-AUC has proven to be a powerful and indispensable method for the multidimensional characterization of nanoparticles directly in solution. The direct correlation of size, shape and optical properties of NPs is possible in a range from 1 nm to 1 μm making it highly relevant for numerous applications in industry and science.

References

1. J. Walter, T. J. Nacken, C. Damm, et al., Determination of the Lateral Dimension of Graphene Oxide Nanosheets Using Analytical Ultracentrifugation. Small, 2015, 11, 814.

2. C. E. Halbig, T. J. Nacken, J. Walter, et al., Quantitative investigation of the fragmentation process and defect density evolution of oxo-functionalized graphene due to ultrasonication and milling. Carbon, 2016, 96, 897.

3. T. Thajudeen, J. Walter, R. Srikantharajah et. al., Determination of the length and diameter of nanorods by a combination of analytical ultracentrifugation and scanning mobility

particle sizer. Nanoscale Horiz., 2017.

4. J. Walter, G. Gorbet, T. Akdas, et al., 2D analysis of polydisperse coreâ€"shell nanoparticles using analytical ultracentrifugation. Analyst, 2017, 142.

507255 A Mobile Device for Nanoparticle Characterization By Wide-Angle Light Scattering (WALS) and Laser-Induced Incandescence (LII)

Abstract:

The combination of the optical measurement techniques Laser-induced incandescence (LII) and Elastic Light Scattering (ELS) allows for an online, non-invasive and comprehensive characterization of nanoparticle systems. With LII the monomer size (primary particle diameter) and particle concentration (volume fraction) are accessible [1], aggregate size (radius of gyration) and aggregate structure (fractal dimension) can be derived from elastically scattered light [2].

We developed and improved a mobile measurement setup based on wide-angle light scattering (WALS) combined with two-color Time-Resoved LII (TiRe-LII) [3]. Conventional approaches for elastic light scattering techniques may suffer from limited angular resolution using only a few detectors at discrete positions or slow measurement rates with a single detector rotating in angular steps around the measurement object. With WALS the scattered light from a small punctual measurement volume is imaged onto a camera-chip using an ellipsoidal mirror. Consequently, scattering diagrams can be obtained from single-shot data over a wide angular range (10°-165°) at a high angular resolution (<0,5°). Here, a continous-wave fiber-laser (2 W maximum output power @ 532 nm) and an industrial CCD-camera were used to acquire scattering data. Evaluation of the scattering data in the Guinier- and Power-Law-regime yields effective values of radius of gyration and fractal dimension. For LII-measurements a pulsed diode-pumped Nd:YAG-laser (15 mJ maximum energy per pulse @ 1064 nm and 100 Hz maximum pulse frequency) is employed. The incandescence signal is detected at two different spectral regions by gated photomultiplier tubes and read-out by an oscilloscope (500 MHz, 4 GaS/s). Primary particle sizes were subsequently derived from an adequate theoretical cooling model considering decay time, particle temperature, ambient temperature and pressure. The volume fraction was determined relating the maximum intensity of the incandescence signal to a known volume fraction of a calibration source.

In practice, scattering data from real aerosols consists of the single scattering signals of aggregates of different morphology and only effective values for aggregate sizes can be determined. Inferring size distributions from the angular distribution of elastically scattered light is mathematically ill-posed as different distributions may lead to similar scatterings signals. Bayesian inference is used as a powerful statistical tool to derive probability densities for recovered size distributions of soot particles from a flat-flame burner (McKenna type). In another experiment, silica particles are synthesized from a liquid precursor and sintered afterwards. With increasing sintering temperature the fractal-like aggregates become more spherical and the scattering behavior shifts from the Rayleigh-Debye-Gans approximation for Fractal Aggregates (RDG-FA) towards Mie-theory. Here, a Principal Component Analysis (PCA) is conducted to correlate the scattering signals to the respective theory.

References

[1] H.A. Michelsen, C. Schulz, G.J. Smallwood, S.Will, Prog. Energ. Combust., 51: 2-48 (2015)

[2] C.M. Sorensen, Aerosol Sci. Technol. 35: 648-687 (2001)

[3] F.J.T Huber, M. Altenhoff, S. Will, Rev. Sci. Instr. 87: 053102 (2016)

Complimentary Techniques to Directly Characterize Liposomes: Nanoparticle Tracking Analysis, Dynamic Light Scattering, Electrophoretic Light Scattering, Small and Wide-Angle X-ray Scattering

Ragy Ragheb[1], Joerg Bolze[1], Graham J. Taylor[2], Nima Tamaddoni[2], Ulf Nobbman[1] and Duncan Griffiths[1]

(1) Malvern Panalytical, Westborough, MA, (2) T&T Scientific, Knoxville, TN

Introduction:

Liposomes have been attractive delivery systems for decades due to their composition of natural biological lipids and structural resemblance to cell membranes suggesting metabolic compatibility, low toxicity, overall expectations of biocompatibility and lack of a strong immune response [1]. They comprise of spherical vesicles with an aqueous core enclosed by one or more phospholipid bilayers or lamellae and are frequently classified based on their size, polydispersity, and number of bilayers [2]. Control over these parameters has remained a challenge with most preparation methods and is further accentuated when moving from a laboratory to industrial scale. The desirable size for drug delivery ranges between 50 and 200nm [3]. They are often used in studies of model biological membranes, phase transition and spacing, targeted drug delivery to specific areas of a human body, etc [4]. The extent of liposome instability in various biological fluids such as plasma depends on the relative concentrations, size and lamellarity, lipid composition and incubation temperatures [1]. Thus, biophysical characterization of these systems and their payloads is critical to understanding and optimizing their fabrication and function. This study looks at optimal conditions for extruding liposomes as well as their stability under different conditions. We highlight the limit of detection for fluorescently labeled liposomes. Malvern Panalytical affords a cost-effective suite of direct characterization of these critical parameters. Nanoparticle Tracking Analysis (NTA) from the NanoSight product range measures size, concentration and fluorescence measurements. Dynamic (DLS) and Electrophoretic (ELS) Light Scattering from the Zetasizer product range measures size, zeta potential, and payload stability. Small- and Wide-Angle X-ray scattering (SAXS/WAXS) measurements enable to determine the lamellarity, structures and dimensions of the lipid bilayers.

Materials:

DOPC (1,2-dioleoyl-*sn*-glycero-3-phosphocholine), DPPC (1,2-dipalmitoyl-*sn*-glycero-3-phosphocholine), and cholesterol (ovine wool, >98%) lipids were obtained in powder form from Avanti Polar Lipids (Alabaster, AL). Fluorescent rhodamine-DPPE lipids were obtained in powder form from Biotium (Fremont, CA). All lipids were dissolved initially in chloroform to create stock solutions and stored at -20°C. Buffer reagents (MOPS, NaCl, NaOH) were obtained from Sigma Aldrich.

Methods:

Liposome Preparation and Extrusion. DOPC, DOPC-cholesterol, and Rh-DPPE incorporated lipid mixtures were initially prepared at the desired molar ratios in chloroform, and the chloroform was subsequently evaporated to a film using a gentle nitrogen stream and overnight vacuum. Lipid films were then rehydrated to the desired molar concentration (typically 2mM DOPC) using aqueous buffer (typically 100 mM NaCl and 10 mM MOPS, pH 7.4 which resulted in the formation of Multi Lamellar Vescicle (MLV) suspensions. For DPPC lipids with a transition temperature of 42°C, the MLV suspension was heated after hydration of the lipid powder (single-component and bTLE) or film (multi-component mixtures). MLV suspensions were repeatedly frozen/thawed 5x, unless otherwise noted, and stored at -20°C until extrusion and Small Unilamellar Vesicle (SUV) formation. SUVs were formed by extruding thawed MLV suspensions through hand-held NanoSizer mini-extruders housing polycarbonate track-etched membranes of various pore-sizes including 50 nm, 100 nm, 200 nm, and 400 nm. The single-use NanoSizer kits (extruders, syringes and needles) and reusable extruder heating block were obtained from T&T Scientific (Knoxville, TN). The total number of passes through the extruder was varied as part of this study and is reported for each sample in the results section.

Analysis. Samples were analyzed by a NanoSight NS300 equipped with a high sensitivity Hamamatsu sCMOS camera, 20x objective lens, and a 50 mW green 532 nm laser. Samples were analyzed using NTA 3.2 Build 16 software. NanoSight technology (Malvern Panalytical) calculates size based on the relationship between Brownian motion and hydrodynamic diameter through the Stokes-Einstein equation. Concentration is calculated by particle observation on a frame-by-frame basis by the sCMOS camera (25 frames per second). When recording the video it averages the concentration across all the frames giving an absolute number average. Samples were processed across 3x 60 sec videos.

Samples were further analyzed by a Zetasizer Nano ZSP at 25 °C. Measurements were made with a disposable cuvette (DTS0012) and disposable folded capillary cell (DTS1070) for sizing and zeta potential respectively. The Nano ZSP incorporates non-invasive backscatter (NIBS™) optics for sizing measurements. The detection angle of 173° enables size measurements of concentrated, turbid samples to be made.

Small-and wide-angle X-ray scattering (SAXS/WAXS) data were measured using the Empyrean Nano edition (Malvern Panalytical), a versatile X-ray scattering platform. The experimental setup uses the line focus of a Cu X-ray tube, a focusing X-ray mirror, an evacuated beam path and a hybrid pixel array detector. The samples were filled in a 1 mm quartz capillary and measured at a liposome concentration of 25 mg/ml. DPPC and DOPC liposomes, before and after extrusion, were investigated with the SAXS/WAXS technique under variation of sample temperature. Extrusion was done by repeatedly passing the liposome samples through an extruder with a pore size of 100 nm.

Discussion:

NTA, DLS, and ELS provided a comprehensive characterization of liposomes across various conditions. Both NanoSight and Zetasizer are similar because they both rely on the Brownian motion of and light scattering from the particle. Both utilize the Stokes-Einstein equation and relate diffusion to size (hydrodynamic diameter). NTA produces a number-based size distribution from a particle-by-particle measurement, while DLS produces an intensity-based distribution from an ensemble measurement. This is further exemplified in the set of experiments shown here.

NTA and DLS confirmed that approximately 11 passes were needed to reach the target pore size (Figure 1). During the initial passes, DLS reflected the larger particles that were present because of their higher scattering intensity. Once the larger particles were removed, passes quickly reflected the pore size of the extruder. NTA displayed an initially larger standard deviation in the data because the particle sizes varied more by number. Neither original lipid concentration nor freeze-thaw cycles had a discernable effect on extruded sizes through different pore sizes (Figures 2 and 4). However, DOPC liposomes subjected to freeze-thaw cycles did have noticeably higher initial concentration when increasing the number of extrusion passes (Figure 4). Robustness of the Nanosizer Extruder by T&T Scientific was demonstrated by the stable size and concentration over 20 mL (11 passes/mL) of extruded DOPC liposomes (Figure 3). There was a noticeable difference in resulting concentrations of liposomes subjected to step-down extrusion vs extrusion directly at different pore sizes (Figure 5). This provided milder conditions and higher retention of liposomes. Increasing the percentage of cholesterol in DOPC further displaced head groups in the lipid bilayer causing an increase in the zeta potential (Figure 6).

Fluorescence measurements on Rh-DPPE incorporated into DOPC showed that lower percentages could be further measured to determine the true limit of detection (Figure 7). The number of fluorescent molecules per vesicle was estimated from geometrical considerations, assuming a liposome size of 130 nm and a lipid headgroup area of $0.6 nm^2$. Rhodamine dye conjugated to DOPC vesicles gave a photo-stable and bright fluorescence signal when excited with the 532 nm laser and using a 565 nm long-pass filter before the sCMOS camera of the NanoSight NS300. Comparison was made between the concentration measured under fluorescence mode and under scatter mode. Since the sample was pure, all particles present and measured in scatter mode were assumed to be liposomes that should be labeled with the dye. This efficiency % is the ratio of the two results. As expected, with increasing amounts of dye, the fluorescence efficiency increases. Scatter and fluorescence size distribution profiles match; one example shown in Figure 7. 100% efficiency is seen with as little as 0.25% dye which is equivalent to 113 fluorophores/particle. More than 0.25% dye in a sample resulted in more than 100% efficiency. This indicated some measurement difficulties that could be optimized with further work, including investigation of dye aggregation or difficulties in scatter mode measurements as potential causes. Decreasing % labeling efficiency with higher mol% Rh-DPPE indicates potential fluorophore quenching.

The SAXS data from the liposomes prior to extrusion (see fig. 8) show several equidistant Bragg peaks, as a consequence of the lipid bilayer stacking in these multilamellar vesicles. The repeat distances in the lamellar stacks, as deduced from the peak positions, were 6.4 nm for DPPC and 6.5 nm for DPPC. The peak widths are similar for the two samples, which suggests that the number of lamellae in the stacks is comparable. However, more higher order peaks are observed for DPPC, which indicates a better long-range order in the lamellar stack. In the WAXS data a distinct diffraction peak at the scattering angle 2θ of 20.9° is seen in case of DPPC, whereas no such peak is present in the data measured from DOPC. The peak is associated with the ordering of the alkyl chains and indicates that DPPC is in the ordered gel phase with a Bragg spacing of 0.42 nm, whereas DOPC is in the disordered fluid phase. This finding is consistent with the tabulated gel-to-fluid transition temperatures, which are +41°C for DPPC but -17°C for DOPC [5].

SAXS and WAXS measurements on DDPC at different temperatures ranging between 5 and 70°C are shown in fig. 9 and the deduced Bragg spacings are given in Table 1.The lamellar repeat distances and the alkyl chain ordering did not change significantly between 5°C and 30°C, At 39°C, which is just slightly below the gel-to-fluid phase transition temperature of DPPC, pronounced changes were observed in the SAXS data: the diffraction peaks are shifted to smaller scattering angles and become significantly broader. This indicates an increased spacing and a reduced order in the lamellar stack. At 43 °C, just above the phase transition temperature, the WAXS peak disappears, confirming the presence of the fluid phase. Furthermore, with increasing temperature the peaks in the SAXS region shift back to the higher angles, indicating a reduction of the lamellar repeat distance. The interesting behavior that was observed just below the transition temperature, indicates the onset of some restructuring in the multi-lamellar vesicles, which points to the occurrence of a transient 'ripple phase'.

The background-corrected SAXS data (see fig. 10) of the extruded DPPC liposomes, measured at 20°C, show the characteristic features from the scattering of single lipid bilayers: a broad hump, followed by some higher oscillations. The sharp Bragg peaks, as observed on the same sample prior to extrusion, are no more present. This confirms that upon extrusion the multi-lamellar vesicles were transformed to (almost) unilamellar vesicles. Two very broad Bragg peaks that are still visible (arrows in fig. 10) on top of the broad hump due to the lipid bilayer structure, indicate that the vesicles are not strictly unilamellar, but still contain a small number (around 2-4) of lipid bilayers. A similar behavior was observed for the extruded DOPC vesicles (not shown), From the extruded DPPC vesicles a WAXS peak, though significantly broadened (see fig. 11), could still be observed at T = 20°C, which again points to the formation of the gel phase. The absence of such peak in case of the extruded DOPC vesicles confirms that these are in the fluid phase.

The modeling of the electron density profile across the lipid bilayer as well as a more detailed study of the interesting behavior of DPPC close to the gel-fluid phase transition temperature are still pending.

Conclusion:

A broad range of characterization information and combination of both NanoSight and Zetasizer systems helped further optimize fabrication and understand the function of liposomes. NTA provided number-based high resolution sizing, accurate distribution profiles, concentration (particles/mL), and fluorescence measurements. DLS provided excellent reproducibility, mean size and PDI measurements over a broad range and non-invasive trend analysis. ELS provided zeta potential as a functionality and stability metric of particles. Furthermore, SAXS data gave insight in the lipid bilayer stacking as a function of sample temperature, whereas the WAXS data allowed to investigate the alkyl chain packing.

References:

[1] Woodle MC. Advanced Drug Delivery Reviews 16 (1995) 249-265.
[2] Patil YP, Jadhav S. Chemistry and Physics of Lipids 177 (2014) 8-18.
[3] Harashima H, Sakata K, Funato K, Kiwada H. Pharmaceutical Research 11 (1994) 402-406.
[4] Chibowski E, Szczes A. Adsorption 22 (2016) 755-765.
[5] www.avantilipids.com[4] reference to the liposome preparation method described

Acknowledgements:

We would like to thank our collaborators Dr's. Graham Taylor and Nima Tamaddoni at T&T Scientific (Knoxville, TN) for working with us to produce this data. They provided the liposomes and their Nanosizer extruders. NTA, DLS, and ELS analysis was completed at their facility.

Figures:

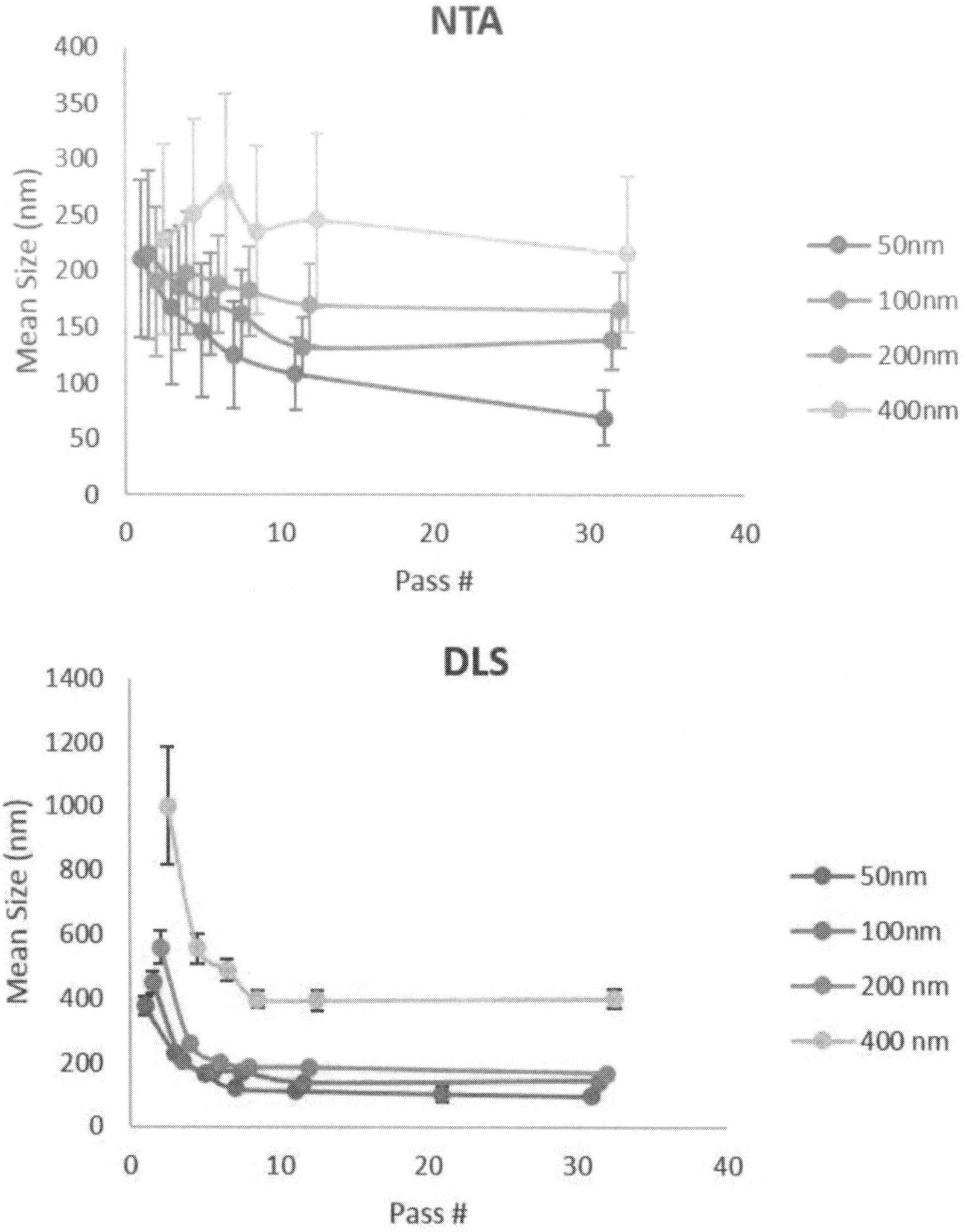

Figure 1. DOPC, 2 mg/mL, size decrease by increased number of extrusion passes, for various extruder pore-sizes.

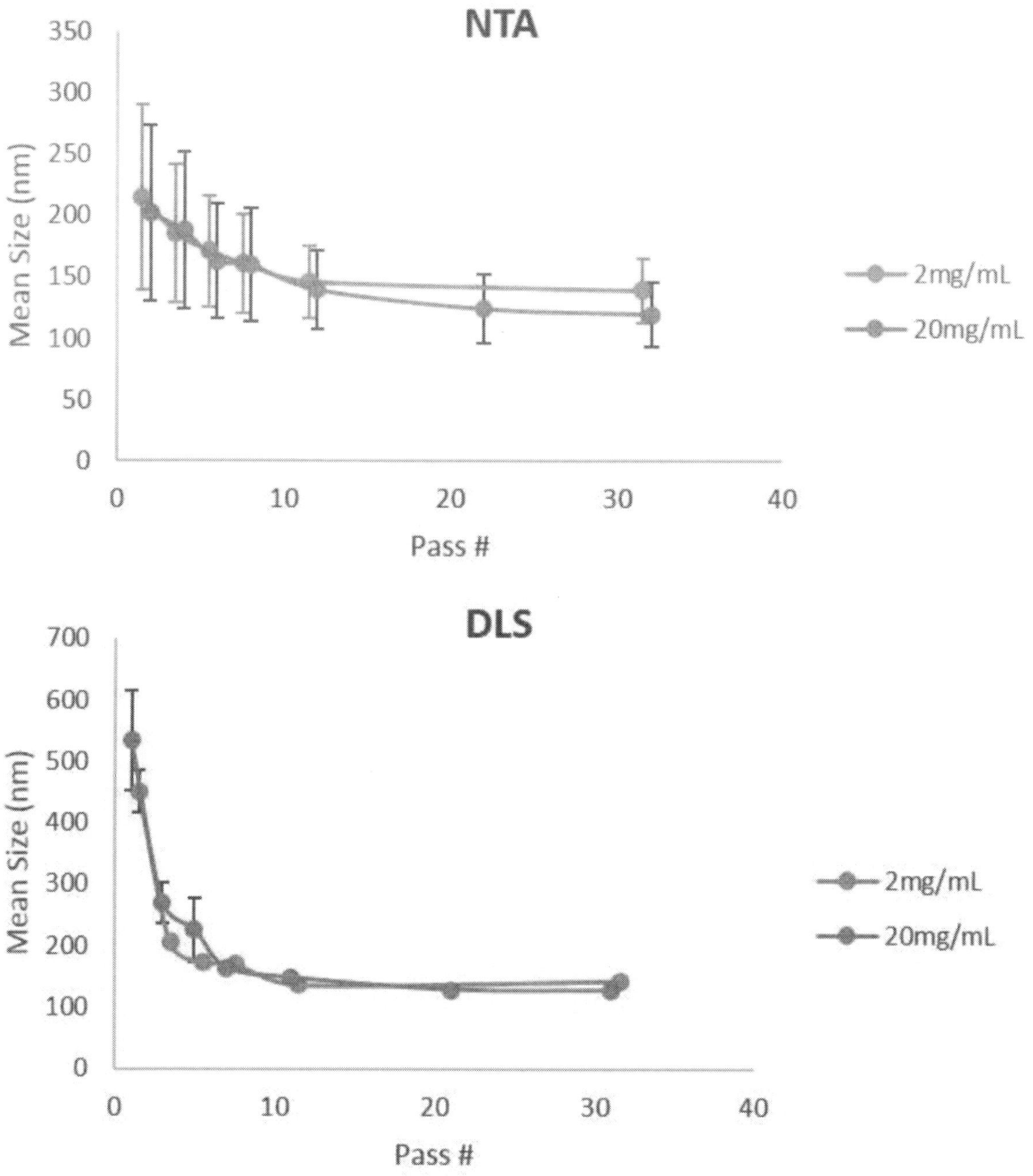

Figure 2. Effect of varying lipid concentration on extrusion sizing by pass # through 100nm pore size.

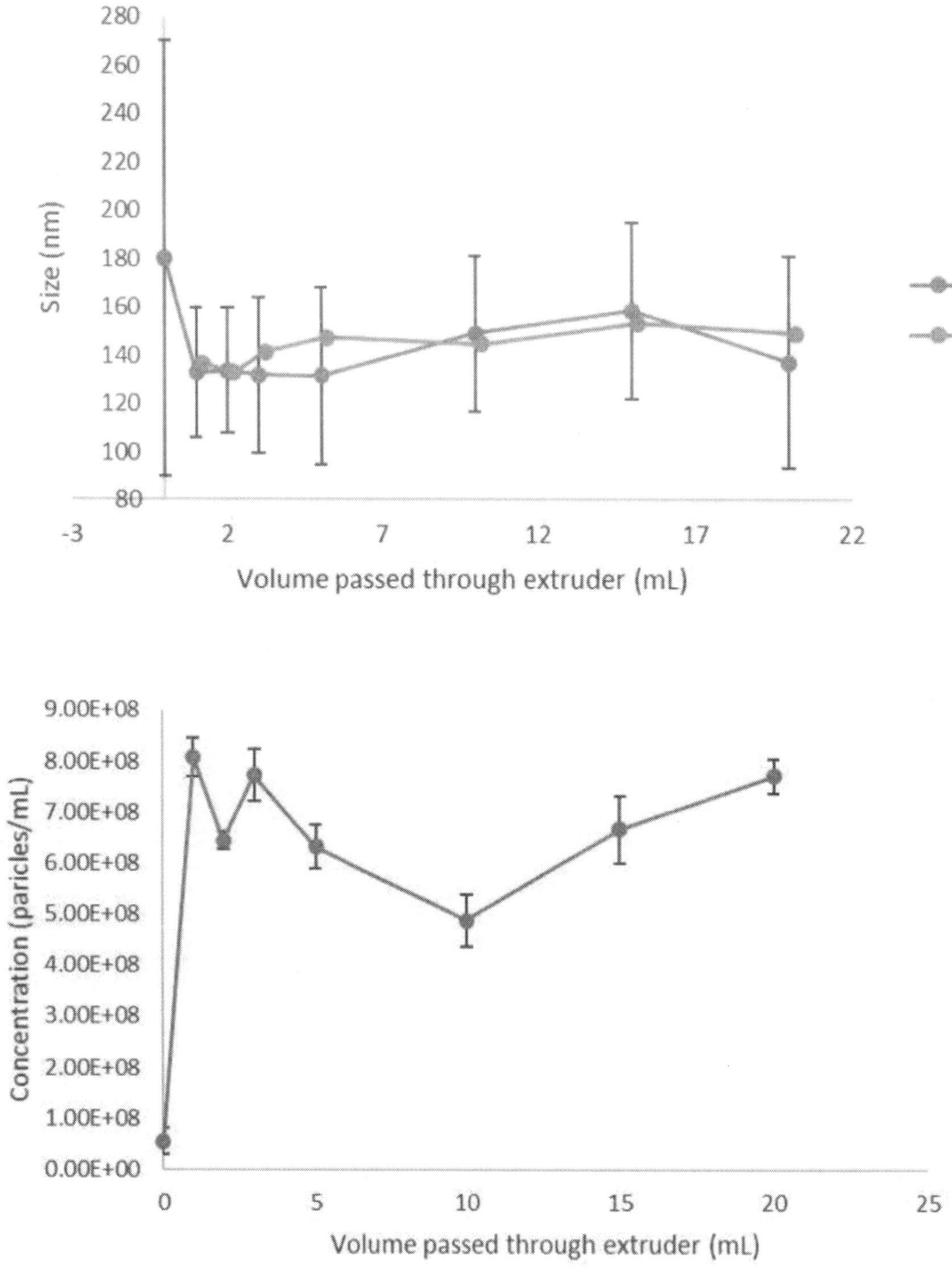

Figure 3. Testing limit of track-etched membrane through 100 nm pore size (each mL passed 11x). Membranes show no performance degradation after >200 passes.

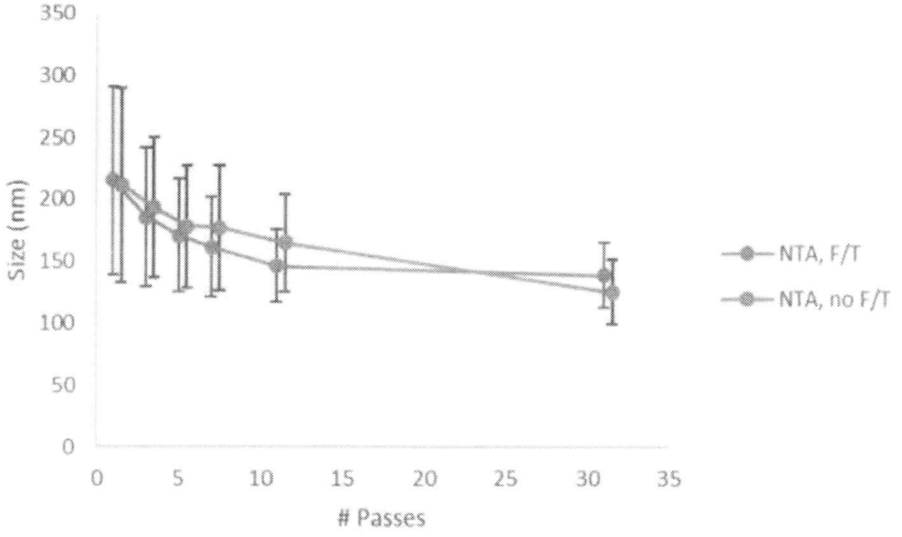
Size (nm)
Passes
NTA, F/T
NTA, no F/T

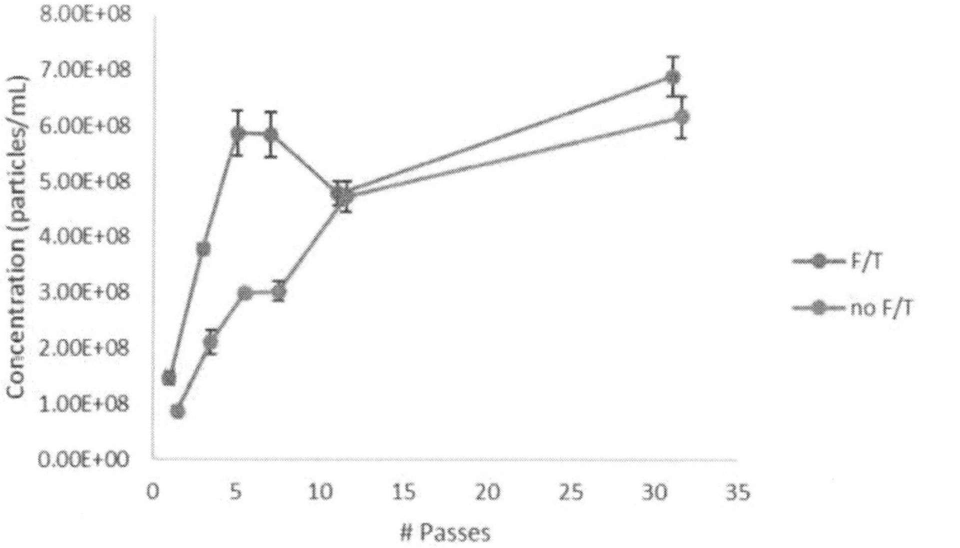
Concentration (particles/mL)
Passes
F/T
no F/T

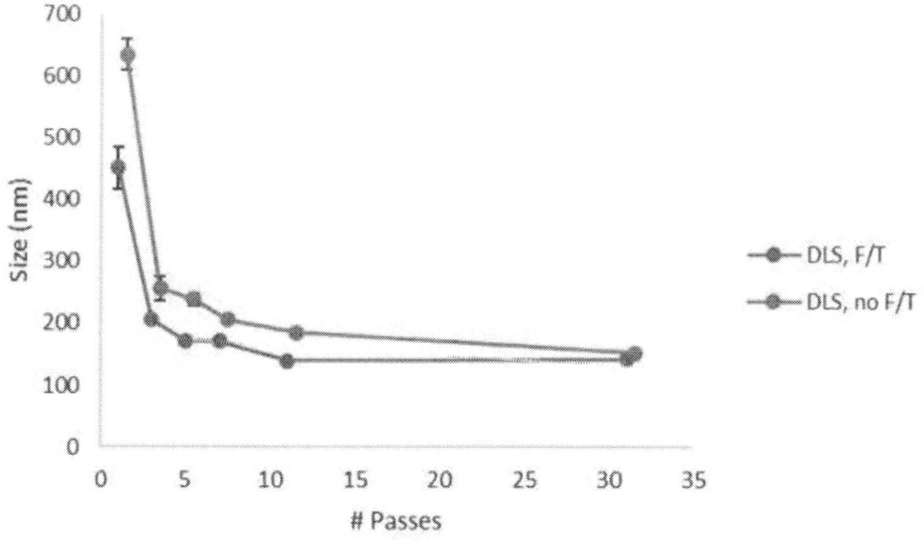
Size (nm)
Passes
DLS, F/T
DLS, no F/T

Figure 4. Effect of Freeze / Thaw cycles (5x) on particle size and concentration as extruded through 100nm pores.

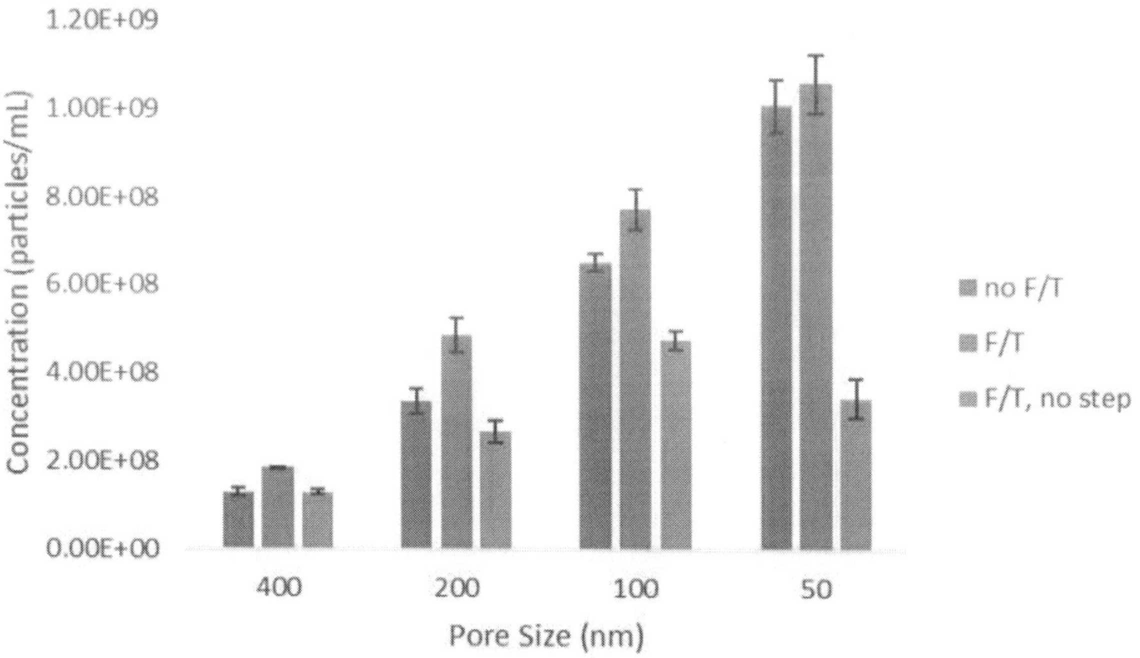

Figure 5. Effect of both freeze/thaw cycles and step-down extrusion on concentration (11x passes/pore size) – NTA.

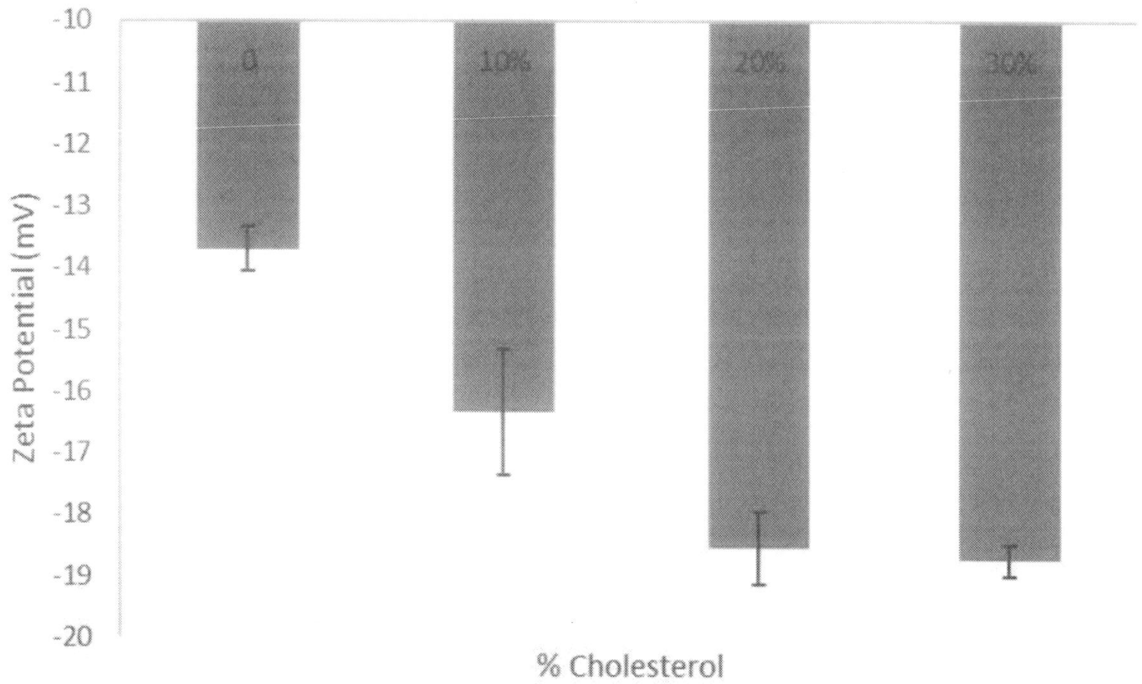

Figure 6. Effect of % Cholesterol on zeta potential (1mM NaCl +0.1mM MOPS) with diffusion barrier method.

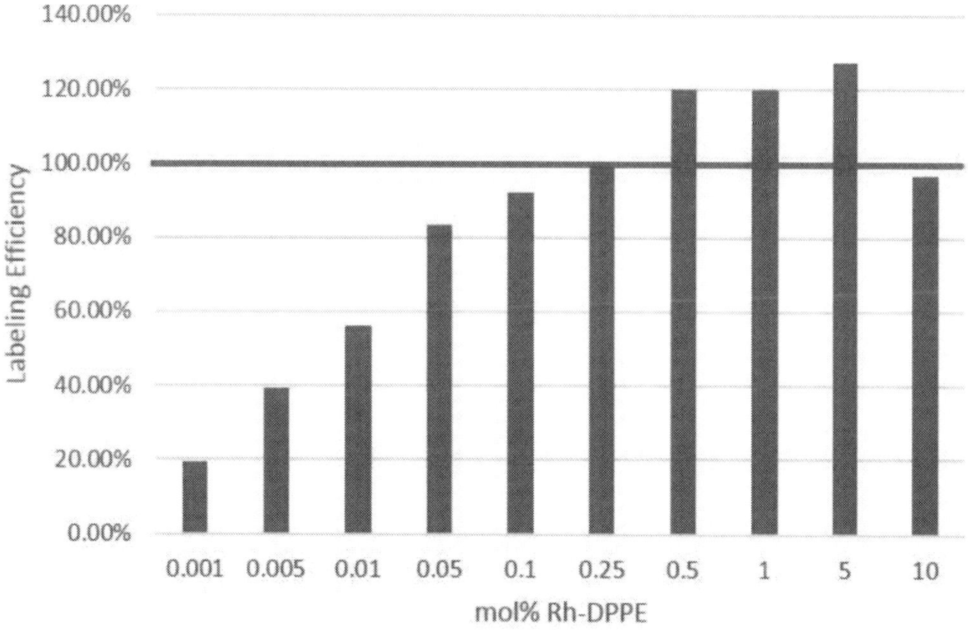

Figure 7. 2 mg/mL DOPC with varying mol% Rh-DPPE Fluorescence-mode concentration efficiency of Rhodamine labeled liposomes as a function of increasing dye loadings. (Fluoresence efficiency obtained with NTA as Fluorescence mode concentration divided by Scatter mode concentration)

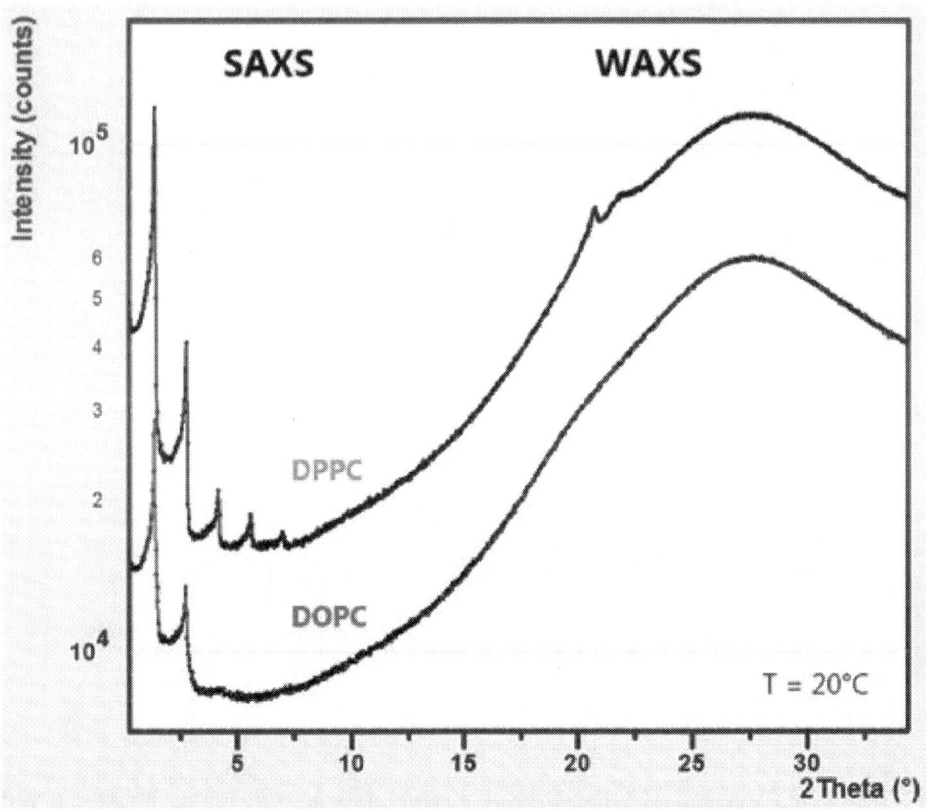

Figure 8. SAXS/WAXS results on DPPC and DOPC liposomes, prior to extrusion. The sample temperature was 20°C. The curves were vertically offset for clarity. The data collection time was 10 min.

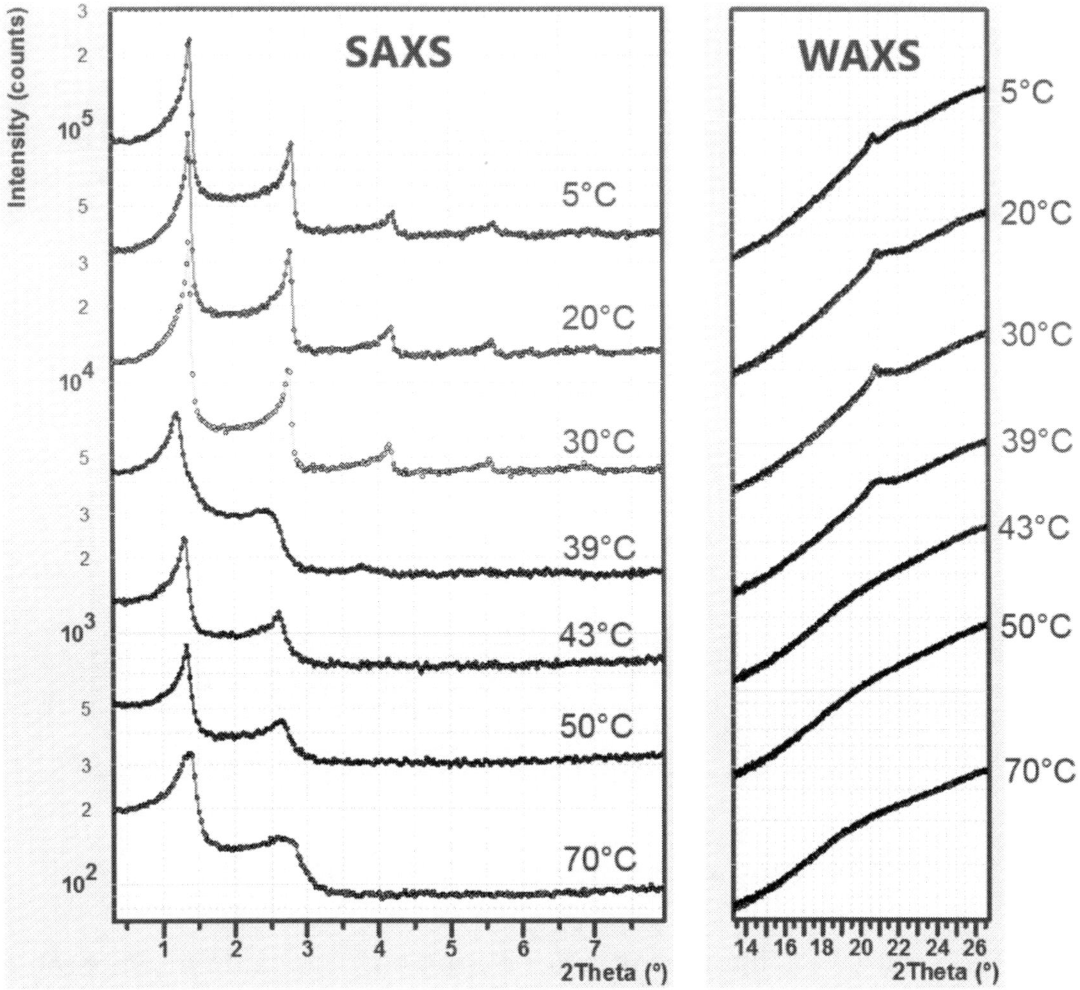

Figure 9. SAXS and WAXS results on DPPC liposomes, prior to extrusion. Data were collected at different temperatures. The curves were vertically offset for clarity.

T [°C]	d1 [nm] from SAXS data	d2 [nm] from WAXS data
5	6.3	0.43 (gel phase)
20	6.4	0.42 (gel phase)
30	6.4	0.43 (gel phase)
39	7.3	0.43 (transient ripple phase)
43	6.7	- (liquid phase)
50	6.6	- (liquid phase)
70	6.3	- (liquid phase)

Table 1: Structural parameters of DPPC liposomes, as a function of sample temperature. d1: Bragg spacing due to lamellar repeat distance; d2: Bragg spacing due to alkyl chain ordering in the gel phase.

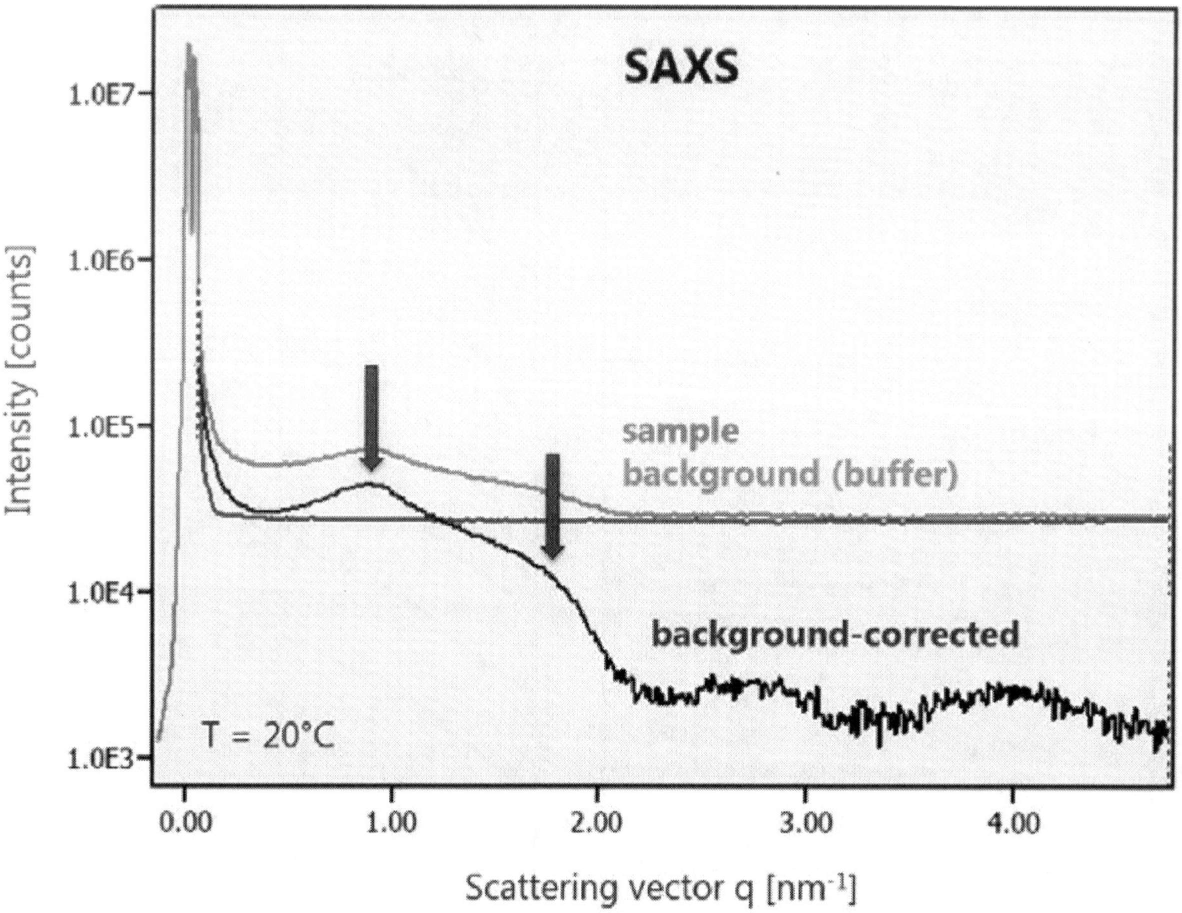

Figure 10. SAXS results on DPPC liposomes, after extrusion. The sample temperature was 20°C. Here Intensities are given as a function of the scattering vector q ($q = 4\pi/\lambda \sin(2\theta/2)$, where λ denotes the wavelength of radiation and 2θ the scattering angle). Also shown are background-corrected SAXS data, where the scattering background, mesaured from the capillary filled with the pure dispersion medium, was subtracted. Measurement time: 60 min. Arrows highlight Bragg peaks due to lipid bilayer structure.

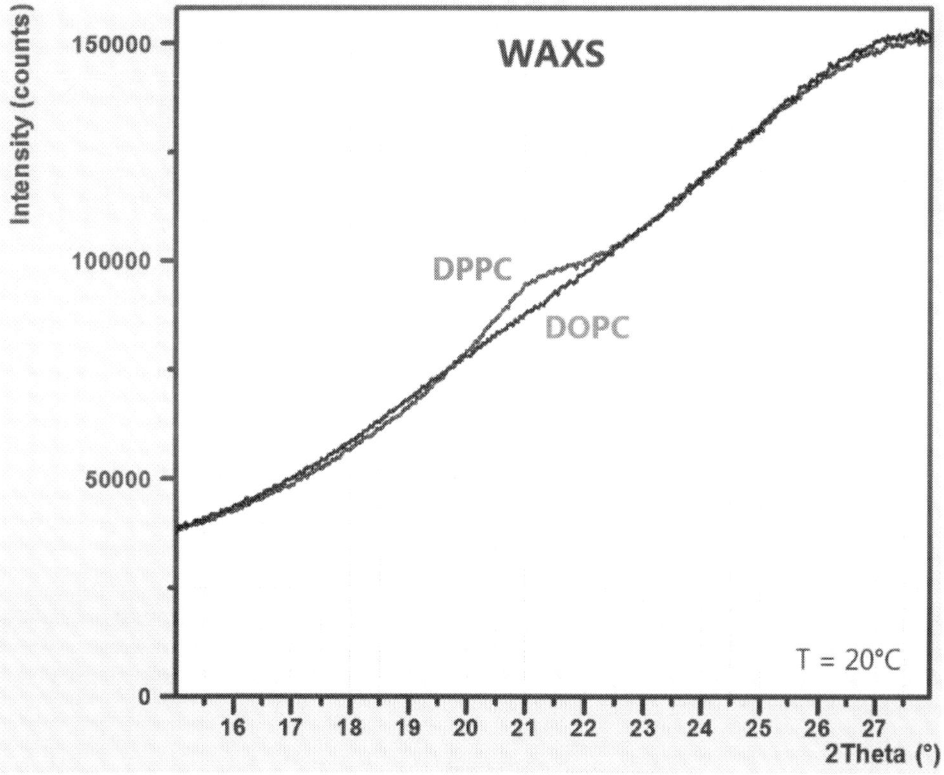

Figure 11. **WAXS results on DPPC and DOPC liposomes, after extrusion. The sample temperature was 20°C. The angular range probes alkyl chain packing within the lipid bilayer, the DPPC peak indicates a gel phase, whereas its absence in DOPC supports a fluid phase.**

508083 Multiwavelength Analytical Ultracentrifugation â€“ Simultaneous Analysis of Hydrodynamic and Optical Properties

Abstract:

Product design is probably the most important paradigm of modern particle technology. By selectively changing the size, shape and composition of the material at hand, characteristic properties can be tailored. The analytical challenge is to fully characterize property functions in dependency of size, shape and structure of the particles at hand. Techniques that are used for the characterization of particle ensembles such as light scattering, laser diffraction, UV-VIS spectroscopy or ultrasound absorption spectroscopy require deconvolution techniques which are often limited for polydisperse samples.

However, analytical ultracentrifugation equipped with an UV-Vis multiwavelength detector (MWL-AUC) is a powerful tool for multidimensional characterization of polydisperse nanoparticles. AUC allows the simultaneous analysis of hydrodynamic and optical properties without the requirement for purification, as it combines UIV/VIS spectroscopy with fractionation in a centrifugal field [1]. It is thus possible to determine distribution of particles by measuring complete spectra from 240 to 1000 nm using a CCD spectrometer at each radial position with high temporal resolution. From the measured turbidity data, concentration profiles may be deduced by the Beer-Lambert law. Subsequently, sedimentation coefficient distributions and particle size distributions (PSD) can be calculated. Our recent improvements on the instrumentation as well as acquisition software and data evaluation allowed us to significantly enhance the possibilities of MWL-AUC as an outstanding characterization method for nanoparticulate systems [1,2]. Optical properties can be directly correlated with size information in polydisperse dispersions. Therefore extinction spectra of any species present within a mixture can be extracted or information on optical properties can be used to deconvolute PSDs comprising chemically distinct particles.

The possibility to adjust the centrifugal force over a wide range enables to characterize particles with largely varying sizes from below 1 nm up to almost 1 µm using the same instrument. This huge dynamic range is accessible via a single run using gravitational sweep experiments [3]. This procedure allows dealing with strong polydispersity, as large particles are sedimenting at low centrifugal fields while smaller particles are characterized at higher rotor speeds. Based on the size and wavelength dependency of light scattering, high dynamic range particle size evaluation is possible, while simultaneously determining optical properties of the individual fractions in a mixture.

In order to illustrate the new possibilities of nanoparticle characterization with MWL-AUC we will report on several material systems including semiconductor quantum dots, silica nanoparticles, carbon allotropes and gold nanorods. These results demonstrate that MWL-AUC is a powerful and highly accurate technique, which provides the unique capability to correlate size and optical properties in just one experiment.

References

1. J. Walter, K. Löhr, E. Karabudak, et al., Multidimensional Analysis of Nanoparticles with Highly Disperse Properties Using Multiwavelength Analytical Ultracentrifugation.
ACS Nano, 2014, 8, 8871.

2. J. Walter, P. J. Sherwood, W. Lin et al., Simultaneous Analysis of Hydrodynamic and Optical Properties Using Analytical Ultracentrifugation Equipped with Multiwavelength Detection. Analytical Chemistry 2015, 87 (6), 3396-3403.

3. J. Walter; W. Peukert, Multidimensional Analysis of Nanoparticles with Highly Disperse Properties Using Multiwavelength Analytical Ultracentrifugation.
Nanoscale 2016, 8 (14), 7484-7495.

508197 In Situ Deformation and Breakage of Sol-Gel Derived Oxide Particles inside a SEM

Abstract:

Knowledge of mechanical properties and particle breakage behavior is of fundamental importance for many particle related processes and the application of particles. Although many (in situ) studies have been dedicated to materials' size dependent mechanical characterization over the past decade, particles as free standing structures, however, have widely been omitted [1]. Important questions include the size dependent deformation behavior of particles and small structures confined in one- or two dimensions and the corresponding structure-propertycorrelations at these small scales. Within this account, structural and mechanical characterization under compression will be discussed for sol-gel derived micron sized spherical SiO_2 and TiO_2 [2-5] particles for the a size range of 200 nm to 4 μm. Mechanical compression of the individual particles was performed by a custom-made scanning electron microscope (SEM) supported manipulation device [1]. The internal structural of the sol-gel particles was tuned in a wide range by thermal annealing. For the amorphous silica particles the degree of internal cross-linking and hydroxylation was changed systematically towards vitreous silica. The changes of the internal structure are directly reflected by the mechanical properties [2]. For fully densified and de-hydroxylated vitreous silica particles a clear brittle-to-ductile transition is found in the size range of 500 – 800 nm, i.e. with increasing particle size the plasticity of the particles decrease and the stable crack propagation and brittle fracture becomes the predominant failure mode [3]. By ex situ Raman spectroscopy the observed plasticity is found to be accommodated by structural densification [4]. In the TiO_2 system the influence of crystallinity was studied: Single-phase nanocrystalline (nc) anatase or rutile particles were obtained by annealing [5]. For the amorphous and the nc particles a significant plastic deformation behavior accompanied by crack initiation occurring at high deformations were observed. The crack propagation is found to follow presumably grain boundaries. The corresponding fracture stresses are Weibull distributed.

References:

[1] S. Romeis, J. Paul, M. Ziener, W. Peukert, A novel apparatus for in situ compression of submicron structures and particles in a high resolution SEM, Rev Sci Instrum 83 (2012).

[2] S. Romeis, J. Paul, M. Hanisch, V.R.R. Marthala, M. Hartmann, R.N. Klupp Taylor, J. Schmidt et al., Correlation of enhanced strength and internal structure for heat-treated submicron Stöber silica particles, Part Part Syst Char 31 (2014) 664–674.

[3] J. Paul, S. Romeis, M. Mačković, V. Marthala, P. Herre, T. Przybilla, M. Hartmann et al., In situ cracking of silica beads in the SEM and TEM - effect of particle size on structure-property correlations, Powder Technol. 270 (2015) 337–347.

[4] S. Romeis, J. Paul, P. Herre, D. de Ligny, J. Schmidt, W. Peukert, Local densification of a single micron sized silica sphere by uniaxial compression, Scripta Mater. (2015) 84–87.

[5] P. Herre, S. Romeis, M. Mačković, T. Przybilla, J. Paul, J. Schwenger, B. Torun et al., Deformation behavior of nanocrystalline titania particles accessed by complementary in situ electron microscopy techniques, J. Am. Ceram. Soc. 330 (2017) 556.

508626 Size and Polydispersity of a Nanoparticle Reference Material By Dynamic Light Scattering

Abstract:

Nanotechnology continues to gain importance in both fundamental and applied sciences. A key parameter of anything "nano" is defined by the typical dimensions encountered, and subsequently the need to measure size. Of the various techniques available to characterize nanoparticle size (such as atomic force [AFM], transmission electron [TEM], or scanning electron microscopy [SEM], and also small angle X-ray scattering [SAXS]), we concentrate on dynamic light scattering [DLS] in this study.

The method itself is a relatively quick, cuvette-based ensemble measurement that extracts information about the diffusion coefficient of the nanoparticles from correlation analysis of the underlying signal fluctuations. DLS is also known as photon correlation spectroscopy [PCS] or quasi-elastic light scattering [QELS]. Due to the nature of the technique, the data are "low resolution" with an inherent polydispersity for even perfect standard materials.

We investigate this polydispersity as well as the other key characteristics of the analysis (z-average cumulant size, intensity distribution, volume distribution) at the practical example of a Nanoparticle reference standard from the National Institute of Standards & Technology. The NIST Reference Material RM8011 contains citrate-stabilized colloidal gold nanoparticles of nominal 10nm diameter.

The experimentally determined standard deviations of the key measurement parameters such as mean size and overall polydispersity may serve as guidance for researchers sizing non-standard samples of unknown nanoparticles that tend to be more polydisperse and show larger statistical variances. These results set a bench mark reference point for a 'best and smallest in class' reference nano material near 10nm diameter.

508180 A Comprehensive Brownian Dynamics Based Forward Model for Analytical (Ultra) Centrifugation

Abstract:

Analytical ultracentrifugation (AUC) is considered to be the gold standard for particle characterization in colloidal systems. Together with Analytical centrifugation (AC) techniques, it covers a wide particle size range from sub 1 nanometer to a few microns. Considerable advances have been made over the last decade in the characterization of biomolecules as well as nanoparticles using A(U)C. SEDFIT and Ultrascan have proved to be very reliable and robust in calculating the sedimentation coefficient distributions from centrifugation data [1]. It is possible to retrieve information with and without the effect of particle diffusion in the colloidal systems. Recently, we had extended direct boundary modelling for analyzing the data obtained from Lumisizer [2]. A Brownian dynamics (BD) based tool, originally developed by Diez et al., was improved upon to study the effect of neglecting diffusion while extracting information from the centrifugation data [3].

Currently, there is appreciable interest in using A(U)C techniques for characterizing non-spherical nanoparticles as well as retrieving the optical properties of nanoparticles [4]. Equally interesting is the use of ramp speed data to analyze multimodal distributions and size distributions with larger standard deviations. Development of new tools for sedimentation coefficient analysis (and thereby the size analysis) requires robust forward models that can generate numerical solutions to the Lammâ€™s equation. BD model represents a viable option for this. The model is based on the motion of particles in a centrifugal field and the change in concentration of the particles can be monitored as a function of time and radial space.

The model can be extended to account for different cell geometries, size distribution of particles as well as multimodal particle size distributions. A major advantage is in generating solution of Lamm equation for ramp speed data, where the rotor speed varies as a function of time. Overall, BD simulations function as an alternative tool to numerical Lamm equation solvers and in certain cases performs as the only viable forward model. The features of ramp speed (along with initial acceleration phase) and multimodal size distributions have also been incorporated. The forward model is also equipped to handle light scattering based on Mieâ€™s theory [5, 6].

It is well known that the sedimentation of macromolecules and particles in the AUC correlates with concentration effects [7]. Hence, transport properties of such analytes are always related to their spatial distribution. As interparticulate interactions alter the sedimentation behavior, it is necessary to consider non-ideality phenomena behavior when analyzing AUC or AC data of concentrated systems [8]. As nanoparticulate systems are rarely monodisperse in size, available software tools for analyzing A(U)C data reach their limits when considering non-ideality. BD is a promising approach towards theoretical description of microscopic inter-particle interactions, since its algorithm allows tracking the trajectory of individual particles. Concentration dependency of the sedimentation and diffusion coefficient has been implemented and tested in our latest BD model. For monodisperse systems, we it can be shown that profiles obtained by BD are in excellent agreement with well-established Lamm equation solvers. Moreover, simulations for polydisperse systems including non-ideal concentration effects are currently being incorporated.

References

[1] http://www.analyticalultracentrifugation.com/default.htm,

http://www.ultrascan.uthscsa.edu/

[2] J. Walter, T. Thajudeen, S. Süß, et al., Nanoscale, 2015, 7 (15), 6574-6587.

[3] A. I. Diez, A. Ortega, G. de la Torre, BMC Biophys. 2011, 4, 6.

[4] J. Walter, P. J. Sherwood, W. Lin, et al., Anal. Chem. 2015, 87 (6), 3396-3403.

[5] J. Walter, K. Löhr, E. Karabudak, et al., ACS Nano 2014, 8 (9), 8871-8886.

[6] T. Thajudeen, J. Walter, M. Uttinger et al., Part. & Part. Sys. Charact., 2016,34, 1600229.

[7] S. E. Harding, P. Johnson, Biochem. J. 1985, 231, 543â€"547.

[8] A. Solovyova; P. Schuck; L Costenaro; C. Ebel, Biophys., J. 2001, 81, 1868-1880.

AUTHOR INDEX

AUTHOR INDEX

AUTHOR INDEX

AUTHOR INDEX

AUTHOR INDEX

AIChE
120 Wall Street, FL 23
New York, NY 10005-4020

ISBN 978-1-5108-6966-0

CURRAN ASSOCIATES INC.
proceedings.com